T0408251

BIOACTIVE COMPOUNDS OF MEDICINAL PLANTS

Properties and Potential for Human Health

Innovations in Plant Science for Better Health: From Soil to Fork

BIOACTIVE COMPOUNDS OF MEDICINAL PLANTS

Properties and Potential for Human Health

Edited by

Megh R. Goyal, PhD, PE

Ademola O. Ayeleso, DTech

Apple Academic Press Inc.
3333 Mistwell Crescent
Oakville, ON L6L 0A2
Canada

Apple Academic Press Inc.
9 Spinnaker Way
Waretown, NJ 08758
USA

International Standard Book Number-13: 978-1-77188-648-2 (Hardcover)
International Standard Book Number-13: 978-1-315-14747-5 (eBook)

Library and Archives Canada Cataloguing in Publication

Bioactive compounds of medicinal plants : properties and potential for
human health / edited by Megh R. Goyal, PhD, PE, Ademola Olabode Ayeleso, DTech.

(Innovations in plant science for better health : from soil to fork)
Includes bibliographical references and index.
Issued in print and electronic formats.
ISBN 978-1-77188-648-2 (hardcover).--ISBN 978-1-315-14747-5 (PDF)

1. Plant bioactive compounds. 2. Medicinal plants. 3. Materia medica,
Vegetable. I. Goyal, Megh Raj, editor II. Ayeleso, Ademola Olabode, editor
III. Series: Innovations in plant science for better health

QK898.B54B56 2018 615.3'21 C2018-903205-7 C2018-903206-5

Library of Congress Cataloging-in-Publication Data

Names: Goyal, Megh Raj, editor. | Ayeleso, Ademola Olabode, editor.
Title: Bioactive compounds of medicinal plants : properties and potential for
human health / editors, Megh R. Goyal, Ademola Olabode Ayeleso.
Description: Toronto ; New Jersey : Apple Academic Press, 2018. | Series:
Innovations in plant science for better health : from soil to fork |
Includes bibliographical references and index.
Identifiers: LCCN 2018024233 (print) | LCCN 2018024978 (ebook) | ISBN
9781315147475 (ebook) | ISBN 9781771886482 (hardcover : alk. paper)
Subjects: | MESH: Plants, Medicinal | Phytotherapy | Diabetes Mellitus--drug
therapy | Plant Extracts--therapeutic use | Africa
Classification: LCC RS164 (ebook) | LCC RS164 (print) | NLM QV 770 HA1 | DDC
615.3/21--dc23
LC record available at https://lccn.loc.gov/2018024233

Apple Academic Press also publishes its books in a variety of electronic formats. Some content that appears in print may not be available in electronic format. For information about Apple Academic Press products, visit our website at **www.appleacademicpress.com** and the CRC Press website at **www.crcpress.com**

CONTENTS

ABOUT SENIOR EDITOR-IN-CHIEF

Megh R. Goyal, PhD
Retired Professor in Agricultural and Biomedical Engineering, University of Puerto Rico, Mayaguez Campus Senior Acquisitions Editor, Biomedical Engineering and Agricultural Science, Apple Academic Press, Inc.
E-mail: goyalmegh@gmail.com

Megh R. Goyal, PhD, PE, is a Retired Professor in Agricultural and Biomedical Engineering from the General Engineering Department in the College of Engineering at the University of Puerto Rico–Mayaguez Campus; and Senior Acquisitions Editor and Senior Technical Editor-in-Chief in Agriculture and Biomedical Engineering for Apple Academic Press, Inc. He has worked as a Soil Conservation Inspector and as a Research Assistant at Haryana Agricultural University and Ohio State University.

During his professional career of 45 years, Dr. Goyal has received many prestigious awards and honors. He was the first agricultural engineer to receive the professional license in Agricultural Engineering in 1986 from the College of Engineers and Surveyors of Puerto Rico. In 2005, he was proclaimed as "Father of Irrigation Engineering in Puerto Rico for the Twentieth Century" by the American Society of Agricultural and Biological Engineers (ASABE), Puerto Rico Section, for his pioneering work on micro irrigation, evapotranspiration, agroclimatology, and soil and water engineering. The Water Technology Centre of Tamil Nadu Agricultural University in Coimbatore, India, recognized Dr. Goyal as one of the experts "who rendered meritorious service for the development of micro irrigation sector in India" by bestowing the *Award of Outstanding Contribution in Micro Irrigation*. This award was presented to Dr. Goyal during the inaugural session of the National Congress on "New Challenges and Advances in Sustainable Micro Irrigation" on March 1, 2017, held at Tamil Nadu Agricultural University. On August 1 of 2018, ASABE bestowed on him Netafim Award for Advancements in Micro irrigation.

As a prolific author and editor, he has written more than 200 journal articles and textbooks and has edited over 55 books. He received his BSc degree in engineering from Punjab Agricultural University, Ludhiana, India; his MSc and PhD degrees from Ohio State University, Columbus; and his Master of Divinity degree from Puerto Rico Evangelical Seminary, Hato Rey, Puerto Rico, USA.

ABOUT CO-EDITOR

Ademola O. Ayeleso
Senior Lecturer, Department of Biochemistry, Adeleke University, Nigeria
E-mail: ademola.ayeleso@gmail.com

Ademola Ayeleso, DTech, is a Senior Lecturer in the Department of Biochemistry at Adeleke University, Ede, Nigeria. Dr. Ayeleso is working on the use of medicinal plants, plant products, and synthesized compounds in the management of diabetes mellitus, which has become a major threat to the lives of people in both urban and rural areas throughout the world.

He has presented research papers at different international conferences. Dr. Ayeleso obtained his BTech degree in Biochemistry at the Federal University of Technology, Akure, Nigeria. He received his MSc degree in Biochemistry at the University of Port Harcourt, Port Harcourt, Nigeria, and his doctoral degree in Biomedical Technology at the Cape Peninsula University of Technology, Bellville, South Africa. He has worked as a Postdoctoral Research Fellow in some universities in South Africa. Dr. Ayeleso is a prolific writer with over 20 publications to his credit. He is presently the Head of Department (Biochemistry) and Director of the Office of Research, Grant and Linkages at Adeleke University, Ede, Nigeria.

LIST OF CONTRIBUTORS

Yapo Guillaume Aboua
Researcher, Faculty of Health and Wellness Sciences, Cape Peninsula University of Technology, P.O. Box 1906, Bellville 7535, South Africa, Mobile: +278-27029043, E-mail: abouay@cput.ac.za

Gbenga Anthony Adefolaju
Senior Lecturer, Department of Preclinical Sciences, University of Limpopo, Private Bag x1106, Sovenga 0727, South Africa, Tel.: +271-52683281, Mobile: +277-86462774, E-mail: gbenga.adefolaju@ul.ac.ac.za, adefolajugbenga@gmail.com

Johnson Akinwumi Adejuyitan
Lecturer 1, Department of Food Science and Engineering, Ladoke Akintola University of Technology, P.M.B. 4000, Ogbomoso, Oyo State, Nigeria, Mobile: +234-8105049457, E-mail: jaadejuyitan@lautech.edu.ng

Oluyemisi Elizabeth Adelakun
Senior Lecturer, Department of Biological Sciences, Covenant University at Ota – Ogun State, Nigeria and Department of Food Science and Engineering, Ladoke Akintola University of Technology, P.M.B 4000, Ogbomoso, Oyo State, Nigeria, Mobile: +234-8030713796, E-mail: Olu_yemisi2000@yahoo.com; oeadelakun@lautech.edu.ng, oluyemisi.adelakun1@covenantuniversity.edu.ng

Anthony Jide Afolayan
Research Professor, Medicinal Plants and Economic Development (MPED) Research Centre, Department of Botany, University of Fort Hare, Private Bag X1314, Alice 5700, South Africa, Tel.: +27-406022321, Fax: +27-866282409, E-mail: aafolayan@ufh.ac.za

Ayodele Jacob Akinyemi
Senior Lecturer, Department of Biochemistry, Afe Babalola University Ado-Ekiti, Private Mail Bag 5454, Nigeria, Mobile: +234-8035866170, E-mail: akinyemiaj@abuad.edu.ng; ajakinyemi2010@yahoo.co.uk

Ademola Olabode Ayeleso
Senior Lecturer, Department of Biochemistry, Adeleke University, Ede, Osun State, Nigeria, Mobile: +2348144556529, E-mail: ademola.ayeleso@gmail.com

Adeniyi Bola
Professor, Department of Pharmaceutical Microbiology, Faculty of Pharmacy, University of Ibadan, P.O. Box 22346, Ibadan, Oyo State Nigeria, Mobile: 08-053039343, E-mail: baadeniyi@yahoo.com

Islamiyat Folashade Bolarinwa
Lecturer 1, Department of Food Science and Engineering, Ladoke Akintola University of Technology, P.M.B. 4000, Ogbomoso, Oyo State, Nigeria, Mobile: +234-9030771688, E-mail: islamiyat202@yahoo.co.uk; ifbolarinwa@lautech.edu.ng

Nicole Brooks
HOD, Department of Wellness Sciences, Faculty of Health and Wellness Sciences, Cape Peninsula University of Technology, 7535 Cape Town, South Africa, Mobile: +272-14603436, E-mail:brooksn@cput.ac.za

Matthew Cheesman
Principal Researcher, School of Pharmacy and QUM Network, Gold Coast Campus, Griffith University, Southport, Queensland 4222, Australia. Phone: +61-755529230,
E-mail: m.cheesman@griffith.edu.au

Ian Edwin Cock
Principal Researcher and Research Group Leader, Environmental Futures Research Institute and School of Natural Sciences, Griffith University, 170 Kessels Rd, Nathan, Queensland 4111, Australia. Phone: +61-737357637, E-mail: I.Cock@griffith.edu.au

Nwanekwu Kenneth Emeka
Lecturer, Department of Microbiology, Faculty of Science, Imo State University, P.M.B. 2000, Owerri, Imo State, Nigeria, Tel: 08-065831258, E-mail: kennwanekwu@yahoo.com

Benedict Abiola Falana
Lecturer, Department of Anatomy, Faculty of Basic Medical Sciences, College of Health Sciences, Osun State University, Osogbo, Nigeria, Mobile: +234-8067033706,
E-mail: benedict.falana@uniosun.edu.ng

Mahady Gail
Research Professor, Department of Pharmacy Practice, University of Illinois, Chicago, IL, 60608, U.S.A., Mobile: 1-3129961669, E-mail: mahady@ucl.edu

Mediline Goboza
Nutrition and Chronic Diseases Research Group, Oxidative Stress Research Centre, Department of Biomedical Sciences, Faculty of Health and Wellness Sciences, Cape Peninsula University of Technology, P.O. Box 1906, Bellville 7535, South Africa, Mobile: +277-87770768,
E-mail: medgoboza@gmail.com

Victoria Adaora Jideani
Professor, Department of Food Science and Technology, Faculty of Applied Science, Cape Peninsula University of Technology, Symphony Way, Off Robert Sobukwe Road, P.O. Box 1906, Bellville 7535, South Africa, Mobile: +277-88391615, E-mail: jideaniv@cput.ac.za

Prisca Kachepe
Nutrition and Chronic Diseases Research Group, Oxidative Stress Research Centre, Department of Biomedical Sciences, Faculty of Health and Wellness Sciences, Cape Peninsula University of Technology, P.O. Box 1906, Bellville 7535, South Africa, Mobile: +277-19541439,
E-mail: kachepepriscak@yahoo.com

Mutiu Idowu Kazeem
Lecturer, Department of Biochemistry, Lagos State University, P.M.B 0001, Lagos, Nigeria, Mobile: +234-8030622000, E-mail: mikazeem@gmail.com

Emmanuel Mukwevho
Associate Professor, Department of Biological Sciences, North-West University, Private Bag X2046, Mmabatho, 2735, South Africa, Tel.: +271-83892854, E-mail: emmanuel.mukwevho@nwu.ac.za

Yvonne Yeukai Murevanhema
Nutrition and Chronic Diseases Research Group, Oxidative Stress Research Centre, Department of Biomedical Sciences, Faculty of Health and Wellness Sciences, Cape Peninsula University of Technology, P.O. Box 1906, Bellville 7535, South Africa, Mobile: +277-20437156,
E-mail: yymurevanhema@mail.com

Anthony Mwakikunga
Senior Lecturer, Anatomy Department, College of Medicine, Blantyre, Malawi,
Mobile: +265-885328924, E-mail: mwakikungaanthony@gmail.com

Esther Emem Nwanna
PhD Candidate and Research scientist, Department of Biochemistry, Federal University of Technology (FUTA), P.M.B 704 Ondo State, Nigeria, Mobile: +234-8068062480, E-mail: esthernwanna@gmail.com

Ganiyu Oboh
Professor, Functional Foods and Nutraceuticals Unit, Department of Biochemistry, Federal University of Technology Akure, Private Mail Bag 704, Akure 340001, Nigeria, Mobile: +234-7031388644, E-mail: goboh2000@yahoo.com

Oluwatosin Ogunkelu
Department of Biochemistry, Lagos State University, P.M.B 0001, Lagos, Nigeria, Mobile: +234-7037479241, E-mail: slygirl006@yahoo.com

Oluwafemi Omoniyi Oguntibeju
Professor, Nutrition & Chronic Diseases Research Unit (Phytomedicine & Diabetes), Oxidative Stress Research Centre, Department of Biomedical Sciences, Faculty of Health and Wellness Sciences, Cape Peninsula University of Technology, P.O. Box 1906, Bellville 7535, South Africa, Mobile: +277-11400428, E-mail: oguntibejuo@cput.ac.za

Olukemi Abimbola Okediran
Graduate Student and Research Assistant, Integrated Science, Department Federal college of Education Oyo State, Nigeria, Mobile: +234-8033644916, E-mail: khemmydot@yahoo.com

Folorunso Adewale Olabiyi
Nutrition and Chronic Disease Research Group, Oxidative Stress Research Centre, Department of Biomedical Sciences, Faculty of Health and Wellness Sciences, Cape Peninsula University of Technology (CPUT), Bellville 7535, South Africa, E-mail: folabiyi@gmail.com

Wilfred Mbeng Otang
Senior Lecturer, School of Biology and Environmental Sciences, Faculty of Agriculture and Natural Sciences, University of Mpumalanga, Mbombela Campus, Private bag X11283, Nelspruit 1200, Mpumalanga, South Africa, Tel.: +27-33257274, E-mail: wilfredotang5@yahoo.com

Adeoye Oluwole Oyewopo
Senior Lecturer, Department of Anatomy, Faculty of Basic Medical Sciences, College of Health Sciences, University of Ilorin, Ilorin, Kwara State, Nigeria, Mobile: +234-8033925431, E-mail: oyewopo.ao@unilorin.edu.ng

Maria Rosa Chitolina Schetinger
Professor, Postgraduate Program in Biological Sciences and Biochemical Toxicoloy, Natural Sciences Exatas, Federal University of Santa Mara – University Campus (Centro de Ciências Naturais e Exatas, Universidad de Federal de Santa Maria, Campus Universitário), Camobi, Santa Maria, RS CEP 97105-900, Brazil, Mobile: +55- 5596757775, E-mail: mariachitolina@gmail.com

Toyin Dorcas Udje
Nutrition & Chronic Diseases Research Unit (Phytomedicine & Diabetes), Oxidative Stress Research Centre, Department of Biomedical Sciences, Faculty of Health and Wellness Sciences, Cape Peninsula University of Technology, Bellville, 7535 Cape Town, South Africa, Mobile: +277-89876608, E-mail: toyudoc@yahoo.com, 215299787@mycput.ac.za

LIST OF ABBREVIATIONS

AA	arachidonic acid
AAS	atomic absorption spectrophotometry
AC	adenyl cyclase
ACE	angiotensin-1 converting enzyme
Ach	acetylcholine
Ache	acetylcholinesterase
AD	Alzheimer's disease
ADA	adenosine deaminase
AGEs	advanced glycation end products
AHE	artemisia herba-alba extract
ANOVA	analysis of variance
AR	aldose reductase
ATCC	American Type Culture Collection
ATP	adenosine triphosphate
BChE	butyrylcholinesterase
BGN	bambara groundnut
BP	blood pressure
CAT	catalase
CC	*Carum carvi* L.
CCl_4	carbon tetrachloride
CE	capillary electrophoresis
CKD	chronic kidney disease
CML	carboxymethyl-lysine
CMV	cytomegaloviruses
CNS	central nervous system
COX	cyclooxygenase
CPC	centrifugal partition chromatography
CPH	cyclophosphamide
CPSO	cold-pressed pomegranate seed oil
CS	*Capparis spinosa*
CVD	cardiovascular disease

DAD	diode array detector
DAG	diacylglycerol
DCMM	dichloromethane-methanol
DKD	diabetic kidney disease
DM	diabetes mellitus
DMSO	dimethysulphoxide
DN	diabetic nephropathy
DNA	deoxyribonucleic acid
DNPH	dinitrophenyl hydrazine
DNSA	dinitrosalicylic acid
ED	erectile dysfunction
ENOS	endothelial nitric oxide synthase
ESI	electron spray ionization
ESRD	end stage renal disease
ETC	electron transport chain
FRAP	ferric ion reducing antioxidant power
FRIN	Forest Research Institute of Nigeria
GAPDH	glyceraldehyde-3-phosphate dehydrogenase
GC	gas chromatography
GC-FID	Gas Chromatography Coupled with Flame Ionization Detector
GI	gastrointestinal
GPC	gel permeation chromatography
GPX	glutathione peroxidase
GR	glutathione reductase
GST	glutathione transeferase
HCV	hepatitis C virus
HDL	high density lipoprotein cholesterol
HER2	human epidermal growth factor receptor-2
HOCl	hypochlorous acid
HPGxP	hydroperoxide glutathione peroxidase
HPLC	high performance liquid chromatography
HPV	human papillomavirus
HSCCC	high speed counter current chromatography
IDDM	insulin dependent diabetes mellitus
IL-1	interleukin-1

IRS	insulin receptor substrates
LAB	lactic acid bacteria
LC	liquid chromatography
LDL	low-density lipoprotein
LPO	lipid peroxidation
LPS	lipopolysaccharide
MAO	monoamine oxidase
MCP-1	monocyte chemotactic protein -1
MDA	malondialdehyde
MDA	mass drug administration
MECC	micellar-electrokinetic capillary chromatography
MFC	minimum fungicidal concentration
MI	myocardial infarction
MIC	minimum inhibitory activity
MS	mass spectrometry
NADPH	nicotinamide adenine dinucleotide phosphate
NAFLD	non-alcoholic fatty liver disease
NCDs	non-communicable diseases
NFκB	nuclear factor kappa-light-chain enhancer of activated beta cells
NIDDM	non-insulin-dependent diabetes mellitus
NMR	nuclear magnetic resonance spectroscopy
NO	nitric oxide
NSAIDs	non-steroidal anti-inflammatory drugs
NTPDase	nucleoside triphosphate phosphohydrolase
NUS	neglected and underutilized species
ORAC	oxygen radical absorbance capacity
OS	oxidative stress
PA	*Phyllanthus amarus*
PBGS	porphobilinogen synthase
PC	paper chromatography
PCR	Polymerase chain reaction
PD	Parkinson's disease
PDA	photo diode array
PFJ	pomegranate fermented juice
PFLE	pomegranate flower extract

PJ	pomegranate juice
PK	protein kinase
PKC	protein kinase C
PS	phosphatidylserine
PUFAs	polyunsaturated fatty acids
RAGE	receptor of advanced glycation end products
RAS	renin-angiotensin aldosterone system
RMS	reactive molecular species
RNS	reactive nitrogen species
ROS	Reactive oxygen species
RSS	reactive sulfur species
SD	standard deviation
SDA	sabouraud dextrose agar
SDH	sorbitol dehydrogenase
SFC	supercritical fluid chromatography
SMCS	S-methyl cysteine sulphoxide
SOD	superoxide dismutase
STZ	streptozotocin
TBARS	thiobarbituric acid reactive species
TCA	trichloroacetic acid
TCM	traditional Chinese medicine
TEAC	trolox equivalent antioxidant capacity
TKs	tyrosine kinases
TLC	thin layer chromatography
TNF-α	tumor necrosis factor
TTAE	talinum triangulare aqueous extract
UAE	urine albumin excretion
VA	vernonia amygdalina
VBA	vitalboside A
VCEAC	vitamin C equivalent antioxidant capacity
VEGF	vascular endothelial growth factor
WHO	World Health Organization
ZP	Zona Pellucida

PREFACE 1 BY MEGH R. GOYAL

"25 grams of soy protein a day,
as part of a diet low in saturated fat and cholesterol,
may reduce the risk of heart disease." [USFDA 21CFR101.82]
—Ramabhau Patil, PhD

"Let food be thy medicine and medicine be thy food—
Who said anything about medicine? Let's eat—
That your food should be better than the medicine."
—attributed to Hippocrates
(https://en.wikipedia.org/wiki/Hippocrates)

Quality and nutritional value of foods are highly dependent on environment, agricultural practices, production conditions, and consumer preferences, which all may provide different effects on human health. One of the main challenges of food science and technology is to optimize food production to have a minimal environmental footprint, to lower production costs, and to improv quality and nutritional value.

Therefore, we introduce this book volume *Bioactive Compounds of Medicinal Plants: Properties and Potential for Human Health,* under the book series *Innovations in Plant Science for Better Health: From Soil to Fork.* This book covers mainly the current scenario of the research and case studies especially under conditions of the African continent on: (1) Bioactive compounds in medicinal plants and health benefits: phytochemistry and antifungal activity of *Desmodium adscendens* root extracts/bioactive compounds in plants and their antioxidant capacity, plants of the genus *Syzygium* (Myrtaceae)—review on ethnobotany/medicinal properties and phytochemistry; (2) Medicinal plants/plant products and health promotion: *Talinum triangulare*—hepatic antioxidant gene expression profile in carbon tetrachloride-induced rat liver injury, review on potential of seeds and value added products of bambara groundnut: antioxidant/ anti-inflammatory and anti-oxidative stress, antioxidant and antimicrobial activity of *Opuntia aurantiaca Lindl*, protective effect of *Gongrenema latifolium* on cyclophosphamide-induced stress on male wistar rats brain in vivo; (3) Medicinal plants and management of diabetes mellitus: review of therapeutic potentials of selected medicinal plants in the management of diabetes mellitus and potential of medicinal activities of *Anchomanes difformis* in the treatment of diabetes mellitus and other disease conditions, screening of different extracts of *Ageratum conyzoides* for inhibition of diabetes-related enzymes, potential of *Catharanthus roseus* and *Punica granutum* in the management and treatment of diabetes mellitus; (4) Medicinal plants and management of hypertension: review on ginger/turmeric supplemented diet as a novel dietary approach for management of hypertension: a review.

This book volume sheds light on the potential of medicinal plants from the African continent for human health for different technological aspects, and it contributes to the ocean of knowledge on food science and technology. We hope that this compendium will be useful for the students and researchers as well as those working in the food, nutraceutical, and herbal industries.

The contributions by the cooperating authors to this book volume have been most valuable this compilation. Their names are mentioned in each chapter and in the list of contributors. We appreciate their patience with our editorial skills. This book would not have been written without the valuable cooperation of these investigators, many of whom are renowned scientists who have worked in the field of food engineering and food science throughout their professional careers.

I am glad to introduce Dr. Ademola O. Ayeleso (co-editor of this book), who is Senior Lecturer, Department of Biochemistry, Adeleke University, Ede, Osun State, in Nigeria. With experience in Medicinal Plant Research, Dr. Ayeleso brings his expertise and innovative ideas in this book. Without his support and leadership qualities as editor of this book volume and extraordinary work on *Bioactive Compounds of Medicinal Plants*, readers will not have this quality book.

We will like to thank editorial staff, Sandy Jones Sickels, Vice President, and Ashish Kumar, Publisher and President at Apple Academic Press, Inc., for making every effort to publish the book when the diminishing water resources are a major issue worldwide. Special thanks are due to the AAP Production Staff for typesetting the entire manuscript and for the quality production of this book.

We request that readers offer their constructive suggestions that may help to improve the next edition.

We express our admiration to our families and colleagues for their understanding and collaboration during the preparation of this book volume. As an educator, there is a piece of advice to one and all in the world: *Permit that our almighty God, our Creator, provider of all and excellent Teacher, feed our life with Healthy Food Products and His Grace; and Get married to your profession.*

—Megh R. Goyal, PhD, PE
Senior Editor-in-Chief

PREFACE 2 BY ADEMOLA AYELESO

Plants are known to contain compounds (also referred to as phytochemicals) in the leaves, stems, flowers, and fruits that can help to promote human health. Therefore, medicinal plants/plant products are drawing the attention of researchers/policymakers because of their demonstrated beneficial effects against diseases with high global burdens such as: diabetes, hypertension, cancer, and neurodegenerative diseases (Alzheimer's disease and Parkinson's disease). The side effects associated with conventional medicine have awakened the interest of researchers to explore the medicinal plants as alternative or complementary medicine.

In Africa, medicinal plants play a major role in the traditional health care system due to accessibility and affordability. In some parts of Africa, traditional medicine is still a fundamental part of the health care system because of the availability of an endless list of medicinal plants. People also seek treatments for diseases using natural sources because modern medicine has not been able to provide remedy. These plants were discovered by traditional healers to have active properties against certain diseases mostly by chance or by testimonies from other users. Hence, there is a need for substantial scientific evidence in terms of efficacy, dosage, and safety in order for traditional herbs to have a place in modern medicine.

This book presents scientific reports on the medicinal values of different plants against diseases. It aims to further encourage the development of plant-based drugs through innovative and ground-breaking research studies and, thus, will help to promote the health and economic well-beings of people around the world. The understanding of the therapeutic values of these plants will also help to improve their sustainability, as people and governments will be encouraged to preserve and conserve the plants for future generations. The book covers the phytochemistry and health-promoting potential of the plants against different ailments, such as diabetes, hypertension, and microbial infections. Some of the mechanisms by which these plants exert their beneficial effects are also reported.

I thank Dr. Megh R. Goyal for his leadership qualities and for inviting me to join his team.

—Ademola Olabode Ayeleso, DTech

WARNING/DISCLAIMER

PLEASE READ CAREFULLY

The goal of this volume, *Bioactive Compounds of Medicinal Plants: Properties and Potential for Human Health*, is to offer current knowledge on medicinal plants under African conditions for human health and to provide the world community with information on how to use this know-how efficiently.

The editors, the contributors, and the publisher have made every effort to make this book as accurate as possible. However, there still may be grammatical errors or mistakes in the content or typography. Therefore, the contents in this book should be considered as a general guide and not a complete solution to address any specific situation.

The editors and publisher shall have neither liability nor responsibility to any person, organization, or entity with respect to any loss or damage caused, or alleged to have caused, directly or indirectly, by information or advice contained in this book. Therefore, the purchaser/reader must assume full responsibility for the use of the book or the information therein.

The mention of commercial brands and trade names are only for technical purposes. It does not mean that a particular product is endorsed over another product or equipment not mentioned. The editors and publisher do not have any preference for a particular product.

All weblinks that are mentioned in this book were active at the time of publication. The editors and the publisher shall have neither liability nor responsibility if any of the weblinks are inactive at the time of reading of this book.

INNOVATIONS IN PLANT SCIENCE FOR BETTER HEALTH: FROM SOIL TO FORK BOOK SERIES

Series Editor-in-Chief:
Dr. Hafiz Suleria
Honorary Fellow at the Diamantina Institute,
Faculty of Medicine, The University of Queensland (UQ), Australia
email: hafiz.suleria@uqconnect.edu.au

The objective of this new book series is to offer academia, engineers, technologists, and users from different disciplines information to gain knowledge on the breadth and depth of this multifaceted field. The volumes will explore the fields of phytochemistry, along with its potential and extraction techniques. The volumes will discuss the therapeutic perspectives of biochemical compounds in plants and animal and marine sources in an interdisciplinary manner because the field requires knowledge of many areas, including agricultural, food, and chemical engineering; manufacturing technology along with applications from diverse fields like chemistry; herbal drug technology; microbiology; animal husbandry; and food science; etc. There is an urgent need to explore and investigate the innovations, current shortcomings, and future challenges in this growing area of research.

We welcome chapters on the following specialty areas (but not limited to):

- Alternative and complementary medicines
- Ethnopharmacology and ethnomedicine
- Extraction of bioactive molecules
- Food and function
- Food processing waste/by products: management and utilization
- Functional and nutraceutical perspectives of foods
- Herbals as potential bioavailability enhancer and herbal cosmetics
- Importance of spices and medicinal and functional foods
- Marine microbial chemistry
- Marine phytochemistry
- Natural products chemistry
- Nutritional composition of different foods materials
- Other related includes nuts, seed spices, wild flora, etc.
- Phytomedicinal properties of different plants

- Phytopharmaceuticals for the delivery of bioactives
- Phytopharmacology of plants used in metabolic disorders
- Traditional plants used as functional foods

ABOUT THE BOOK SERIES EDITOR-IN-CHIEF

Dr. Hafiz Suleria is an eminent young researcher in the field of food science and nutrition. Currently, he is an Honorary Fellow at the Diamantina Institute, Faculty of Medicine, The University of Queensland (UQ), Australia. Before joining the UQ, he worked as a lecturer in the Department of Food Sciences, Government College University Faisalabad, Pakistan. He also worked as a Research Associate in a PAK-US Joint Project funded by the Higher Education Commission, Pakistan, and Department of State, USA, with the collaboration of the University of Massachusetts, USA, and the National Institute of Food Science and Technology, University of Agriculture, Faisalabad, Pakistan.

Dr. Suleria's major research focus is on food science and nutrition, particularly in screening of bioactive molecules from different plant, marine, and animal sources, using various cutting-edge techniques, such as isolation, purification, and characterization. He also did research work on functional foods, nutraceuticals, and alternative medicine. He has published more than 60 peer-reviewed scientific papers in different reputed/impacted journals. He is also in collaboration with more than five universities where he is working as a co-supervisor/special member for PhD and postgraduate students and also involved in joint publications, projects, and grants.

Expertise: Food Science and Nutrition, Functional Foods, Bioactive Molecule Research, Marine Bioactives, Nutraceuticals, Medicinal Plants, Natural Product Chemistry, and Drug Discovery

BOOKS IN THE SERIES

- **Bioactive Compounds of Medicinal Plants: Properties and Potential for Human Health**
  Editors: Megh R. Goyal, PhD, and Ademola O. Ayeleso, DTech

- **Plant- and Marine-Based Phytochemicals for Human Health: Attributes, Potential, and Use**
  Editors: Megh R. Goyal, PhD, and Durgesh Nandini Chauhan, MPharm

- **Human Health Benefits of Plant Bioactive Compounds: Potentials and Prospects**
  Editors: Megh R. Goyal, PhD, and Hafiz Ansar Rasul Suleria, PhD

BOOK ENDORSEMENTS

In this book, an assemblage of recent findings on the therapeutic potentials of several medicinal plants/natural products is presented. The editors have selected chapters made of original research and updated reviews from experienced authors who are specialists in the field. Readers are assured of interesting discoveries and an insight into the future position of medicinal plants/natural products in modern therapeutics. I recommend the book.

—Pius Fasinu, PhD
Pharmacologyist, National Center for Natural Products Research
School of Pharmacy, University of Mississippi, USA

This reference book is a dependable, broad and extremely educational presentation of recent progress in various vital up-and-coming areas of medicinal plants and health. It has exceptional chapters written by erudite scholars. Therefore, this book can serve as a reference point for both students and lecturers in all higher institutions.

—Bashiru Olaitan Ajiboye, PhD
Biochemistry Program, Department of Chemical Sciences,
Afe Babalola Programme, Ado-Ekiti, Ekiti State, Nigeria

OTHER BOOKS ON AGRICULTURAL & BIOLOGICAL ENGINEERING FROM APPLE ACADEMIC PRESS, INC.

Management of Drip/Trickle or Micro Irrigation
Megh R. Goyal, PhD, PE, Senior Editor-in-Chief

Evapotranspiration: Principles and Applications for Water Management
Megh R. Goyal, PhD, PE, and Eric W. Harmsen, Editors

Book Series: Research Advances in Sustainable Micro Irrigation
Senior Editor-in-Chief: Megh R. Goyal, PhD, PE

Volume 1: Sustainable Micro Irrigation: Principles and Practices
Volume 2: Sustainable Practices in Surface and Subsurface Micro Irrigation
Volume 3: Sustainable Micro Irrigation Management for Trees and Vines
Volume 4: Management, Performance, and Applications of Micro Irrigation Systems
Volume 5: Applications of Furrow and Micro Irrigation in Arid and Semi-Arid Regions
Volume 6: Best Management Practices for Drip Irrigated Crops
Volume 7: Closed Circuit Micro Irrigation Design: Theory and Applications
Volume 8: Wastewater Management for Irrigation: Principles and Practices
Volume 9: Water and Fertigation Management in Micro Irrigation
Volume 10: Innovation in Micro Irrigation Technology

Book Series: Innovations and Challenges in Micro Irrigation
Senior Editor-in-Chief: Megh R. Goyal, PhD, PE

- Micro Irrigation Engineering for Horticultural Crops: Policy Options, Scheduling and Design

- Micro Irrigation Management: Technological Advances and Their Applications
- Micro Irrigation Scheduling and Practices
- Performance Evaluation of Micro Irrigation Management: Principles and Practices
- Potential of Solar Energy and Emerging Technologies in Sustainable Micro Irrigation
- Principles and Management of Clogging in Micro Irrigation
- Sustainable Micro Irrigation Design Systems for Agricultural Crops: Methods and Practices
- Engineering Interventions in Sustainable Trickle Irrigation: Water Requirements, Uniformity, Fertigation, and Crop Performance

Book Series: Innovations in Agricultural & Biological Engineering
Senior Editor-in-Chief: Megh R. Goyal, PhD, PE

- Dairy Engineering: Advanced Technologies and Their Applications
- Developing Technologies in Food Science: Status, Applications, and Challenges
- Engineering Interventions in Agricultural Processing
- Engineering Practices for Agricultural Production and Water Conservation: An Inter-disciplinary Approach
- Emerging Technologies in Agricultural Engineering
- Flood Assessment: Modeling and Parameterization
- Food Engineering: Emerging Issues, Modeling, and Applications
- Food Process Engineering: Emerging Trends in Research and Their Applications
- Food Technology: Applied Research and Production Techniques
- Modeling Methods and Practices in Soil and Water Engineering
- Processing Technologies for Milk and Dairy Products: Methods Application and Energy Usage
- Soil and Water Engineering: Principles and Applications of Modeling
- Soil Salinity Management in Agriculture: Technological Advances and Applications
- Technological Interventions in the Processing of Fruits and Vegetables
- Technological Interventions in Management of Irrigated Agriculture
- Engineering Interventions in Foods and Plants

- Technological Interventions in Dairy Science: Innovative Approaches in Processing, Preservation, and Analysis of Milk Products
- Novel Dairy Processing Technologies: Techniques, Management, and Energy Conservation
- Sustainable Biological Systems for Agriculture: Emerging Issues in Nanotechnology, Biofertilizers, Wastewater, and Farm Machines
- State-of-the-Art Technologies in Food Science: Human Health, Emerging Issues and Specialty Topics
- Scientific and Technical Terms in Bioengineering and Biological Engineering
- Engineering Practices for Management of Soil Salinity: Agricultural, Physiological, and Adaptive Approaches
- Processing of Fruits and Vegetables: From Farm to Fork

PART I

BIOACTIVE COMPOUNDS IN MEDICINAL PLANTS AND HEALTH BENEFITS

CHAPTER 1

PHYTOCHEMISTRY AND ANTIFUNGAL ACTIVITY OF *DESMODIUM ADSCENDENS* ROOT EXTRACTS

NWANEKWU KENNETH EMEKA, ADENIYI BOLA, and MAHADY GAIL

CONTENTS

1.1 INTRODUCTION

Medicinal plants have long been used for centuries as healthcare products. Traditional medicine of many communities of the world, however, may have originated from the age-long link with nature and natural products, of which plants and plant materials form a greater part [7]. A number of people rely on herbs for the treatment of various diseases ranging from common cold to other infectious diseases caused by viruses, bacteria, protozoa, and fungi, which are endemic in the communities [5]. of particular concern in this study is community-based fungal infections. Among these infections, dermatomycoses commonly called ringworm and candidiasis are the most common mycoses in the rural communities. These infections are widespread and are difficult to control most of the times with commonly available antibiotics, thereby necessitating the need for alternative treatments [9]. *Desmodium adscendens* (a member of the family Fabaceae) is a perennial leguminous plant growing in the form of a multibranched shrub up to 50 cm in height in the equatorial and circum-equatorial zones of Central and South America and Africa [3]. It grows in small areas of the Andes at elevations of 1000 to 3000 m above sea level and produces numerous light purple flowers and green fruits in small, bean-like pods. It is indigenous to many tropical countries and grows in open forests, pastures along roadsides, and, like many weeds, just about anywhere the soil is disturbed. Today, many tribes in many parts of the world use *D. adscendens* as a medicine for various purposes. A tea of the plant is given for treating nervousness, and it is used in baths to treat vaginal infections [8], promote lactation in women, and treat wounds, sores, malaria, diarrhea, and ovarian and uterine problems [10]. Most of the antimicrobial studies carried out on this plant have focused on leaves and were mostly against bacterial pathogens.

This chapter is aimed at screening the root extracts of *D. adscendens* (Strongback tick-trefoil) for its antifungal properties against fungal pathogens with the goal of developing alternative treatment options.

1.2 MATERIALS AND METHODS

1.2.1 PLANT MATERIALS

The plant samples used in this research were bought from local traders at Oje market, Ibadan, Nigeria. All samples were identified, authenticated, and stored at the Forestry Research Institute of Nigeria (FRIN) herbarium,

Ibadan, Nigeria, with voucher number FHI 108301. They were air dried and ground to powder.

1.2.2 MICROORGANISMS

The microorganisms in this research were clinical strains of *Candida valida, Candida albicans, Candida pseudotropicalis, Candida tropicalis, Candida glabrata, Candida krusei, Trychophyton rubrum, Trychophyton tonsurans, Trychophyton interdigitale*, and *Epidermophyton floccosum* that were obtained from the University College Hospital and American Type Culture Collection (ATCC) strains of *Candida parapsilosis* ATCC 22011, *C. krusei* ATCC 6825, and *C. albicans* ATCC 90029.

1.2.3 EXTRACTION OF PLANT ACTIVE COMPONENTS

The air-dried grinded root samples weighing 600 g were Soxhlet extracted with hexane and methanol solvents in succession for 6 hours each according to the method by Adeniyi et al. [1]. The extracts were concentrated by evaporation under vacuum until dry and then stored in the refrigerator at 4°C for further use.

1.2.4 PHYTOCHEMICAL SCREENING

The dried grinded plant samples were subjected to phytochemical analysis to determine the presence of secondary metabolites such as saponins, tannins, alkaloids, cardenolides, and anthraquinones. Standard procedures were employed in the screening process as described by Sonibare et al. [13].

1.2.5 ANTIFUNGAL SCREENING

The antifungal screening of the plant extracts was carried out using the agar diffusion technique as described by Adeniyi et al. [2]. Sabouraud dextrose agar (SDA) plates were seeded with 0.2 mL of 1/100 dilution of an overnight culture of the fungal isolates and left to stand. A standard cork borer of 6 mm diameter was used to bore uniform wells on the agar surface, and 1000 μg/mL concentration of the test extract was filled in each well. The antifungal antibiotics griseofulvin, tioconazole, ketoconazole, and nystatin (250 μg/mL) were also added to the agar wells as control. The plates were left to stand for an hour to allow for sufficient diffusion of the extract into the medium before incubation. All plates seeded were incubated at 25°C for 48 hours. At the end of the incubation period, diameters of the zone of inhibition observed were measured in millimeters.

1.2.6 DETERMINATION OF MINIMUM INHIBITORY CONCENTRATION (MIC) AND MINIMUM FUNGICIDAL CONCENTRATION (MFC)

The minimum inhibitory concentration (MIC) of the active plant extracts was determined using the broth dilution method as described by Obiukwu and Nwanekwu [12]. The extracts were serially diluted using Mueller Hinton broth to get varying concentrations of 1000, 500, 250, 125, and 62.5 μg/mL in different test tubes. The tubes were subsequently inoculated with 0.1 mL suspension of the test organism and incubated at 25°C for 48 hours. The least concentration of the extract at which no detectable microbial growth was observed was recorded as the MIC.

Furthermore, from the tubes showing no visible detectable microbial growth, 0.1 mL of the content was plated out onto the surface of an agar medium and then incubated for 24 to 48 hours at 25°C. The minimum fungicidal concentration (MFC) was taken as the lowest concentration without growth on the agar plate. Fungicidal activity of the plant extracts with time (kill kinetics) was determined with the viable counting technique.

1.3 RESULTS

The phytochemical screening of the powdered plant samples to determine the presence of secondary metabolites revealed the presence of tannins, saponins, alkaloids, and anthraquinones (Table 1.1).

Table 1.2 presents the results of the antifungal screening test and the MIC and MFC of the plant extract and the antibiotic ketoconazole against the test fungal isolates. The plant extracts showed antifungal activity against all fungal isolates except *C. valida* and *C. glabrata*, with the best activity exhibited against *C. albicans*. This is evident in the highest inhibition zone diameter recorded at 25 mm. A comparative analysis showed that the plant extract was less active against the standard strains and dermatophytes than

TABLE 1.1 Phytochemical Composition of *Desmodium adscendens* Root extracts

Metabolites	Saponins	Tannins	Alkaloids	Cardenolides	Anthraquinone
Results	Yes	Yes	Yes	No	Yes

Yes = Present; No = Absent.

TABLE 1.2 Antifungal Activity of *Desmodium adscendens* Extracts Against Fungal Species Showing Inhibition Zone Diameter (mm), MIC (µg/mL) and MFC (µg/mL)

Test organism	Diameter zone of inhibition (mm)	MIC (µg/mL)	MFC (µg/mL)	Antibiotic control (Ketoconazole) (mm)
C. albicans	25	250	500	12
C. krusei	21	250	1000	-R
C. glabrata	–	–	–	-R
C. tropicalis	22	250	500	-R
C.pseudotropicalis	19	125	500	-R
C. valida	–	–	–	–
C. parapsilosis ATCC 22019	15	125	500	-R
C. krusei ATCC 6825	18	125	500	-R
C.albicans ATCC90029	15	125	250	-R
T.rubrum	17	250	500	-R
T. interdigitales	17	250	1000	12
T.tonsurans	18	250	1000	20
E. floccosum	18	125	1000	-R

against the clinical strains. The MIC of the plant extract ranged from 125 to 250 μg/mL and the MFC ranged from 250 to 1000 μg/mL.

Four conventional antifungal drugs (ketoconazole, griseofulvin, tioconazole, and nystatin) were used as control drugs. of these four drugs, only ketoconazole showed activity against some of the fungal species, while the remaining drugs were inactive.

The kill kinetics of the active plant extracts against the organisms are presented in Figures 1.1 and 1.2. These figures show a plot of the percentage number of surviving cells of the test organism with time of exposure. This analysis revealed a reduction in the number of viable organisms with an increase in exposure time.

1.4 DISCUSSION

The extracts had varying degrees of antifungal activities against the tested fungi. This was confirmed by the varying diameter zones of inhibition of the extract on the fungal pathogens in the in vitro test. The plant extract possessed plant metabolites such as tannins, saponins, alkaloids, and anthraquinones, which have been shown to possess antimicrobial properties [4, 6]. In this study, the antifungal activity of the

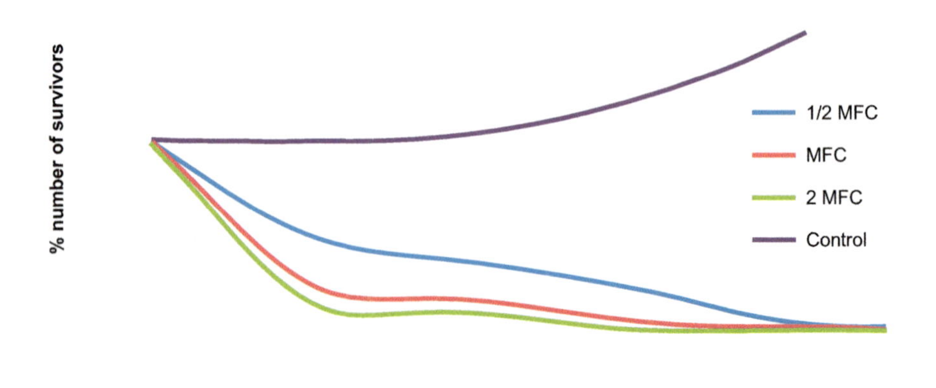

FIGURE 1.1 Kinetic kill curve of *D. adscendens* extracts against *T. rubrum.*

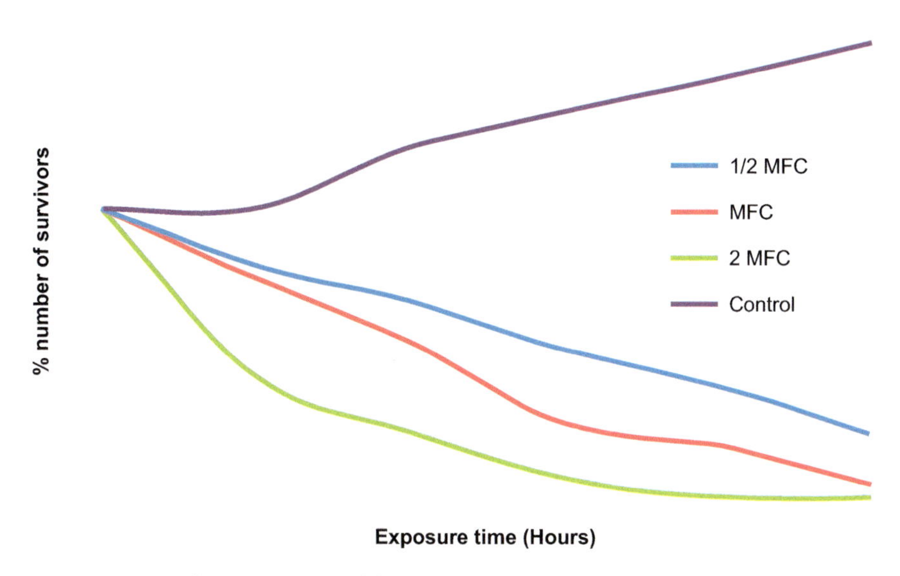

FIGURE 1.2 Kinetic kill curve of *D. adscendens* extracts against *C. tropicalis*.

extract was observed to be enhanced with an increase in the concentration of the extract. Furthermore, the antifungal properties reported in this chapter are in agreement with those reported by Muanda et al. [11] and Rastogi et al. [14], who at different times studied the antimicrobial and ethnopharmacological properties of the *D. adscendens* plant. They reported that the plant extract demonstrated considerable antimicrobial activity against common microbial pathogens. The antifungal activity demonstrated by *D. adscendens* gives credence to its traditional medicinal use for the treatment of venereal sores, vaginal discharge, painful urination, and other sexually transmitted infections [8, 14]. The higher activity observed against the clinical strains of the test organisms is noteworthy as it indicates that the plant extracts could be effectively applied in clinical use in infection management. The results of the comparative study between the test extract and the available antibiotics is of significance as the test organisms showed high levels of resistance to these conventional drugs, thus buttressing the significance of this study as a search for alternative therapy to these infections.

1.5 CONCLUSIONS

As stated earlier, many people in the developing nations lack access to essential medicine and thus rely on herbal therapy. However, while herbal remedies provide cheap and available treatment of infections, care must be taken to ensure that appropriate regulations are implemented and adhered to, so as to maximize their inherent benefits while reducing any adverse reactions.

1.6 SUMMARY

The phytochemical properties and antifungal activity of the methanol extracts of *D. adscendens* roots were investigated against common fungal pathogens implicated in various mycoses. These fungal species included *C. valida, C. pseudotropicalis, C. tropicalis, C. glabrata, C. krusei, C. albicans, Trychophyton rubrum, T. tonsurans, T. interdigitale,* and *E. floccosum* and standard typed strains of *C. parapsilosis* ATCC 22011, *C. krusei* ATCC 6825, and *C. albicans* ATCC 90029. The preliminary phytochemical screening revealed the presence of tannins, saponins, alkaloids, and anthraquinones. The extract was active against most of the test organisms except *C. valida* and *C. glabrata*, with a high activity exhibited against *C. albicans*. The zone of inhibition diameter ranged from 15 to 25 mm. The MIC and MFC values of the extracts ranged from 25 to 125 μg/mL and 25 to 500 μg/mL, respectively. The antifungal drugs griseofulvin, tioconazole, and nystatin used as controls were inactive against the organisms except ketoconazole, which was active against only four of the organisms tested. Thus, this confirms the antifungal potential of *D. adscendens* and its capability of being a good replacement for the conventional drugs as an alternative if properly processed.

ACKNOWLEDGMENTS

This work was financially supported by a grant from the University of Ibadan Senate research grant number SRG/COM/2006/10A.

KEYWORDS

- **antibiotics**
- **antifungal**
- **ATCC**
- **candidiasis**
- **dermatomycoses**
- *Desmodium adscendens*
- **Fabaceae**
- **medical plant**
- **metabolites**
- **methanol**
- **MFC**
- **MIC**
- **pathogens**
- **phytochemistry**

REFERENCES

1. Adeniyi, B. A., Odelola, H. A., & Oso, B. A., (1996). Antimicrobial potentials of *Diospyros mespiliformis (Ebenaceae). Afr. J., Med. Sc., 25*, 221–224.
2. Adeniyi, B. A., Fong, H. H., S., Pezzuto, J. M., Luyengi, L., & Odelola, H. A., (2000). Antibacterial activity of diospyrin, isodiospyrin and bisisodiospyrin from *Diospyros piscatoria* (Girke) *[Ebenaceae]. Phytotherapy Research, 14*, 112–117.
3. Addy, M. E., (1990). Effects of *Desmodium adscendens* fraction on contractions of respiratory smooth muscles. *J. Ethnopharmacol, 29*(3), 325–335.
4. Addy, M. E., (1992). Some secondary plant metabolites in *Desmodium adscendens* and their effects on arachidonic acid metabolism. Prostaglandins leukotrienes. *Essent. Fatty Acids, 47*(1), 85–91.
5. Barreto, G. S., (2002). Effect of butanolic fraction of *Desmodium adscendens* on the anococcygeus of the rat. *Braz. J. Biol., 62*(2), 223–230.
6. Ghosal, S., & Bhattacharya, S. K., (1972). Desmodium alkaloids, part II: Chemical and pharmacological evaluations of *Desmodium gangeticum. Planta. Medica., 22*(4), 434–440.
7. James, A. D., (2007). Returning to our medicinal roots. *Mother Earth News*, 26–33.

8. Leslie, T., (2005). *The Healing Power of Rainforest Herbs*. Tropical plant database. Carson city, NV, 89701.
9. Marples, M. J., & Di Menna, M. E., (1956). The incidence of *Candida albicans* in Dunedin, New Zealand. *J. Path. Bact., 64*, 497.
10. Muanda, F. N., Bouayed, J., Djilani, A., Yao, C. Y., Soulimani, R., & Dicko, A., (2010). Chemical composition and, cellular evaluation of the antioxidant activity of *Desmodium adscendens* leaves. *ECAM,* 1–9.
11. Muanda, F. N., Soulimani, R., & Dicko, A., (2011). Study on biological activities and chemical composition of extracts from *Desmodium adscendens* leaves. *Journal of Natural Products, 4*, 100–107.
12. Obiukwu, C. E., & Nwanekwu, K. E., (2010). Evaluation of the antimicrobial potentials of 35 medicinal plants from Nigeria. *International Science Research Journal, 2*, 48–51.
13. Sonibare, M. A., Soladoye, M. O., Oyedokun, O. E., & Sonibare, O. O., (2009). Phytochemical and antimicrobial studies of four species of Cola Schott and Endl. (*sterculiaceae*). *Afr. J. Traditional, Complementary and Alternative Medicines, 6*(4), 518–525.
14. Rastogi, S., Pandey, M. M., & Rawat, A. K., (2011). An ethno medicinal, phytochemical and pharmacological profile of *Desmodium gangeticum* (L.) DC and *Desmodium adscendens* (Sw.) DC. *J. Ethnopharmacol., 136*(2), 283–296.

CHAPTER 2

BIOACTIVE COMPOUNDS IN PLANTS AND THEIR ANTIOXIDANT CAPACITY

OLUYEMISI ELIZABETH ADELAKUN,
ISLAMIYAT FOLASHADE BOLARINWA,
and JOHNSON AKINWUMI ADEJUYITAN

CONTENTS

2.1 INTRODUCTION

Plant species provide a large variety of bioactive compounds, and they act not only as a source of food but also as medicinal and supplementary healthcare products. The consumption of plant foods has been linked with reduction in the risk of developing chronic diseases such as cancer, diabetes, obesity, and cardiovascular diseases [2]. Regular intake of these plant foods in our daily life has been reported to have the potential of preventing free radical-mediated degenerative diseases [59]. The capacity of some plant-derived foods to reduce the risk of chronic diseases has been associated partly to the occurrence of secondary metabolites (phytochemicals) that have been shown to exert a wide range of biological activities. Although these metabolites may have low potency as bioactive compounds when compared with pharmaceutical drugs, when they are consumed regularly and in significant amounts as part of the diet, they may have a noticeable long-term physiological effect [24].

The bioactive compounds derived from the plant secondary metabolism have a clear therapeutic potential by contributing toward antioxidant activities. Generally, antioxidants are an important class of compounds that when present at low concentrations relative to an oxidizable substrate, significantly delay, retard, or inhibit oxidation of that substrate [31]. Phenolic antioxidants in particular are compounds that act as terminators of free radicals [19]. Free radicals have extremely high chemical reactivity due to the presence of unpaired electrons, which explains why and how they inflict damage on cells [48]. Free radicals have been reported to be the cause of several diseases such as liver cirrhosis, atherosclerosis, cancer, and diabetes, while bioactive compounds that elicit high free radical scavenging potential could play a vital role in ameliorating these disease processes [9, 27, 37, 61]. Thus, antioxidants can play an important role to protect the human body against damage induced by reactive oxygen species (ROS) and oxidative stress (OS) [65].

ROS, which are the natural byproducts of mitochondrial respiration and other cellular processes, have high chemical reactivity. Therefore, when they occur in excess of normal needs in the cell, they may damage the cell's structural and functional integrity. They usually do this either by directly modifying cellular DNA, proteins, and lipids or by initiating chain reactions that can cause extensive oxidative damage of these critical molecules [13, 32, 56]. Beneficial effects of ROS in low concentration involve defense against microbial pathogens [70]. In healthy individuals, the generation of ROS is well balanced by the counterbalancing act of antioxidant defenses. However, if there is an imbalance between the ROS produced and the antioxidant status of an individual, a process referred to as OS occurs [32, 43]. Plant foods serve as sources of a wide variety of dietary antioxidants such as vitamins C and E, carotenoids, flavonoids, and other phenolic compounds.

This chapter reviews potential of bioactive compounds in plants and their antioxidant properties under African conditions.

2.2 REACTIVE OXYGEN SPECIES AND OXIDATIVE STRESS

Free radicals are defined as atoms, molecules, or ions possessing unpaired electrons, which are very unstable and active toward chemical reactions with other molecules [31, 41]. Free radicals are derived from three elements, namely oxygen, nitrogen, and sulfur, thus creating ROS, reactive nitrogen species (RNS), and reactive sulfur species (RSS).

- ROS include free radicals like the superoxide anion ($O_2^{\cdot-}$), hydroperoxyl radical ($HO_2^{\cdot}$), hydroxyl radical (·OH), nitric oxide (NO), and other species like hydrogen peroxide (H_2O_2), singlet oxygen ($^1O_2$), hypochlorous acid (HOCl), and peroxynitrite ($ONOO^-$). RNS derive from NO by reacting with $O_2^{\cdot-}$ and forming $ONOO^-$. RSS are easily formed by the reaction of ROS with thiols [41].
- The ROS represent the most important class of radical species generated in living systems [4]. ROS are formed continuously as byproducts of normal metabolic reactions, and their formation can be accelerated by accidental exposure to occupational chemicals like pesticides. Figure 2.1 shows some of the causes of generation of free radicals, their possible cellular targets, and their consequences [26]. When an individual is healthy, the ROS generated is well balanced by the counterbalancing act of antioxidants. However, when there is an imbalance between ROS generation and the antioxidant status of that individual, a condition termed OS occurs [26, 43].

2.3 ANTIOXIDANTS AND REACTIVE OXYGEN SPECIES

According to Cornelli [16], an antioxidant is a product that inhibits oxidation in vitro and reduces OS in vivo. Antioxidant scavenging activity is the first line of defense against ROS that are generated through physiological

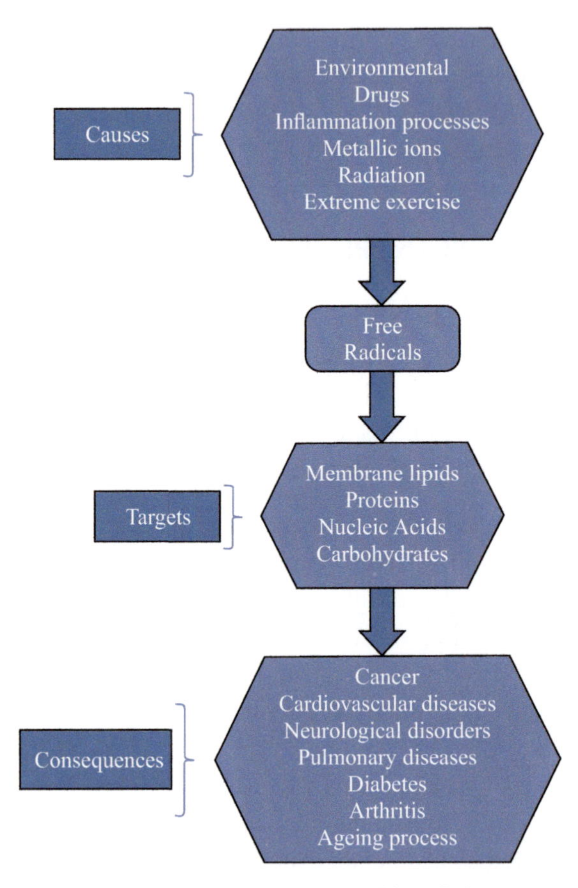

FIGURE 2.1 Some causes of oxidative stress, possible cellular targets, and conditions that are associated with oxidative stress.

processes and can also protect against these ROS by their chain-breaking properties. In this type of protection, antioxidants react with the reactive species. The rate of the reaction of an antioxidant with the reactive species determines the effectiveness of that antioxidant [8].

The adverse effects of ROS are represented by OS that can arise due to a lack of antioxidant defense or by an increase in oxidative processes in the body. The OS occurs when there is an imbalance between the generation of ROS and the antioxidant defense systems in the body, so that the latter become overwhelmed [36]. Many different illnesses (such as cardiovascular disease, cancer, and neurological and endocrinological disorders) have been related to OS, which can be either a cause or a consequence of the disease [16]. Fat-containing food materials can also deteriorate from oxidation, thereby reducing their nutritional quality; synthetic or natural antioxidants can be used in order to prevent such deterioration [40]. Many researchers have demonstrated other benefits of antioxidants, which have been reviewed by Wootton-Beard and Ryan [71], Yoshihara et al. [73], and Moure et al. [49]. Antioxidants can prevent loss of pigments and improve the stability of pigments from red beet juice in the food industry, protect and stabilize aroma compounds in food, and can also be applied in oral pharmaceutical adjuncts to limit infertility [68] and as cosmetic compositions.

2.4 ANTIOXIDANT DEFENSE SYSTEMS

Production of ROS occurs during normal cell metabolism and after exposure to various physical and chemical agents. These are counteracted by antioxidant enzyme systems such as [5] glutathione peroxidase (GPX), glutathione reductase (GR), glutathione transferase (GST), and catalase (CAT) and by nonenzymatic antioxidant systems in the organism. Under these circumstances, antioxidant enzymes in cells act as a defense against ROS by transforming them into inactive and harmless compounds. Another important antioxidant defense enzyme against OS in cells is superoxide dismutase (SOD). The superoxide ion can react with nitric oxide (NO) to produce peroxynitrite, which is an oxidant and more toxic than superoxide. The SOD turns superoxide into the less harmful H_2O_2, and then, CAT decomposes H_2O_2 [69].

Although the human body is equipped with antioxidant defense enzyme systems mentioned above, which deactivate highly reactive free radicals, when these antioxidant defense enzymes are overwhelmed, antioxidant nutrients (found in plant foods) are needed to combat these free radicals and turn them into harmless or waste products that can then be eliminated through the liver. These antioxidant nutrients are therefore functional components of food, which can offer additional health benefits [52].

2.5 BIOACTIVE COMPOUNDS FROM PLANTS

Generally, bioactive compounds of plants are defined as secondary metabolites that cause pharmacological or toxicological effects in human and animals; they can be identified and characterized from extracts of plant roots, stem, bark, leaves, flowers, fruits, and seeds [10]. The qualitative and quantitative studies of bioactive compounds from plant materials depend mostly on the selection of proper extraction methods. These methods are usually affected by common factors such as: the matrix properties of the botanical source, solvent, temperature, pressure, and time [33]. Bioactive plant compounds can be classified according to the type of extraction [11]:

- Hydrophilic or polar compounds (e.g., phenolic acids, flavonoids, organic acids, sugars); and
- Lipophilic or nonpolar compounds (e.g., carotenoids, alkaloids, terpenoids, fatty acids, tocopherols, steroids).
- They can also be classified according to their distribution in nature:
- Narrowly distributed (simple phenols, pyrocatechol, aldehydes);
- Widely distributed (flavonoids, phenolic acids); and
- The least abundant polymers (tannin and lignin).

2.6 PHENOLIC COMPOUNDS

Phenolic compounds are part of the important group of secondary metabolites produced by plants. They are characterized by at least one aromatic ring (C-6) that bears one or more hydroxyl groups [45]. Biosynthesis of phenolic compounds in plants through aromatic amino acids commences via the shikimate pathway. The significance of this pathway is demonstrated by the fact that, under normal growth conditions, 20% of the carbon fixed by plants flows through this route. The aromatic amino acids phenylalanine, tyrosine, and tryptophan are formed via this pathway, which are later utilized for protein synthesis or transformed via phenylpropanoid metabolism to secondary metabolites such as phenolic compounds [22] (Figure 2.2).

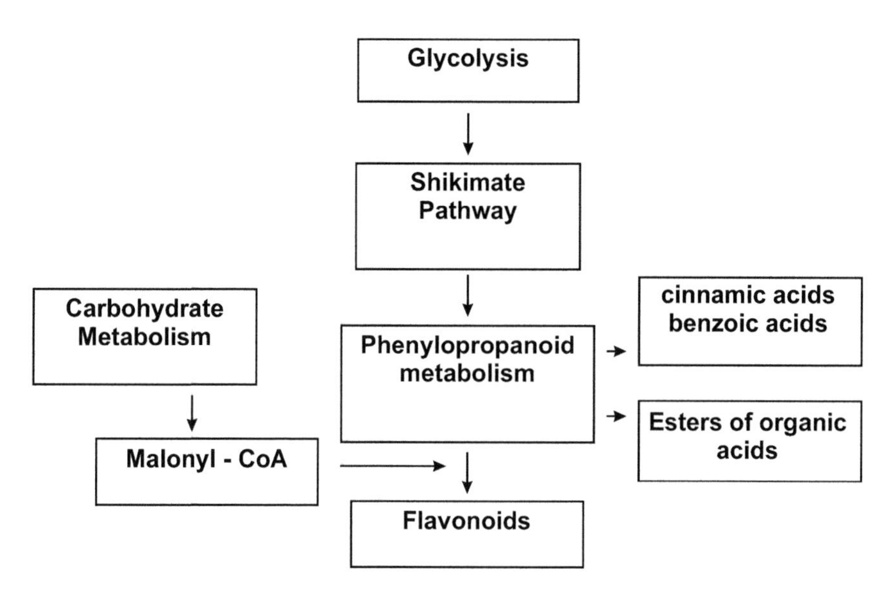

FIGURE 2.2 Biosynthesis pathways leading to the formation of main groups of phenolic compounds.

Phenolic compounds are divided into several groups and are based on the number of constitutive carbon atoms in conjunction with the structure of the basic phenolic skeleton [45]. Figures 2.3–2.7 show the structures of some of major groups of phenolic compounds.

Flavonoids are the most common and widely distributed group of plant phenolic compounds. Their general structure, as phenylpropanoid derivatives, consists of a 15-carbon skeleton with two benzene rings linked via a heterocyclic pyran ring (Figures 2.3–2.6). They comprise a large class of plant secondary metabolites with relevant action on the plant defense system and are reported as important constituents of the human diet [28]. The chemical nature of flavonoids depends on their structural class, according to the degree of methylation, hydrogenation, and hydroxylation and other substitutions and conjugations. Flavonoids can be divided into a variety of classes such as flavones, flavanols, flavonols, flavanonols, flavanones, isoflavones, or anthocyanins (Figures 2.3–2.6) [38, 60] with different antioxidant, antibacterial, antiviral, and anticancer activities (Figure 2.4) [28, 38].

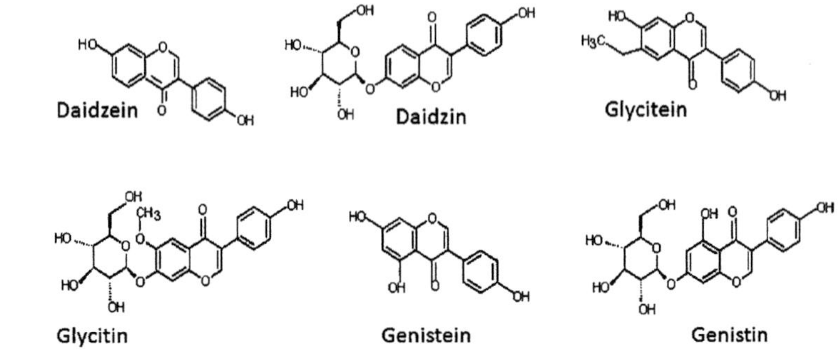

Daidzein Daidzin Glycitein

Glycitin Genistein Genistin

ISOFLAVONES

FIGURE 2.3 Examples of naturally occurring flavonoids: Isoflavones.

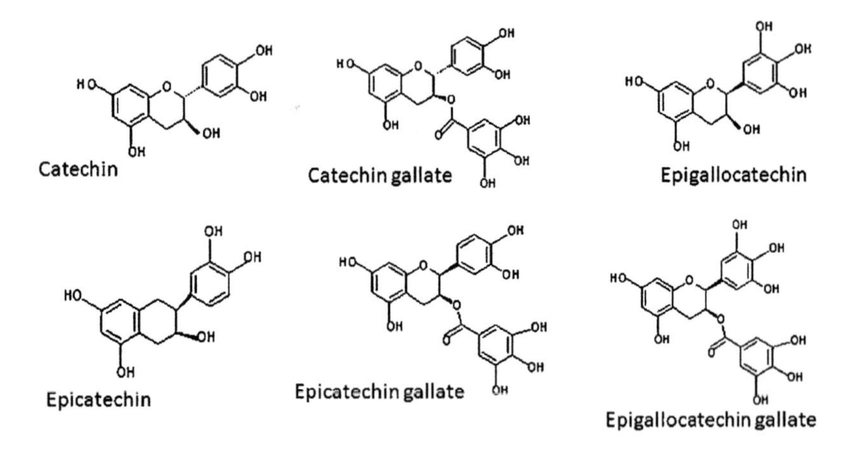

FLAVANOLS

FIGURE 2.4 Examples of naturally occurring flavonoids: Flavanols.

When added to food products, flavonoids are responsible for preventing fat oxidation and protecting vitamins and enzymes, while also contributing to food color and taste (Figure 2.5).

Phenolic acids have a benzene ring, a carboxylic group, and one or more hydroxyl and/or methoxyl groups and are usually divided in two groups: (i) benzoic acids, and (ii) hydroxycinnamic acids. The benzoic acids have seven carbon atoms (C6-C1), while hydroxycinnamic acids have nine carbon atoms (C6-C3), but the most common types found in vegetables have

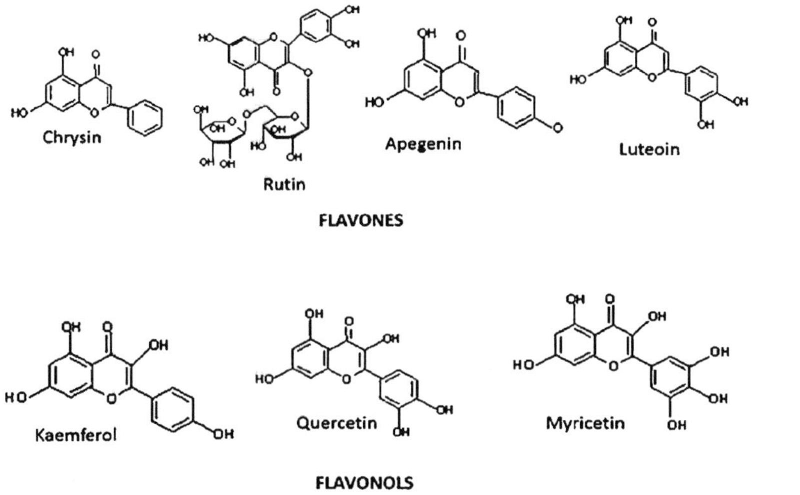

Chrysin
Rutin
Apegenin
Luteoin
FLAVONES
Kaemferol
Quercetin
Myricetin
FLAVONOLS

FIGURE 2.5 Examples of naturally occurring flavonoids: Flavones and Flavonols.

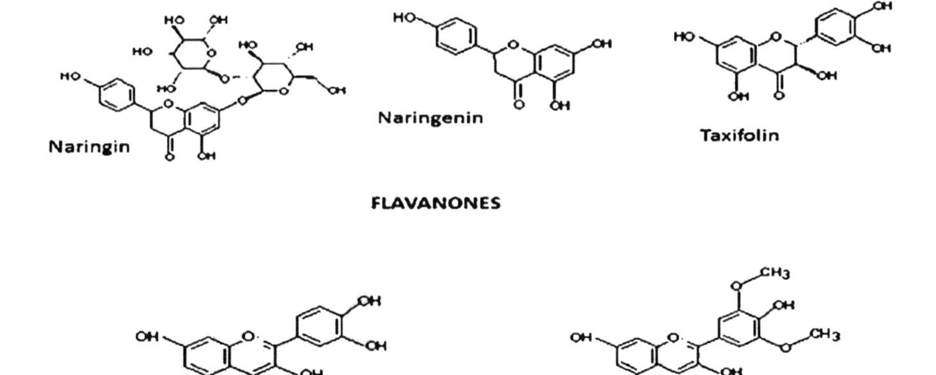

Naringin
Naringenin
Taxifolin
FLAVANONES
Cyanidin
Malvidin
ANTHOCYANIDINS

FIGURE 2.6 Examples of naturally occurring flavonoids: Flavanones and Anthocyanidins.

seven carbon units [72] (Figure 2.7), where they are rarely found in the free form (commonly present as esterified compounds). Phenolic acids may be about one-third of the phenolic compounds in the human diet, where these substances have a high antioxidant activity [11, 28].

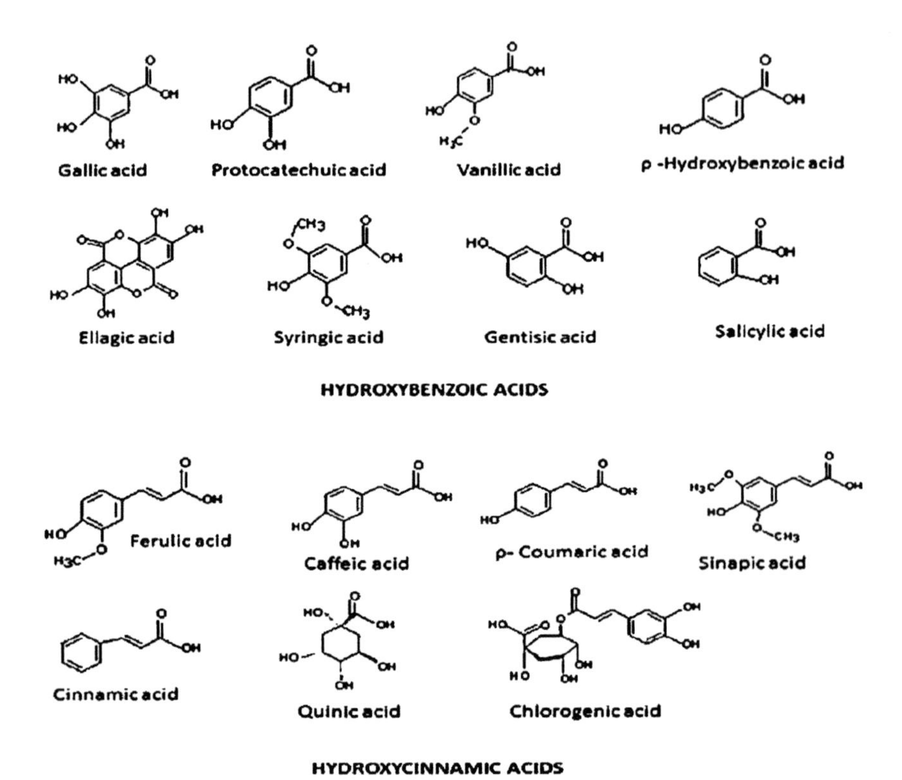

FIGURE 2.7 Examples of naturally occurring phenolic acids: Hydroxybenzoic and Hydroxycinnamic acids.

Tannins are another well-known example of phenolic compounds with intermediate to high molecular weight. These compounds are usually classified into two major groups: (i) hydrolysable tannins, and (ii) nonhydrolyzable tannins or condensed tannins [15]. Hydrolysable tannins have a center of glucose or a polyhydric alcohol partially or completely esterified with gallic acid or hexahydroxydiphenic acid, forming gallotannin and ellagitannins [54]. Condensed tannins are polymers of catechin and/or leucoanthocyanidin that constitute the main phenolic fraction responsible for the characteristic astringency of vegetables. These substances are polymeric flavonoids that form the anthocyanidin pigments [63].

2.7 MEASUREMENT OF ANTIOXIDANT CAPACITY OF BIOACTIVE COMPOUNDS

Plants provide a large variety of bioactive compounds with substantial applications in the area of health and food and are good sources of phenolic compounds with nutritional and therapeutic applications. There is a strong correlation between antioxidant activity and the total phenolic content in the plants, which suggests that phenolic compounds could be the major contributor of their antioxidant capacity. Phenolic compounds are widely distributed in plants, usually with a higher concentration in leaves and green stems. These compounds are considered natural defense substances, and their concentration in each plant may be influenced by several factors such as physiological variations, environmental conditions, geographic variation, genetic factors, and evolution [42].

Phenolic compounds form an important class of compounds, which serve to inhibit the oxidation of material of both commercial and biological importance [51]. Phenolic compounds do not have intrinsic antioxidant activity unless substitution at either the *ortho*- or *para*-position increases the electron density at the hydroxyl group and lowers the O–H bond energy. The antioxidant efficiency of phenolic acids has been related to the number of hydroxyl groups in the molecule and also to their hydrogen radical donating abilities [46]. For instance, in the hydroxycinnamic acids group (Figure 2.7), the presence of electron donating groups on the benzene ring of ferulic acid (3-methoxy and 4-hydroxyl) contributes to its property of terminating free radical chain reactions. Plant foods provide a wide variety of dietary antioxidants. Health benefits of phenolic compounds are reported to be delivered by several mechanisms such as:

- Free radical scavenging;
- Protection and regeneration of other dietary antioxidants (i.e., vitamin E); and
- Chelating of pro-oxidant metal ions.

The species and levels of phenolic compounds vary dramatically among the plant materials. Phenolic compounds with different structures or levels are likely to have different functional properties. Analyzing the antioxidant capacity of phenolic compounds in plant materials before their health promoting properties can be adequately studied is therefore necessary. Analysis of phenolic compounds in plant samples is difficult because of variation in their structure and lack of appropriate standards [21]. Table 2.1 shows principle methods used for measuring antioxidant capacity in vitro and their respective mechanisms.

2.8 BIOACTIVE COMPOUNDS FROM LEAVES OF *VERNONIA AMYGDALINA* L. AND *OCIMUM GRATISSIMUM* AND THEIR HEALTH BENEFITS

Vernonia amygdalina (VA) leaves are called African bitter leaf and are also called as *ewuro* (Figure 2.8) by the Yoruba speaking tribe of Nigeria; these leaves are harvested throughout the year. It is mostly used as a medicinal plant and possess antimalarial [1, 44, 53], antimicrobial [4, 23], and anticancer activities [35]. VA is a vegetable plant used throughout tropical Africa for both food and traditional treatment of diseases such as infertility, diabetes, gastrointestinal problems, and sexually transmitted diseases [25]. Nutritional and phytochemical evaluations have revealed the presence of high levels of antioxidant vitamins (A, C, E, and riboflavin), mineral elements (Fe, Se, Zn, Cu, Cr, and Mn) and phytocompounds (polyphenols, flavonoids, and tannins) in the leaves of VA [7, 34].

Atangwho et al. [6] studied the antioxidant and antidiabetic properties of the sequential extracts of *VA* based on the chemical composition of most effective antidiabetic extract. Using DPPH (2,2-diphenyl-1-picryl-hydrazyl) and ABTS (2,2'-azino-bis(3-ethylbenzothiazoline-6-sulphonic acid) radical scavenging as well as FRAP (Ferric ion reducing antioxidant power) assays, they reported that the extracts showed a consistent dose-dependent trend of potent antioxidant activity in different solvents (water extract > methanol extract > chloroform extract > petroleum ether extract). After a 14-days administration in diabetic rats, the chloroform extract recorded the highest blood (23.5%) and serum (21.4%) glucose-lowering effects ($P < 0.05$). GC–MS analysis of the chloroform extract revealed high levels of linoleic acid (4.72%), α-linolenic acid (10.8%), and phytols (12.0%). Their study revealed the beneficial effect of VA in human nutrition and its potential role in the management of diabetes.

TABLE 2.1 Principle Methods for Measuring Antioxidant Activities *In-vitro* and Their Respective Mechanisms

Method	Mechanism	Reaction	Measurement	Ref.
Ability to chelate transition metals (Cu^{2+})	Chelation of Cu^{2+}	Complexation reaction of Cu^{2+} with pyrocatechol violet to generate a colored product. The presence of antioxidants decreases the formation of the Cu^{2+}-pyrocatechol complex, reducing the intensity of the color.	Reduction in the absorbance at 620 nm	[64]
Ability to chelate transition metals (Fe^{2+})	Chelation of Fe^{2+}	Complexation reaction of Fe^{2+} with ferrozine, generating a colored product. The presence of antioxidants decreases the formation if the Fe^{2+}-ferrozine complex, reducing color intensity.	Reduction in the absorbance at 562 nm	[50]
ABTS	ABTS capture	(2,2′-azino-bis(3-ethylbenzothiazoline-6-sulfonic acid)) is stabilized in the presence of hydrogen-donating free radicals, changing the color from dark green to light green.	Reduction in the absorbance at 734 nm	[29]
DPPH	DPPH capture	DPPH radical (2,2-diphenyl-1-picryl-hydrazyl) reacts with hydrogen-donating antioxidants, changing the color from violet to yellow.	Reduction in the absorbance at 517 nm	[62]
FRAP	Iron reducing power	In the presence of electron-donating antioxidants, the Fe^{3+}-TPTZ [2,4,6-Tripyridyl-S-Triazine] complex is reduced to Fe^{2+}-TPTZ, changing the color from light blue to dark blue.	Increase in the absorbance at 593 nm	[55]
ORAC	Peroxyl radical capture	The peroxyl radical, generated from the breakdown of AAPH [2,2′-Azobis(2-amidinopropane) dihydrochloride] in the presence of atmospheric oxygen, reacts with a fluorescent indicator to produce a non-fluorescent product. In the presence of antioxidants, the fluorescence is maintained.	Reduction in fluorescence (excitation at 485 nm and emission at 520 nm)	[20]

TABLE 2.1 (Continued)

Method	Mechanism	Reaction	Measurement	Ref.
TBARS	Quantification of lipid peroxidation products	Reaction of thiobarbituric acid with hydroperoxide decomposition products. Malonaldehyde is the main compound quantified. Absorbance and antioxidant activity are inversely proportional.	Increase in the absorbance at 532 nm	[58]

FIGURE 2.8 *Vernonia amygdalina* (African bitter leaf) plant.

Erasto et al. [23] studied the bioactive sesquiterpene lactones from the leaves of VA. Phytochemical analysis of the leaves of VA yielded two known sesquiterpene lactones (vernolide and vernodalol). These two compounds exhibited significant bactericidal activity against five gram-positive bacteria while lacking efficacy against gram-negative strains. In the antifungal test, vernolides exhibited high activity with LC_{50} values of

0.1, 0.2, 0.3, and 0.4 mg/mL against *Penicillium notatum*, *Aspergillus flavus*, *Aspergillus niger*, and *Mucor hiemalis*, respectively, and vernodalol showed moderate inhibition against *A. flavus*, *P. notatum*, and *A. niger* with LC_{50} values of 0.3, 0.4, and 0.5 mg/mL, respectively. Both compounds were ineffective against *Fusarium oxysporum*, a microbe known to be highly resistant to chemical agents. However, the antimicrobial results of their study correspond positively with the claimed ethno-medical uses of the leaves of VA in the treatment/management of various infectious diseases.

Ocimum gratissimum L. (Og) is a medicinal plant and is called *efinrin* (Figure 2.9) by Nigerian people, and it is popularly known as African basil. This plant has a strong-smelling aromatic flavor. For medicinal properties, in southwestern Nigeria, Og is mainly known for its antimicrobial activities against bacteria causing diarrhea [3]. It is a traditional herb with a wide array of phytochemical constituents and has been reported for its diverse physiological properties. Its medicinal uses include vasorelaxation and anti-inflammatory, antimycotoxicogenic, and antioxidant activities [17, 18, 39, 57]. Og contains phytochemicals such as flavonoids, polyphenols, and volatile compounds like eugenol, thymol, and geraniol [67]. Identification of xanthomicrol, cirsimaritin, vicenin-2, luteolin 7-O-glucoside, quercetin 3-O-glucoside, quercetin 3-O-rutinoside, and kaempferol-3-O-rutinoside as principal flavonoids in Og has been reported [30].

Venuprasad et al. [66] studied the phytochemical components of Og by LC-ESI–MS/MS and its antioxidant and anxiolytic effects. Their work revealed the presence of various bioactive compounds such as polyphenols, flavonoids, and fatty acids. The antioxidant activity of Og was verified by an array of in vitro antioxidant assays, which exhibited a strong free radical scavenging activity. A 70% ethanolic fraction of Og leaf showed protective effects against H_2O_2/AAPH-induced damage to macromolecules such as DNA, proteins, and lipids.

Identification of chicoric acid as a hypoglycemic agent from Og leaf extract in a biomonitoring in vivo study was reported by Casanova et al. [12]. Four phenolic substances were identified: L-caftaric acid [1], L-chicoric acid [2], eugenyl-β-D-glucopyranoside [3], and vicenin-2 [4]. Their results were the first to describe the hypoglycemic activity of chicoric acid in an animal model of diabetes mellitus. In their study, a

FIGURE 2.9 *Ocimum gratissimum* (African basil) plant.

correlation was established between the hypoglycemic activity of Og and its chemical composition by means of a biomonitored fractionation. Chicoric acid, the active substance identified, was described as not likely to be the only compound responsible for the hypoglycemic activity. However, their results suggested a possible interaction between chicoric acid and a substance present in the precipitate from the aqueous fraction, probably tartaric acid. The antioxidant and cytoprotective activity of Og extracts against H_2O_2-induced toxicity in human HepG2 cells was also reported by Chiu et al. [14]. Their findings indicated that aqueous Og extracts exert antioxidant and protective activities against OS induced by H_2O_2 on human HepG2 cells; secondary metabolites pharmacological or toxicological effects phenolic compounds reactive free radicals antioxidant nutrients

2.9 SUMMARY

Bioactive compounds in plants are secondary metabolites that cause pharmacological or toxicological effects in human and animals. Plants provide a large variety of bioactive compounds and are good sources of phenolic compounds that possess antioxidant capacity. The human body has antioxidant enzyme defense systems that deactivate highly reactive free radicals, but when these antioxidant defense enzymes are overwhelmed, antioxidant nutrients (found in plant foods) are needed to combat these free radicals, turning them harmless or waste products that can then be easily eliminated through the liver. Regular intake of plant foods has the potential of preventing free radical-mediated degenerative diseases. Phenolic compounds are widely distributed in plants, usually with a higher concentration in leaves and green stems. The leaves of two traditional medicinal plants (*Vernonia amygdalina* and *Ocimum gratissimum*) have the potential to ameliorate various degenerative diseases due to the presence of inherent bioactive compounds.

KEYWORDS

- **ABTS**
- **antioxidant nutrients**
- **antioxidants**
- **bioactive compound**
- **degenerative diseases**
- **DPPH**
- **FRAP assays**
- **H_2O_2**
- **HepG2 cells**
- **medicinal plant**
- *Ocimum gratissimum*
- **oxidative stress**
- **pharmacological effect**

- **phenolic compound**
- **radical scavenging**
- **reactive free radicals**
- **secondary metabolites**
- **toxicological effects**
- ***Vernonia amygdalina***

REFERENCES

1. Abos, A. O., & Raseroka, B. H., (2003). In vivo antimalarial activity of *Vernonia amygdalina*. *British Journal of Biomedical Science*, *60*(2), 89–91.
2. Adams, S. M., & Standridge, J. B., (2006). What should we eat? Evidence from observational studies. *Southern Medical Journal*, *99*, 744–748.
3. Adebolu, T. T., & Salau, A. O., (2005). Antimicrobial activity of leaf extracts of *Ocimum gratissimum* L., on selected bacteria causing diarrhoea in southwester Nigeria. *African Journal of Biotechnology*, *4*(7), 682–684.
4. Akinpelu, D. A., (1999). Antimicrobial activity of *Vernonia amygdalina* L., leaves. *Fitoterapia, 70(4),* 432–434.
5. Anderson, D., (1996). Antioxidant defense against reactive oxygen species causing genetic and other damage. *Mutation Research*, *350*, 103–108.
6. Atangwho, I. J., Egbung, G. E., Ahmad, M., Yam, M. F., & Asmawi, M. Z., (2013). Antioxidant versus anti-diabetic properties of leaves from *Vernonia amygdalina* L., growing in Malaysia. *Food Chemistry*, *141*, 3428–3434.
7. Atangwho, I. J., Ebong, P. E., Eyong, E. U., Williams, I. O., Eteng, M. U., & Egbung, G. E., (2009). Comparative chemical composition of leaves of some antidiabetic medicinal plants: *Azadirachta indica*, *Vernonia amygdalina* and *Gongronema latifolium*. *African Journal of Biotechnology*, *8*(18), 4685–4689.
8. Balk, M. J., Bast, A., & Guido Haenen, R. M. M., (2009). Evaluation of the accuracy of antioxidant competition assays: incorrect assumptions with major impact. *Free Radical Biology and Medicine*, *47*, 137–144.
9. Behera, B. C., Verma, N., Sonone, A., & Makhija, U., (2006). Determination of antioxidative potential of lichen *Usnea ghattensis* in vitro. *LWT—Food Science and Technology*, *39*, 80–85.
10. Bernhoft, A., (2010). *Brief Review on Bioactive Compounds in Plants*. The Norwegian Academy of Science and Letters, Oslo, Norway.
11. Bravo, L., (1998). Polyphenols: chemistry, dietary sources, metabolism, & nutritional significance. *Nutrition Reviews*, *56*, 317–333.
12. Casanova, L. M., da Silva, D., Sola-Penna, M., Camargo, L. M., M., Celestrini, D. M., Tinoco, L. W., & Cost, S. S., (2014). Identification of chicoric acid as a hypogly-

cemic agent from *Ocimum gratissimum* leaf extract in a biomonitoring in vivo study. *Fitoterapia, 93*, 132–141.
13. Chan, S. H. H., Tai, M. H., Li, C. Y., & Chan, J. Y. H., (2006). Reduction in molecular synthesis or enzyme activity of superoxide dismutases and catalase contributes to oxidative stress and neurogenic hypertension in spontaneously hypertensive rats. *Free Radical Biology and Medicine, 40*, 2028–2039.
14. Chiu, Y. W., Lo, H. J., Huang, H. Y., Chao, P. Y., Hwang, J. M., Huang, P. Y., Huang, S. J., Liu, J. Y., & Lai, T. J., (2013). The antioxidant and cytoprotective activity of *Ocimum gratissimum* extracts against hydrogen peroxide-induced toxicity in human HepG2 cells. *Journal of Food and Drug Analysis, 21*, 253–260.
15. Chung, K. T., Wong, T. Y., Wei, C. I., Huang, Y. W., & Lin, Y., (1998). Tannins and human health: A review. *Critical Reviews in Food Science and Nutrition, 38*, 421–464.
16. Cornelli, U. M. D., (2009). Antioxidant use in nutraceuticals. *Clinics in Dermatology, 27*, 175–194.
17. Costa, R. S., Carneiro, T. C., Cerqueira-Lima, A. T., Queiroz, N. V., Alcantara-Neves, N. M., Pontes-de-Carvalho, L. C., Velozo-Eda, S., Oliveira, E. J., & Figueiredo, C. A., (2012). *Ocimum gratissimum* L., and rosmarinic acid, attenuate eosinophilic airway inflammation in an experimental model of respiratory allergy to *Blomia tropicalis*. *International Immunopharmacology, 13*, 126–134.
18. Dambolena, J. S., Zunino, M. P., López, A. G., Rubinstein, H. R., Zygadlo, J. A., Mwangi, J. W., & Kariuki, S. T., (2010). Essential oils composition of *Ocimum basilicum* L., and *Ocimum gratissimum* L., from Kenya and their inhibitory effects on growth and fumonisin production by Fusarium verticillioides. *Innovative Food Science and Emerging Technologies, 11*, 410–414.
19. Da Silva Oliveira, C., Maciel, L. F., Miranda, M. S., & Da Silva Bispo, E., (2011). Phenolic compounds, flavonoids and antioxidant activity in different cocoa samples from organic and conventional cultivation. *British Food Journal, 113*, 1094–1102.
20. Dávalos, A., Gómez-Cordovés, C., Da Silva Bispo, E., & Bartolomé, B., (2004). Extending applicability of the oxygen radical absorbance capacity (ORAC-fluorescein) assay. *Journal of Agricultural and Food Chemistry, 52*, 48–54.
21. De Castro, R. J. S., & Hélia Harumi Sato, H. H., (2015). Biologically active peptides: Processes for their generation, purification and identification and applications as natural additives in the food and pharmaceutical industries. *Food Research International, 74*, 185–198.
22. Diaz, J., Bernal, A., Pomar, F., & Merino, F., (2001). Induction of shikimate dehydrogenase and peroxidase in pepper (*Capsicum annuum* L.) seedlings in response to copper stress and its relation to lignification. *Plant Science, 161*, 179–188.
23. Erasto, P., Grierson, D. S., & Afolayan, A. J., (2006). Bioactive sesquiterpene lactones from the leaves of *Vernonia amygdalina*. *Journal of Ethnopharmacology, 106*, 117–120.
24. Espin, J. C., Garcia-Conesa, M. T., & Tomas-Barberan, F. A., (2007). Nutraceuticals: Facts and fiction. *Phytochemistry, 68*, 2986–3008.
25. Farombi, E. O., & Owoeye, O., (2011). Antioxidant and chemo-preventive properties of *Vernonia amygdalina* and *Garcinia biflavonoid*. *International Journal of Environmental Research and Public Health, 8*, 2533–2555.

26. Ferreira, I. C. F. R., Barros, L., & Abreu, R. M. V., (2009). Antioxidant in mushrooms. *Current Medicinal Chemistry, 16,* 1543–1560.
27. Gerber, M., Boutron-Ruault, M. C., Hercberg, S., Riboli, E., Scalbert, A., & Siess, M. H., (2002). Food and cancer: state of the art about the protective effect of fruits and vegetables. *Bull. Cancer, 89,* 293–312.
28. Giada, M. L. R., (2013). Food phenolic compounds: main classes, sources and their antioxidant power. In: J. A., M., Gonzalez (Ed.), Oxidative stress and chronic degenerative diseases – A role for antioxidants, *In. Tech.,* pp. 87–112.
29. Gómez-Guillén, M. C., López-Caballero, M. E., López De Lacey, A., Alemán, A., Giménez, B., & Montero, P., (2010). Antioxidant and antimicrobial peptide fractions from squid and tuna skin gelatin. In: Bihan, E. L., & Koueta, N., (Eds.), *Sea By-Products as a Real Material: New Ways of Application.* Kerala Transworld Research Network Signpost, pp. 89–115.
30. Grayer, R. J., Kite, G. C., Abou-Zaid, M., & Archer, L. J., (2000). The application of atmospheric pressure chemical ionization liquid chromatography–mass spectrometry in the chemotaxonomic study of flavonoids: characterisation of flavonoids from *Ocimum gratissimum* var. *gratissimum. Phytochemical Analysis, 11,* 257–267.
31. Halliwell, B., & Gutteridge, J. M. C., (1999). *Free radicals in Biology and Medicine.* 3rd edition, Oxford University Press.
32. Halliwell, B., (1984). Oxygen radicals: a common sense look at their nature and medical importance. *Medical Biology, 62,* 71–77.
33. Hernandez, Y., Lobo, M. G., & Gonzalez, M., (2009). Factors affecting sample extraction in the liquid chromatographic determination of organic acids in papaya and pineapple. *Food Chemistry, 114,* 734–741.
34. Igile, G. O., Oleszek, W., Jurzysta, M., Burda, S., Fanfunso, M., & Fasanmade, A. A., (1994). Flavonoids from *Vernonia amygdalina* and their antioxidant activities. *Journal of Agricultural and Food Chemistry, 42,* 2445–2448.
35. Izevbigie, F. B., (2003). Discovery of water-soluble anticancer agents (edotides) from a vegetable found in Benin City Nigeria. *Experimental Biology and Medicine (Maywood), 228(3),* 293–298.
36. Katsiki, N., & Manes, C., (2009). Is there a role for supplemented antioxidants in the prevention of atherosclerosis? *Review, Clinical Nutrition, 28,* 3–9.
37. Kris-Etherton, P. M., Hecker, K. D., Bonanome, A., Coval, S. M., Binkosi, A. E., & Hilpert, K. F., (2002). Bioactive compounds in foods: their role in the prevention of cardiovascular disease and cancer. *American Journal of Medicine, 113,* 71S–88S.
38. Kumar, S., & Pandey, A. K., (2013). Chemistry and biological activities of flavonoids: an overview. *Scientific World Journal,* ID 162750.
39. Lee, M. J., Chen, H. M., Tzang, B. S., Lin, C. W., Wang, C. J., Liu, J. Y., & Kao, S. H., (2011). *Ocimum gratissimum* aqueous extract protects H9c2 myocardiac cells from H_2O_2 induced cell apoptosis through AKT signalling. *Evidence-based Complementary and Alternative Medicine,* http://dx.doi.org/10.1155/2011/578060.
40. Locatelli, M., Gindro, R., Travaglia, F., Coïsson, J., Rinaldi, M., & Arlorio, M., (2009). Study of the DPPH scavenging activity: Development of free software for the correct interpretation of data. *Food Chemistry, 114,* 889–897.
41. Lü, J., Lin, P. H., Yao, Q., & Chen, C., (2010). Chemical and molecular mechanisms of antioxidants: Experimental approaches and model systems. *Journal of Cellular and Molecular Medicine, 14,* 840–860.

42. Martins, S., Mussatto, S. I., Martínez-Avila, G., Montañez-Saenz, J., Aguilar, C. N., & Teixeira, J. A., (2011). Bioactive phenolic compounds: production and extraction by solid-state fermentation. *Review on Biotechnology Advances*, *29*, 365–373.
43. Marubayashi, S., Dohi, K., Ochi, K., & Kawasaki, T., (1985). Role of free radicals in ischemic rat liver cell injury: Prevention of damage by alpha-tocopherol administration. *Surgery*, *99*, 184–191.
44. Masaba, S. C., (2000). The antimalarial activity of *Vernonia amygdalina* L. (Compositae). *Transactions of the Royal Society of Tropical Medicine and Hygiene*, *94*(6), 694–695.
45. Michalak, A., (2006). Phenolic compounds and their antioxidant activity in plants growing under heavy metal stress. *Polish Journal of Environmental Studies*, *15*, 523–530.
46. Miller, N. J., & Rice-Evans, C. A., (1997). The relative contributions of ascorbic acid and phenolic antioxidants to the total antioxidant activity of orange and apple fruit juices and blackcurrant drink. *Food Chemistry*, *60*, 331–337.
47. Miller, D. M., Buettner, G. R., & Aust, S. D., (1990). Transition metals as catalysts of autoxidation reactions. *Free Radical Biology and Medicine*, *8*, 95–108.
48. Mohajeri, A., & Asemani, S. S., (2009). Theoretical investigation on antioxidant activity of vitamins and phenolic acids for designing a novel antioxidant. *Journal of Molecular Structure*, *930*, 15–20.
49. Moure, A., Cruz, J. M., Franco, D., Dominguez, J. M., Sineiro, J., Dominguez, H., Nunez, M. J., & Parajo, J. C., (2001). Natural antioxidants from residual sources. *Food Chemistry*, *72*, 145–171.
50. Nazeer, R. A., & Kulandai, K. A., (2012). Evaluation of antioxidant activity of muscle and skin protein hydrolysates from giant kingfish, *Caranx ignobilis* (Forsskal). *International Journal of Food Science and Technology*, *47*, 274–281.
51. Nikolic, K. M., (2006). Theoretical study of phenolic antioxidants properties in reaction with oxygen-centered radicals. *Journal of Molecular Structure: Theochem*, *774*, 95–105.
52. Oboh G., & Amusan T. V., (2009). Nutritive value and antioxidant properties of cereal gruels produced from fermented maize and sorghum. *Food Biotechnology*, *23*, 17–31.
53. Oboh, G., (2006). Nutritive value and haemolytic properties (in vitro) of the leaves of *Vernonia amygdalina* L., on human erythrocyte. *Nutrition Health*, *182*(2), 151–160.
54. Okuda, T., Yoshida, & Hatano, T., (1995). Hydrolyzable tannins and related polyphenols. *Progress in the Chemistry of Organic Natural Products*, *66*, 1–117.
55. Ou, B., Huang, D., Hampsch-Woodill, M., Flanagan, J. A., & Deemer, E. K., (2002). Analysis of antioxidant activities of common vegetables employing oxygen radical absorbance capacity (ORAC) and ferric reducing antioxidant power (FRAP) assays: A comparative study. *Journal of Agricultural and Food Chemistry*, *50*, 3122–3128.
56. Pang, X. B., Yao, W. B., Yang, X. B., Xie, C., Liu, D., Zhang, J., & Gao, X. D., (2007). Purification, characterization and biological activity on hepatocytes of a polysaccharide from *Flammulina velutipes mycelium*. *Carbohydrate Polymers*, *70*, 291–297.
57. Pires, A. F., Madeira, S. V., Soares, P. M., Montenegro, C. M., Souza, E. P., Resende, A. C., Soares de Moura, R., Assreuy, A. M., & Criddle, D. N., (2012). The role of endothelium in the vaso-relaxant effects of the essential oil of *Ocimum gratissimum*

in aorta and mesenteric vascular bed of rats. *Canadian Journal of Physiology and Pharmacology, 90,* 1380–1385.

58. Raghavan, S., & Kristinsson, H. G., (2008). Antioxidative efficacy of alkali-treated tilapia protein hydrolysates: A comparative study of five enzymes. *Journal of Agricultural and Food Chemistry, 56,* 1434–1441.
59. Sasipriya, G., & Siddhuraju, P., (2012). Effect of different processing methods on antioxidant activity of underutilized legumes, *Entada scandens* seed kernel and *Canavalia gladiata* seeds. *Food and Chemical Toxicology, 50,* 2864–2872.
60. Scalbert, A., & Williamson, G., (2000). Dietary intake and bioavailability of polyphenols. *Journal of Nutrition, 130*(8), 2073S–2085S.
61. Serafini, M., Bellocco, R., Wolk, A., & Ekstrom, A. M., (2002). Total antioxidant potential of fruit and vegetables and risk of gastric cancer. *Gastroenterologia, 123,* 985–991.
62. Sharma, O. P., & Bhat, T. K., (2009). DPPH antioxidant assay revisited. *Food Chemistry, 113,* 1202–1205.
63. Stafford, H. A., (1983). Enzymic regulation of procyanidin biosynthesis: lack of a flav-3-en-3-ol intermediate. *Phytochemistry, 11,* 2643–2646.
64. Theodore, A. E., Raghavan, S., & Kristinsson, H. G., (2008). Antioxidative activity of protein Hydrolysates prepared from alkaline-aided channel catfish protein isolates. *Journal of Agricultural and Food Chemistry, 56,* 7459–7466.
65. Tutour, B. L., (1990). Antioxidative activities of algal extracts. Synergistic effect with vitamin E., *Phytochemistry, 29,* 3759–3765.
66. Venuprasad, M. P., Kandikattu, H. K., Razack, S., & Khanum, F., (2014). Phytochemical analysis of *Ocimum gratissimum* by LC-ESI–MS/MS and its antioxidant and anxiolytic effects. *South African Journal of Botany, 92,* 151–158.
67. Vieira, R. F., Grayer, R. J., Paton, A., & Simon, J. E., (2001). Genetic diversity of *Ocimum gratissimum* L., based on volatile oil constituents, flavonoids and RAPD markers. *Biochemical systematic and Ecology, 29,* 287–304.
68. Visioli, F., & Hagen, T. M., (2011). Antioxidants to enhance fertility: Role of eNOS and potential benefits. *Pharmacological Research, 64,* 431–437.
69. Wang, B. S., Yen, G. C., Chang, L. W., Yen, W. J., & Duh, P. D., (2006). Protective effects of burdock (*Arctium lappa Linne*) on oxidation of low-density lipoprotein and oxidative stress in RAW 264.7 macrophages. *Food Chemistry, 101,* 729–738.
70. Wickens, A. P., (2001). Ageing and the free radical theory. *Respiratory Physiology, 128*(3), 379–391.
71. Wootton-Beard, P. C., & Ryan, L., (2011). Improving public health: The role of antioxidant rich fruit and vegetable beverages. *Food Research International, 44,* 3135–3148.
72. Yang, C. S., Landau, J. M., Huang, M. T., & Newmark, H. L., (2001). Inhibition of carcinogenesis by dietary polyphenolic compounds. *Annual Review of Nutrition, 21,* 381–406.
73. Yoshihara, D., Fujiwara, N., & Suzuki, K., (2010). Antioxidants: Benefits and risks for long-term health. *Maturitas, 67,* 103–107.

CHAPTER 3

PLANTS OF THE GENUS *SYZYGIUM* (MYRTACEAE): A REVIEW ON ETHNOBOTANY, MEDICINAL PROPERTIES, AND PHYTOCHEMISTRY

IAN EDWIN COCK and MATTHEW CHEESMAN

CONTENTS

3.1 INTRODUCTION

Plants have been used as therapeutic agents since the start of civilization. In many developing countries, herbal medicinal systems remain important in the treatment of many ailments. Ayuvedic medicine is still commonly practiced within India with an estimated 85% of Indians still using crude plant preparations for the treatment of wide variety of diseases and ailments [79]. Traditional Chinese medicine (TCM) and African medicinal systems also account for major portion of healthcare in their populations. Even in countries where allopathic/Western medicine is dominant, medicinal plants still make significant contributions. Furthermore, many people are returning to herbal medicine systems due to the perception that natural medicines are often a safer alternative than allopathic drugs. Individuals are also seeking treatments for diseases, which have not yet been remedied by modern medicine.

Many of the prescription drugs currently marketed for a wide variety of ailments were originally isolated from plants and/or are semi-synthetic analogues of phytochemicals. It has been estimated that approximately 25% of all prescription drugs currently in use are of plant origin [122, 194]. Furthermore, 75% of new anticancer drugs marketed between 1981 and 2006 were derived from plant compounds [122]. Traditional plant-based medicines are generally used as crude formulations (e.g., infusions, tinctures and extracts, essential oils, powders, poultices, and other herbal preparations). Modern natural product drug discovery generally focuses on isolating and characterizing the individual phytochemical components with the aim of producing an analogue with increased bioactivity/bioavailability. Such studies have given rise to many useful drugs such as quinine (from *Cinchona* spp.), digoxin (from *Digitalis* spp.) as well as the anticancer drug paclitaxel (from *Taxus brevifolia* Nutt.) and vincristine and vinblastine (from *Vinca rosea*). However, the bioactivities observed for crude extracts are often much enhanced or even entirely different to those seen for the individual components [32, 81]. Crude plant extracts may contain hundreds, or even thousands, of different chemical constituents that interact in complex ways. Often it is not known how an extract works, even when its therapeutic benefit is well established. Thus, phytochemical and mechanistic studies of traditional medicines may be complex, and it is often difficult to assign single compound/bioactivity relationship.

The genus *Syzygium* (Family Myrtaceae) is the largest genus of flowering plants globally, comprising approximately 1200–1800 species [174]. The genus has a wide range, occurring naturally in subtropical and tropical regions of Africa and Madagascar, Asia, and throughout Oceania and the Pacific region. The greatest diversity occurs in Australia and Southeast Asia. It is believed that many species in this region have not yet been correctly categorized taxonomically, and it is likely that new species will be categorized in the future [174]. Most *Syzygium* species range from medium to large evergreen trees and shrubs, many of which have a history of usage in traditional medicinal systems. Some species have culinary uses and are commercially produced for their edible fruits (e.g., *Syzygium jambos*, commonly known as rose apple). These are eaten fresh or used to produce jams and preserves. Other species are used as spices and flavoring agents. For the economically most important species *Syzygium aromaticum* (clove), the unopened flower bud is used as a spice. This plant has also been used as traditional medicine because of its antimicrobial and anesthetic properties [25]. The antimicrobial activity of *S. aromaticum* is well known, and numerous studies have reported on the antibacterial [8] and antifungal [126] activities of oils and extracts from clove. Other *Syzygium* species from Southeast Asia (*S. jambos*) [113] and India (*Syzygium lineare* and *Syzygium cumini*) [50] have also been shown to have antimicrobial activity. Most information available on antimicrobial activities of Australian *Syzygium* species has been anecdotal, although Australian Aborigines are known to have used them as medicinal agents [36].

This chapter summarizes recent research on the medicinal properties, phytochemistry, and therapeutic mechanisms of *Syzygium* species and also highlights future areas of research on the medicinal activities of *Syzygium* species.

3.2 ETHNOPHARMACOLOGY

Members of genus *Syzygium* have been used for a broad range of medicinal purposes by traditional healers from a wide variety of ethnic and cultural groups (Table 3.1). The best documented of these are the species used in traditional Indian medicinal systems, particularly the Ayurveda. Ayurvedic practitioners employ various species of *Syzygium* for a wide variety of medicinal purposes including coughs and colds, diarrhea and

TABLE 3.1 Summary of Origin, Medicinal Properties, and Phytochemical Constituents (where known) for Selected *Syzygium* Species Worldwide

Species	Common name	Origin	Medicinal properties	Known constituents	Ref.
Syzygium alliiligneum Hyland	Mission Beach satinash, onionwood, puddenwood	North eastern Australia	The fruit is mainly used as a nutritious, high antioxidant food.	Unknown.	[36]
Syzygium anisatum (Vicfkery) Craven & Biffen	Ringwood, anise myrtle, aniseed tree	Australia	Potent broad spectrum antibacterial activity	Contains a diversity of terpenoids. Has especially high levels of monoterpenoids. Essential oils produced from the leaves are high in anethole and chavicol.	[23, 36, 197]
Syzygium aqueum Alston	Water cherry, watery rose apple, laulau	Indonesia and Malaysia	Has high antioxidant capacity. Has antihyperglycemic effects so is useful for the treatment of diabetes.	High antioxidant capacity Contains relatively high levels of terpenoids and flavonoids.	[105, 106, 125, 200]
Syzygium aromaticum (L.) Merrill & Perry	Clove	Native to Indonesia, but naturalized widely throughout Asia	Used as an analgesic and anodyne for toothache. Also used as a carminative, a digestive aid and as an antihelmintic. It has good antimicrobial properties. Possible antidiabetes use.	High levels of terpenoids. Especially high in eugenol. Has a high antioxidant capacity.	[47, 97, 130]

Species	Common name	Origin	Medicinal properties	Known constituents	Ref.
Syzygium australe (H.L. Wendl. Ex Link) B .Hyland	Brush cherry, scrub cherry, creek lilly-pilly, creek satinash, watergum	Eastern Australia	Both leaves and fruit have broad spectrum antibacterial properties. Inhibits anthrax. Inhibits cancer cell proliferation.	High antioxidant capacities with particularly high ascorbic acid levels.	[26, 34, 75, 88, 152, 201]
Syzygium cordatum (Hochst.)	Waterbessie, Water berry, undoni, umSwi, umJoni, and hute	Southern Africa including South Africa, Zimbabwe, Mozambique	Used to treat respiratory ailments, tuberculous, stomach complaints, diarrhoea and diabetes. Has antibacterial activity.	Phenolic compounds, triterpenoids.	[38-40, 188]
Syzygium Francisii (F.M. Bailey) L.A.S. Johnson	Giant water gum, rose satinash, Francis water gum	Eastern Australia	Antiseptic properties	High Antioxidant capacity. Contains relatively high levels of terpenoids and Flavonoids.	[26]
Syzygium grande (Wight) Walp.	Sea apple	South East Asia	Antiseptic properties	The flavonoid xyloside and the triterpenoid rhamnoside have been reported in leaf extracts.	[150]
Syzygium cumini (L.) Skeels.	Jambul, jambolan, jamblang jamun, Java plum	Indian sub-continent	Antibacterial antidiabetic (antihyperglycemic), anti-inflammatory properties	High antioxidant capacity Rich in anthocyanins, ellagic acid, isoquercetin kaemferol and myrecetin.	[13, 16, 117, 155]

TABLE 3.1 (Continued)

Species	Common name	Origin	Medicinal properties	Known constituents	Ref.
Syzygium forte (F. Muell.) B.Hyland	White apple, Flaky barked satinash, Brown satinash	North eastern Australia	Antiseptic properties	High antioxidant capacity. Contains relatively high levels of terpenoids and Flavonoids.	[26]
Syzygium guineense Wall.	Bambara	Africa	Good antibacterial and antifungal properties. It has antimalarial and anti-hypertensive properties.	A wide variety of terpenoids have been reported.	[12, 49, 172]
Syzygium jambos L. (Alston)	Rose apple, water apple, wax apple, Malay apple, cloud apple, plum rose	South east Asia	Antimicrobial activity, useful in the treatmnent of diabetes, antidermatotophytic activity. It has anti-inflammatory properties.	Triterpenoids including friedelin, β-amyrin acetate, betulinic acid and lupeol.	[16,48, 91, 113]
Syzygium leuhmannii (F. Muell.) L.A.S. Johnson	Riberry, small leaved lilly-pilly, cherry satinash, cherry alder, clove lilly-pilly	Eastern Australia	Both leaves and fruit have broad spectrum antibacterial properties. Inhibits anthrax. Inhibits cancer cell proliferation.	High antioxidant capacities with particularly high ascorbic acid levels.	[26, 28, 34, 75, 152, 201]
Syzygium lineare Wall.	None known	Indian sub-continent	Used to cool the body and increase stamina. The leaf has insecticidal properties.	2,3-diacetoxy-2-benzyl-4,4,6,6-tetramethyl-1,3-cyclohexanedione has been identified in leaf extracts and has been postulated to contribute the insecticidal properties.	[76]

Species	Common name	Origin	Medicinal properties	Known constituents	Ref.
Syzygium maire (A.Cunn.) Sykes & Garn.-Jones	Swamp Maire	New Zealand	High antioxidant content.	High antioxidant content. Individual contents are poorly characterized.	[64]
Syzygium malaccense (L.) Merr. & L.M. Perry	Malay rose apple, Malay apple, Otaheite cashew	South East Asia, New Guinea, Australia	Bark decoctions were used to treat mouth ulcers. Leaves are used by women with irregular menstruation.	High antioxidant content.	[70]
Syzygium moorei F. Muell.	Coolamon, watermelon tree, durobby, robby	Eastern Australia	Antiseptic properties.	High antioxidant capacity. Contains relatively high levels of terpenoids and flavonoids.	[26]
Syzygium oleosum (F. Muell.) B. Hyland	Blue lilly-pilly	Eastern Australia	Mainly used as a nutritious food due to is antioxidant content.	High antioxidant content. Individual contents are poorly characterized.	[26, 132]
Syzygium paniculatum Gaertn.	Magenta cherry, Magenta lilly-pilly	Eastern Australia	High antioxidant activity. Has antiproliferative activity against pancreatic cancer cells.	Contains significant levels of gallic acid, chlorogenic acid, catechin and epicatechin.	[193]
Syzygium papyraceum B. Hyland	Paperbark satinash	Eastern Australia	Potent antibacterial activity. Kills HepG2 and MDA-MB-231 carcinoma cell growth.	High antioxidant content. Individual contents are poorly characterized.	[26, 154]

TABLE 3.1 (Continued)

Species	Common name	Origin	Medicinal properties	Known constituents	Ref.
Syzygium polyanthum (Wight) Walp.	Indian bayleaf, Indonesian bayleaf	Indochina, Melanesia	Leaf extracts have potent antibacterial activity and can kill Bacillus spp. spores.	Individual contents are poorly characterized.	[96]
Syzygium samarangense (Blume) Merr. & L.M. Perry	Java apple, Semarang rose apple, wax jambu	South East Asia	The flowers are astringent. Used to treat diarrhoea and fever. Has antibacterial activity and antihyperglycemic effects. High antioxidant content.	High antioxidant content. Triterpenoids, tannins, desmethoxymatteucinol, 5-O-methyl-4'-desmethoxymatteucinol, oleanic acid and β-sitosterol.	[156, 166]
Syzygium suboriculare (Benth.) T.G. Hartley & L.M. Perry	Lady satinash, rolypolysatinash, red bush apple	Northern Australia, Papus New Guinea	Used to treat coughs and colds, diarrhoea and dysentery and as a general antiseptic.	High antioxidant capacity. Contains relatively high levels of terpenoids and flavonoids.	[26, 95]
Syzygium wilsonii (F.Muell.) B.Hyland	Powder-puff lilly-pilly, Wilson's satinash	North eastern Australia	Antiseptic properties.	High antioxidant capacity. Contains relatively high levels of terpenoids and flavonoids.	[26]

Ethnopharmacological uses, constituents and common names are derived from search in Google Scholar and PubMed databases.

dysentery, fever, toothache, pain, inflammation, pneumonia, sexually transmitted diseases, wounds, hemorrhages, ulcers, and as a general tonic. of the Ayurvedic *Syzygium* species, *S. aromaticum* and *S. cumini* are arguably most useful, being a component of multiple traditional therapies for the treatment of many ailments (Table 3.1). These will be discussed separately for each medicinal bioactivity.

Similarly, African medicinal systems made use of native *Syzygium* species in the treatment of multiple different disorders. Different species occur in different parts of the continent, and their usage, whilst often widespread, is also often associated with specific cultural/ethnic groups. Thus, a species used by one cultural group for a specific property may have had different therapeutic uses (or no uses) by other groups in different regions of the continent. Perhaps the best known and most useful of the African species is *Syzygium cordatum*. This species is native to southern Africa, extending up to central Africa. The therapeutic properties of *S. cordatum* are widely known, and it is a part of the pharmacopeia for all ethnic groups in the areas in which it grows, including Zulu, Xhosa, Sotho, and Afrikaaner people. Multiple parts of the plant are used for a variety of purposes, with bark and leaves being the most commonly used, although roots are also used to treat some conditions. Decoctions of the leaves, bark, and roots are used in the treatment of respiratory ailments as well as stomach complaints and diarrhea and as an emetic [188]. Many of these conditions are caused by microbial pathogens, and extracts produced from this species have antimicrobial activity [38–40]. It has also been reported that leaf extracts are useful in the treatment of diabetes by reducing blood glucose levels and stimulating hepatic glycogen storage [118].

Despite having the greatest taxonomic biodiversity, less information is available about the ethnobotanical usage of Australian *Syzygium* species. Traditional/indigenous Australian knowledge of plant-based therapeutics was generally passed on orally, with little written ethnopharmacological record. Thus, as Australian Aboriginal culture merged into mainstream society with a corresponding increased reliance on Western medicine, much of this knowledge has been lost. However, several species including *Syzygium anisatum*, *Syzygium australe*, *Syzygium leuhmannii*, and *Syzygium paniculatum* are known to have been used as both food and medicinal plants. Multiple species have been reported to be used to treat coughs and colds, stomach disorders, and diarrhea, although the species and how it was used varies widely between regions and different tribal groups [36, 95]. Many of these conditions are caused by pathogenic infections, and several Australian *Syzygium* species have been reported to have good antimicrobial activity [26, 34, 75, 152, 201]. Ethnobotanical reports also indicate that some species were used for oral care in the treatment of toothache in a similar manner to the use of *S. aromaticum* in traditional Asian therapeutic systems [95]. Whether this is due to the presence of eugenol (as it has been postulated for *S. aromaticum*) [78] is yet to be determined.

3.3 PHYTOCHEMISTRY AND ANTIOXIDANT PROPERTIES

The consumption of foods rich in antioxidants is associated with the decreased incidence of chronic diseases [134]. High antioxidant levels are preventive against the development of degenerative diseases such as cancer [69], cardiovascular diseases [191], neural degeneration [203], and diabetes and obesity [186]. Phenolic compounds are generally strong antioxidants [144]. Their primary action involves the protection of cell constituents against oxidative damage through the scavenging of free radicals, thereby averting their deleterious effects on nucleic acids, proteins, and lipids in cells [144]. Phenolics interact directly with receptors or enzymes involved in signal transduction [116], clearly indicating that they play a specific role in human physiology. Common plant phenolic compounds include flavonoids, tannins, anthocyanins, and gallic acid, all of which are common phytochemical constituents of *Syzygium* species.

Recent research studies have documented the exceptionally high antioxidant content of the Australian species *S. australe* and *S. leuhmannii* [88, 121]. These studies have reported that the fruit of these species have extremely high ascorbic acid levels (0.72 and 1.77 μmol/g *S. australe* and *S. leuhmannii* fruit, respectively). This is approximately 10–25 times higher (g/g) than the ascorbic acid content in blueberries (which were used as a standard). Because of the high ascorbic acid level, much of the interest has been geared toward *S. australe* and *S. leuhmannii*. However, *S. australe* and *S. leuhmannii* fruits also contain many other compounds, which have also contributed to their high antioxidant activities [88, 121]. Whilst many of these compounds are yet to be identified, *S. australe* and *S. leuhmannii* fruit have been shown to contain benzoic acids, flavanols, or flavanones [88]. *S. australe* and *S. leuhmannii* fruit are good sources of gallic acid, malic acid, and oxalic acid [88]. These fruits are also very rich in delphinidin 3,5-diglucoside, peonidin 3,5-diglucoside, petunidin 3,5-diglucoside, malvidin 3,5-diglucoside, cyanidin 3,5-diglucoside, cyanidin 3 glucoside, rutin, quercetin hexoside, quercetin rhamnoside, and kaempferol.

High antioxidant contents have also been reported for several Asian *Syzygium* species. *S. aromaticum* essential oil has antioxidant activity similar to that of α-tocopherol [97]. The authors attributed the high antioxidant activity to high levels of eugenol and eugenyl acetate. Similarly, *S. cumini* fruit infusions [17] and ethanolic extracts [55] also have high antioxidant activity as determined by ABTS, DPPH, TEAC, superoxide, hydroxyl, and peroxyl (ORAC) radical scavenging assays. Aqueous and ethanolic *S. cumini* seed extracts have significant protective effects against hydroxyl radical-induced strand breaks in pBR322 DNA [9]. Furthermore, the authors reported significant inhibition of hepatic lipid peroxidation and increases in GSH levels and GST, SOD, and CAT activity, which are likely to contribute to these protective effects.

3.4 ANTIMICROBIAL ACTIVITY

3.4.1 INHIBITION OF BACTERIAL GROWTH

Many *Syzygium* species have a history of usage to treat medical conditions related to microbial infections, and numerous recent investigations have focused on their antibacterial properties. The antimicrobial properties of the South Asian species used in traditional Indian medicine systems have been the most extensively reported. The growth inhibition properties of *S. aromaticum* have been reported to have antibacterial activity against a wide panel of microbes [5, 136, 137, 147], partially justifying its use in the treatment of a wide range of diseases and medical disorders [47, 97, 130]. *S. aromaticum* is used in several traditional healing systems for the treatment of oral disease and to alleviate toothache pain. The antibacterial activity of this species against oral caries-causing pathogens has been particularly well studied. Essential oils produced from *S. aromaticum* buds (cloves) have been reported to be good inhibitors of *Streptococcus mutans, Staphylococcus aureus*, and *Lactobacillus acidophilus* when screened in a disc diffusion assay [5]. Indeed, zones of inhibition as large as 35 mm were recorded when the oils were used at dilutions consistent with their traditional use. The same study also determined the minimum inhibitory concentration (MIC) for the crude oils and solvent fractions. MIC values indicative of moderate antibacterial activity (approximately 3 mg/mL against *S. mutans*) were determined for the crude clove oil by using an agar diffusion assay. As lower polarity compounds (as would be present in the *S. aromaticum* essential oil) are insoluble in aqueous solutions, they may not diffuse well through agar gels and thus often produce falsely low antibacterial results. Therefore, the use of the liquid dilution assay may have indicated more potent activity. Interestingly, the solvent fractionations of the oil were substantially less potent as bacterial growth inhibitors than the unfractionated oil, with MIC values approximately of higher magnitude. Many traditional medicines require the synergistic interaction of various phytochemical components to potentiate activity. The lower potency of the essential oil fractions may indicate that some essential oil components are required to enhance the activity of other components.

Another study also examined the growth inhibitory properties of *S. aromaticum* essential oil against gram-negative anaerobic periodontal oral pathogens including *Porphyromonas gingivalis*, *Prevotella intermedia*, *Streptococcus mutans,* and *Actinomyces viscosus* [25]. Whilst the MIC's determined for an *S. aromaticum* methanol fraction (>2500 μg/mL) were similar to those reported by Aneja and Joshi [5], the MIC values against *P. gingivalis* and *P. intermedia* indicate substantially more potent growth inhibition (625 and 156 μg/mL, respectively). Furthermore, Cai and Wu [25] also identified a number of compounds in the methanolic fraction and screened them for antibacterial activity. Kaempferol and myricetin were particularly potent inhibitors of *P. gingivalis* and *P. intermedia* growth, each with MIC values of 20 μg/mL. Gallic acid, 5,7-dihydroxy-2-methylchromone 8-C-β-glucopyranoside, biflorin, rhamnocitrin, and oleanolic acid were also potent inhibitors of *P. gingivalis* and *P. intermedia* growth, with MIC values generally in the range 150–1000 μg/mL.

Several other studies have also reported antibacterial activity for crude *S. aromaticum* extracts against multiple gram-negative bacteria including *Camphylobacter jejuni*, *Eneterococcus faecalis*, *Escherichia coli*, *Klebsiella pneumoniae*, *Pseudomonas aeruginosa*, *Proteus vulgaris*, and *Salmonella enteritidis*, as well as the gram-positive bacteria *Bacillus subtilis*, *S. aureus*, and *Staphylococcus epidermidis* [24, 51, 57, 60, 84, 94, 104]. Notably, another study reported that *S. aromaticum* oil inhibited the growth of methicillin-resistant strains of *S. aureus* and *Streptococcus* spp. [195]. A different study reported growth inhibitory activity against multiple strains of *S. epidermidis* and attributed the activity to eugenol [78]. Diluted (1–2%) *S. aromaticum* essential oil significantly inhibited the growth rate of *Listeria monocytogenes* [119]. *S. aromaticum* essential oil components also potentiate the effects of conventional antibiotics [68, 109, 115]. The synergistic interactions between eugenol and ampicillin/gentamycin are particularly noteworthy. The same studies also reported moderate additive combinational effects for eugenol in combination with vancomycin and β-lactam antibiotics.

S. cumini extracts and essential oils have also been reported to have broad-spectrum antibacterial activity. A recent study used agar diffusion assays to screen aqueous and ethanolic *S. cumini* extracts against a broad panel of bacteria, including several medically important human pathogens [135]. That study tested the ability of *S. cumini* extracts to inhibit the growth of *Bacillus cereus*, *Bacillus megaterium*, *Bacillus pumilus*, *Bacillus subtilis*, *Clostridium acetobutylicum*, *Citrobacter freundii*, *Enterobacter aerogenes*, *Enterococcus faecium*, *E. coli*, *K. pneumoniae*, *P. mirabilis*, *Salmonella typhi*, *Salmonella paratyphi*, *S. aureus*, *S. epidermidis*, *Streptococcus faecalis*, *Streptococcus lactis*, *Streptococcus pyogenes* and *Vibrio cholerae*. The ethanolic extract had particularly broad-spectrum antibacterial activity, inhibiting the growth of 14 of the 19 bacterial species tested. Furthermore, large zones of inhibition were reported, indicating potent antibacterial activity. However, only relatively high extract concentrations were tested in that study (10–50 mg/disc). Furthermore, MIC values were not determined, making it impossible to compare the efficacy with other studies. Similar results were also reported in another study, which screened *S. cumini* essential oil [155]. In a previous study [135], MIC values were not determined. However, this study did determine some of the essential oil components and postulated that the antibacterial activity may be due to the terpenoid components. In particular, relatively high levels of pinocarveol, α-terpineol, myrtenol, eucarvone, muurolol, geranylacetone, α-cadinol, and pinocarvone were reported. A recent study reported antibacterial activity for a *S. cumini* essential oil against *E. coli, P. aeruginosa, Neisseria gonorrhoeae, B. subtilis, S. aureus, and E. faecalis* [112]. Whilst that study also did not quantify the potency by determining MIC values, it also highlighted terpenoids as a likely contributor to antibacterial activity. In addition to identifying several of the same compounds as in the study by Shafi et al. [155], this study also identified α-pinene, β-pinene, trans-caryophyllene, 1,3,6-octatriene, δ-3-carene, humulene, and α-limonene as the major constituents.

The antibacterial activity of other Asian *Syzygium* species has also been reported. *S. jambos* leaf extracts have been shown to inhibit the growth of *Aeromonas hydrophilia, Alcaligenes faecalis, B, cereus, C. freundii, E. coli, K. pneumoniae, P. mirabilis. Pseudomonas fluorescens, Salmonella newport, Serratia marcescens, Shigella sonnei, S. aureus, S. epidermidis,* and *S. pyogenes* [10]. The study used disc diffusion susceptibility screening and MIC assays and reported potent inhibition against several bacteria species. Indeed, MIC values between 180 and 800 μg/mL were reported for *A. faecalis, A. hydrophilia, S. aureus* and *B. cereus*. Another study screened *S. jambos* bark extracts against an extended bacterial panel and also reported potent bacterial growth inhibition [48]. That study reported MIC values as low as 0.25 μg/mL against several medically important bacterial pathogens such as: *Enterococcus gallinarum, S. aureus, Staphylococcus cohnii, Staphylococcus hominis, Staphylococcus warneri*, and *Yersinia enterocolitica*. The authors attributed the potent antibacterial activity to high tannin content of bark extracts, but did not identify the individual components. Similarly, *Syzygium jambolanum* seed extracts are potent inhibitors of *B. subtilis, E. coli, K. pneumoniae, P. aeruginosa, S. aureus*, and *Salmonella typhimurium* growth, with MIC values of 62–250 μg/mL [30]. *Syzygium alternifolium* and *Syzygium samarangense* fruit extracts are potent inhibitors of *B. cereus, E. coli, K. pneumoniae, P. aeruginosa*, and *S. aureus* [141].

Several African *Syzygium* species also have potent antibacterial activity. Perhaps the best studied of these species is the Southern African species *S. cordatum*. Several studies have reported the growth inhibitory properties of *S. cordatum* extracts against diarrhea-causing bacteria [108, 165].

Potent growth inhibitory activity was reported for *S. cordatum* stem bark extracts against *S. aureus, Shigella boydii, Shigella dysentery, Shigella flexneri*, and *V. cholerae*, with MIC values of 150–625 μg/mL [108]. Interestingly, the bark extract tested in that study did not affect *E. coli* growth. In contrast, a different study also screened *S. cordatum* bark extract against *E. coli* and reported moderate (1.44 mg/mL) growth inhibitory activity [165]. Another study examined the antibacterial activity of *S. cordatum* bark and leaf extracts against bacteria associated with food poisoning and food spoilage [38]. Broad-spectrum antibacterial activity was reported for both extracts, with similar potency against most bacterial species (generally 150–750 μg/mL). Recent studies have highlighted the potential of *S. cordatum* as growth inhibitors of the bacterial triggers of some autoimmune inflammatory diseases [44] and thus of the initial events of several autoimmune diseases. *S. cordatum* leaf extracts are potent inhibitors of *P. mirabilis* [40] and *K. pneumoniae* growth [39] and thus have potential in the prevention and treatment of rheumatoid arthritis and ankylosing spondylitis, respectively.

Another study used disc diffusion assays to examine the antibacterial activity of *Syzygium guineense* extracts and isolated compounds against *B. cereus, E. coli*, and *S. sonnei* [49]. That study only determined growth inhibition using discs with standard amounts of compound infused and unfortunately did not determine the MIC values (despite referring to these standard discs as MICs); thus, it is difficult to make a comparison with other studies. However, that study is of particular interest as it used a bioactivity driven separation approach to isolate, screen, and identify the inhibitory components. Ten triterpenoids were isolated, and two of these were determined to have significant antibacterial activity. These were identified as arjunolic acid and asiatic acid [49].

Despite having greatest diversity among species, the Australia *Syzygium* species have received less attention, possibly due to the paucity of ethnobotanical records. However, recent studies [75, 88] have reported bacterial growth inhibitory activity for some Australian *Syzygium* species the growth inhibitory properties of *S. australe* and *S. leuhmannii* perhaps have been the most extensively studied. Much of the interest in these species is derived from the very high antioxidant capacity of fruits. A recent study screened *S. australe* and *S. leuhmannii* fruit extracts against *A. hydrophilia*, *A. faecalis*, *B. cereus*, *B. subtilis*, *C. freundii*, *E. aerogenes*, *E. coli*, *K. pneumoniae*, *P. aeroginosa*, *P. fluorescens*, *Salmonella salford*, *S. marcescens*, *S. aureus*, and *Y. enterocolitica* [152]. That study reported broad-spectrum antibacterial activity for extracts of both species. Indeed, the methanolic extracts inhibited the growth of approximately 90% of the bacterial species screened. Another study reported that *S. australe* and *S. leuhmannii* leaf and fruit extracts inhibited the growth of *Bacillus anthracis* [201]. This is a significant finding as this is a spore-forming bacterium and is considered particularly difficult to destroy. Similarly, leaf extracts from both these *Syzygium* species also have potent antibacterial activity against a panel of bacteria [34, 37]. A recent study also examined the antibacterial properties of several other Australian *Syzygium* spp. including *Syzygium forte, Syzygium francissi, Syzygium moorei, Syzygium puberulum* and *Syzygium wilsonii* [26]. All species demonstrated significant antibacterial activity, although they were generally substantially less potent inhibitors of bacterial growth than was reported for the *S. australe* and *S. leuhmannii* leaf and fruit [34, 37, 152].

3.4.2 INHIBITION OF FUNGAL GROWTH

Antifungal activity has been reported for several *Syzygium* species, although less quantitative data are available for the antifungal activity of *Syzygium* extracts compared to antibacterial activity. The antifungal properties *S. aromaticum* have generally been more extensively studied than for the other species. One study reported antifungal activity for *S. aromaticum* essential oil against multiple strains of five *Candida* spp., including the medically important species *C. albicans* (candidiasis) and *C. tropicalis* (involved in the onset of Crohn's disease) [27, 130]. The oil also inhibited the growth of *C. glabrata, C. krusei*, and *C. parapsilosis*, which although not generally pathogenic in most people, may cause nosocomial infections in immunocompromised individuals [59, 131, 184]. The study reported similarly potent antifungal activity for the crude essential oil and for the major component eugenol against *Candida* spp. (300–650 μg/mL). The same study also tested the essential oil and pure eugenol against

Aspergillus flavus, *Aspergillus fumigatus*, and *Aspergillus Niger* and reported similarly potent antifungal activity for both. Furthermore, both the essential oils and pure eugenol were potent inhibitors of the filamentous fungal pathogens *Epidermophyton floccosum* (tinea), *Trichophyton mentagrophytes*, and *Trichophyton rubrum* (athlete's foot, jock itch, ringworm) as well as *Microsporum canis* and *Microsporum gypseum* (tinea). The potency of the essential oil and eugenol against this fungal panel was particularly noteworthy as the majority of the fungi screened were fluconazole-resistant strains, further highlighting their therapeutic potential. That study determined that the *S. aromaticum* essential oil and pure eugenol killed fungal cells by two mechanisms, which include direct induction of lesions in the fungal cell membrane and the induction of a substantial reduction in the production of the fungal membrane component ergosterol [184]. Further studies have also reported antifungal activity against onychomycosis-causing fungi [61] and *Saccharomyces cerevisiae* [28].

Several other Asian *Syzygium* species also have antifungal activity. One study reported antifungal activity of *S. jambolanum* seed extracts against *C. albicans*, *Cryptococcus neoformans*, *Microsporum gypseum, Rhizopus* spp. and several *Aspergillus* and *Trichophyton* species [30]. The *S. jambolanum* seed extracts were particularly potent inhibitors of *C. albicans* and all *Aspergillus* spp., with MIC values approximately 30–60 μg/mL. Similarly, *Syzygium lineare* leaf extracts are good inhibitors of *T. rubrum* growth [51]. In contrast, the same study reported that the *S. lineare* leaf extracts were inactive or had only low activity against *Trichophyton mentagrophytes*, *Trichophyton simii*, *Epidermophyton floccosum, Scopulariopsis spp., A. niger, Botyritis cinereae, Curvularia lunata*, and *C. albicans*. A separate study examined the antifungal activity of *S. alternifolium* and *S. samarangense* fruit extracts [141]. The *S. alternifolium* petroleum ether extract and the methanolic and petroleum ether *S. samarangense* extracts were potent inhibitors of *C. albicans* growth, with MIC values of generally <500 μg/mL.

Some African *Syzygium* species have also antifungal activity. In a study screening South African plants traditionally used by the Venda people for antifungal activity, *S. cordatum* bark extracts were highlighted as amongst the most potent of the traditional medicines tested against *C. albicans* [173]. A further study examined *S. cordatum* leaf extracts against a panel of five *Fusarium* fungal plant pathogens [149]. That study reported that the growth of all fungal species was inhibited by the extract. MIC values of ≥ 7.5 mg/mL indicated only low potency. However, there is some doubt about the validity of the reported results as very high MIC values of approximately 20 mg/mL were reported for the positive control (nystatin), indicating that the assay may have been faulty. Furthermore, that study also reported minimal fungicidal concentrations (MFC) that were substantially lower than the MIC values for the same extracts (MFC = 3.75 mg/mL compared to MIC = 7.5 mg/mL). Thus, it is likely that these results are erroneous. Limited reports are available on the antifungal activity of *Syzygium* spp. from other regions of the world, although leaf extracts of the Australian species *S. australe* have been reported to have low potency for *A. niger* growth inhibitory activity [34]. These extracts were unable to inhibit *C. albicans* growth. The same study also screened *S. leuhmannii* leaf extracts and found them to be completely devoid of antifungal activity for all fungi tested.

3.4.3 INHIBITION OF PROTOZOAL GROWTH

S. aromaticum also has therapeutic potential for the treatment of protozoal diseases. A recent study reported that *S. aromaticum* essential oil and its major constituent (eugenol) inhibit the growth of *Giardia lamblia* [102]. Both the essential oil and pure eugenol displayed similarly potent antigiardial activity (130 and 100 µg/mL, respectively). Furthermore, both the essential oil and eugenol induced significant morphological changes in the Giardia trophozoites, including shape changes, presence of cytoplasmic precipitates and autophagic vesicles, internalization of the flagella and ventral disc, membrane blebs, as well as intracellular and nuclear clearing [102]. These morphological changes were accompanied by decreased ability of the trophozoites to adhere to surfaces and cell lysis. These events occurred rapidly, with the decreased adherence apparent within one hour and lysis of 50% of the trophozoites following five hours exposure to *S. aromaticum* essential oil [102].

S. aromaticum essential oil also has profound effects on the growth of the epimastigote and bloodstream trypomastigote forms of the protozoal parasite *T. cruzi* [151]. As *T. cruzi* is the cause of Chagas disease, *S. aromaticum* may potentially treat the disease. The essential oil was a potent growth inhibitor, with IC_{50} values of 99.5 and 57.5 μg/mL against epimastigotes and trypomastigotes, respectively [151]. The same study also reported substantial morphological changes including rounding of the cell body and alteration of the nucleus, although no changes to the plasma membrane were noted.

Anti-protozoal activity has also been reported for *S. cumini* [45, 145]. A recent study reported significant inhibitory activity of *S. cumini* essential oil against *Leishmania amazonensis*, with an IC_{50} of 60 μg/mL [45]. Interestingly, this oil also had significant molluscicidal activity against *Biomphalaria glabrata* (an intermediate host in *Leishmania* spp. lifecycle), with an LC_{50} of 90 μg/mL. Thus, *S. cumini* essential oil may be valuable in combating the neglected tropical diseases leishmaniasis and schistosomiasis via multiple mechanisms. The same study used GC-MS analysis to determine the major components of the oil to be α-pinene, (Z)-β-ocimene, and (E)- β-ocimene. However, an LC_{50} of 175 μg/mL was determined in toxicity tests against *Artemia nauplii*. As an LC_{50} <1000 μg/mL has been defined as toxic in this assay [145], this may affect the usefulness of *S. cumini* essential oil for this purpose. A follow-up study from the same group evaluated the effects of the crude essential oil and the major component (α-pinene) on the promastigote and anexic amastigote forms of the protozoa and reported that both the essential oil and α-pinene were potent inhibitors of *L. amazonensis* proliferation [45]. The anti-*Leishmania* effects were determined to be mediated by immunomodulatory activity. Both the essential oil and α-pinene induced increase in phagocytic and lysosomal activity and mediated increase in NO levels [45].

Methanolic stem bark extracts of the African species *S. giuneense* have been reported to have antimalarial activity in a murine model [62]. Significant suppression of *Plasmodium berghei* ANKA strain was observed in early stage assays. Indeed, a 400 mg/kg body weight dose suppressed nearly 50% of the parasites growth. Similar results were determined in curative and parasite suppression tests. While there are few reports of direct antiplasmoidal activity for other *Syzygium* spp., *S. aromaticum* has mosquito repellent [168] and larvicidal properties [168] and thus may be useful in controlling other phases of *Plasmodium* spp. lifecycle.

3.4.4 ANTIVIRAL ACTIVITY

The antiviral activity of *Syzygium* species has been less extensively studied. *S. aromaticum* fruit extract has been reported to have potent antiviral activity against hepatitis C virus (HCV), with greater than 90% inhibition of HCV protease activity at 100 μg/mL [72]. Similarly, an aqueous extract of *S. aromaticum* was reported to be a potent inhibitor of murine and human cytomegaloviruses (CMV) in vivo, with an IC_{50} value of 7.6 μg/mL determined using a plaque reduction assay [204]. A different study also reported antiviral activity for an *S. aromaticum* extract against herpes virus and identified the antiviral compound as eugenin [175]. Furthermore, isolated compounds present in *S. aromaticum* oils inhibit viral replication. In particular, eugenol is effective against herpes simplex-1 (HSV-1) and HSV-2 viruses [19].

3.5 ANTICANCER ACTIVITY

The anticancer properties of the Asian species *S. aromaticum* and *S. cumini* have been most extensively studied. Numerous studies have reported cytotoxic properties of these species against a broad panel of carcinoma cell lines of various anatomical origins. One study examined antiproliferative activity of *S. aromaticum* essential oil extracts against HeLa (cervical cancer), MCF-7 and MDA-MB-231 (breast cancer), DU-145 (prostate cancer), and Te-13 (esophageal cancer) cell lines and reported that with the exception of DU-145 cells, the proliferation of all cell lines was significantly inhibited [52]. The same study also determined that the antiproliferative activity was generally due to apoptosis, with maximal cell death occurring within 24 hours at 300 μL extract /mL of cells. Furthermore, that study also examined the phytochemistry of the active extracts and reported that eugenol and β-caryophyllene were the major constituents, accounting for 79% and 13% of the crude *S. aromaticum* essential oil, respectively. Notably, the crude essential oil and the extracts were not significantly toxic for normal human peripheral blood lymphocytes.

Further studies have reported potent cytotoxicity for *S. aromaticum* leaf, bark, and stem extracts against MCF-7 human estrogen-dependent breast cancer and MDA-MB-231 estrogen independent breast cancer cell lines, with IC_{50} values as low as 30 μg/mL [3]. Furthermore, the authors reported that the cytotoxicity was due to the induction of apoptosis. The extracts contained high levels of betulinic acid, although the authors postulated that other components were responsible for the cytotoxic activity. *S. aromaticum* flower-bud (clove) extracts also prevented the induction of lung cancer by benzo(a)pyrene in vivo in a murine model [16]. Treatment of mice with a clove infusion significantly reduced the number of proliferating cells and increased the number of apoptotic cells. Mechanistic studies also revealed an upregulation of the expression of the pro-apoptotic proteins p53 and Bax concurrent with a downregulation of the antiapoptotic protein Bcl-2. The extract also downregulated the expression of COX-2, cMyc, Hras. and several growth-promoting proteins. Furthermore, the expression and activation of caspase 3 was apparent at very early stage of carcinogenesis. A different study assessed the ability of *S. aromaticum* fruit extracts and flavonoids isolated from these extracts for protective activity against gastric cancer formation [14]. The study showed that the *S. aromaticum* flavonoids kaempferol-3-O-β-D-glucopyranoside and kaempferol-3-O-α-L-rhamnopyranoside induced significant cytotoxicity. Interestingly, kaempferol-3-O-α-L-rhamnopyranoside had high bonding energy on human epidermal growth factor receptor-2 (HER2). Further *S. aromaticum* flavonoid (5-hydroxy-7,4'-dimethoxy-6,8-di-C-methylflavone) lacked cytotoxic affects, but had cytostatic activity against AGS cells by inhibiting cell cycle progression at the G2/M phase.

Because multiple parts of *S. cumini* are used therapeutically to treat cancer, the fruit has attracted interest among researchers. Several studies have reported potent antiproliferative activity for *S. cumini* fruit extracts against HeLa and SiHa cervical carcinoma cells [18], A549 human non-small cell lung carcinoma cells [7] and U251 (brain tumor cell line), and HEPG2 (hepatocarcinoma cells) and H460 (lung carcinoma cell line) [120]. Aquil et al. [7] and Nazif [120] postulated that anthocyanins were responsible for the antiproliferative activity. They. [7] identified cyanidin, delphinidin, malvidin, peonidin, petunidin, and pelargonidin as the major anthocyanin fruit components. Nazif [120] identified glycoside derivatives of these compounds, such as pelargonidin-3-O-glucoside, pelargonidin-3,5-O-diglucoside, cyanidin-3-O-malonyl glucoside, and delphinidin-3- O-glucoside. *S. cumini* seed extracts also have antiproliferative activity against multiple carcinoma cell lines [120]. A recent study reported that *S. cumini* seed extracts inhibited the proliferation of a panel of human carcinoma cell lines including MCF-7 (breast cancer), A2780 (ovarian carcinoma), PC-3 (prostate carcinoma), and H460 (non-small cell lung carcinoma) [202]. A different study demonstrated that similar seed extracts also inhibit skin carcinogenesis in vivo in a murine model [127].

Further studies have reported that other Asian *Syzygium* species also inhibit cancer cell proliferation. The anticancer properties of *S. samarangense* against CHO-AA8 Chinese hampster ovary, MCF-7 and SKBR-3 human breast cancer carcinoma cells [4], and SW480 human colon cancer cells [166] have been reported. Both studies determined potent antiproliferative activity of these extracts. Indeed, IC_{50} values of 10–60 μg/mL were determined for *S. samarangense* fruit pulp extract and some semi-purified fractions of the extract cells [166]. Amor et al. [4] reported considerably more potent antiproliferative activities for compounds purified from the fruit pulp extract (≤0.2 nM). These potencies are especially promising compared to the IC_{50} values of other conventional cancer chemotherapeutic drugs against several cell lines [1.14 μg/mL (4.8 μM) for pure cisplatin; 2.5–7.5 nM for paclitaxel [98]]. Different cell lines and assay systems were used by Amor et al. [4] compared to these other studies, and this may account for these differences in the results. However, the IC_{50} values reported by Amor et al. [4] are particularly noteworthy as they indicate that the isolated *S. samarangense* fruit pulp compounds are at least 10–20 times more potent anticancer drugs than cisplatin and paclitaxel.

Interestingly, research studies by Amor et al. [4] and Simirgiotis et al. [166] attributed the anticancer properties to chaclones. The methylated chalcones 2'-hydroxy-4',6'-dimethoxy-3'-methylchalcone, 2',4'-dihydroxy-6'-methoxy-3',5'-dimethylchalcone, 2',4'-dihydroxy-6'-methoxy-3'-methylchalcone,2',4'-dihydroxy-6'-methoxy-3'-methyldihydrochalcone, and 2',4'-dihydroxy-6'-methoxy-3',5'-dimethyldihydrochalcone were identified as the highly potent purified compounds by Amor et al. [4]. Simirgiotis et al. [166] identified the same chalcones as well as the quercetin glycosides reynoutrin, hyperin, myricitrin, quercitrin, quercetin, and guaijaverin.

Fewer studies have examined the anticancer properties of the African and Australian *Syzygium* species. Indeed, while several papers reported that selected African *Syzygium* species attenuate oxidative stress in cell lines [42, 43], the authors of this chapter were unable to find any studies that specifically reported anticancer activity of the African *Syzygium* species. While there is also a dearth of anticancer studies for the Australian species, one study has reported that *S. australe* and *S. leuhmannii* leaf and fruit extracts were potent inhibitors of carcinoma cell growth [75]. Indeed, the study reported IC_{50} values of 27 and 172 μg/mL against HeLa cervical and CaCo2 colorectal carcinoma cell lines, respectively. Furthermore, the study screened multiple extracts and correlated the antiproliferative activity with extracts with high antioxidant capacity. However, that study did not identify the active components, nor did it evaluate the anti-proliferative mechanism.

The anticancer properties of the other *Syzygium* species remain largely unexplored. However, a defining feature of the genus is that the fruit (and often the leaves) possess high antioxidant activities. The induction of cellular oxidative stress has been linked with the etiology and progression of several types of cancer [22, 182]. Thus, the high antioxidant contents of many *Syzygium* species may inhibit cancer formation and/or progression. Chromosome instability is also a common feature of many of the cancers that have been linked with oxidative stress, suggesting that increased oxidative stress may contribute to development of genetic instability. Oxidative stress leading to genetic instability would result in the emergence of new tumor phenotypes. In such populations, a decrease in apoptosis but an increase in tumor growth and subsequent tumor progression are observed. Consumption of high levels of antioxidants (as are characteristic of *Syzygium* species) may block this genetic instability and thus block tumorigenesis.

The anticancer properties of the other *Syzygium* species remain largely unexplored. However, a defining feature of the genus is that the fruit (and often the leaves) possess high antioxidant activities. The induction of cellular oxidative stress has been linked with the etiology and progression of several types of cancer [22, 182]. Thus, the high antioxidant contents of many *Syzygium*species may inhibit cancer formation and/or progression. Chromosome instability is also a common feature of many of the cancers that have been linked with oxidative stress, suggesting that increased oxidative stress may contribute to development of genetic instability.

Oxidative stress leading to genetic instability would result in the emergence of new tumor phenotypes. In such populations, a decrease in apoptosis but an increase in tumor growth and subsequent tumor progression are observed. Consumption of high levels of antioxidants (as are characteristic of *Syzygium* species) may block this genetic instability and thus block tumorigenesis.

Most of the currently used anticancer agents prescribed for the treatment of established tumors (e.g., doxorubicin, daunorubicin, mitomycin C, etoposide, cisplatin, arsenic trioxide, ionizing radiation, photodynamic therapy) depend exclusively or in part on the production of ROS for cytotoxicity. Sensitivity of tumor cells to oxidative stress and/or apoptosis may affect treatment success [46, 146]. Studies on the antioxidant/ pro-oxidant effects of extracts from plant species from other genuses have demonstrated that the ability of a plant extract to exert antioxidant activity depends on multiple factors. *Aloe vera* antioxidant components for example may function as either an antioxidant or an oxidant, with their action being dependent upon their concentration [35]. The *A. vera* anthraquinone aloe emodin exerts antioxidant behavior at lower concentrations, but acts as a pro-oxidant at high concentrations. In contrast, a different *A. vera* anthraquinone (aloin) has an antioxidant effect at higher concentrations, but has pro-oxidant effects at low concentrations [35]. *A. vera* extracts and components may act as either antioxidants or as oxidants, depending on different levels of the various constituents, and on their ratios. Thus, although many *Syzygium* species have very high antioxidant contents, it is possible that the individual components may act as either antioxidants or as oxidants and thus may also be effective in the treatment of cancer as well as in its prevention at different concentrations.

Similar pro-oxidant effects have been reported for other antioxidant phytochemicals including flavonoids [138] and tannins [167], which are present in relatively high concentrations in many *Syzygium* species (Table 3.1). The pro-oxidant/antioxidant effect of plant extracts is due to the balance between the free radical scavenging activities and reducing power of their phytochemical components [35]. The antioxidant vitamin ascorbic acid is present in high levels in several *Syzygium* species. While ascorbic acid has well-characterized antioxidant bioactivities, it also acts as a pro-oxidant at high concentrations due to the greater reducing power of ascorbic acid than its free radical scavenging activity [77]. In the presence of transition metal ions, ascorbic acid will function as a reducing agent, thus reducing the metal ions. In this process, it is converted to a pro-oxidant. Therefore, high dietary intake of ascorbic acid in individuals with high iron levels (e.g., premature infants) may result in unexpected health effects due to the induction of oxidative damage to susceptible biomolecules [66].

ROS-based tumor therapy would cause tumor regression, if the tumor cells are not apoptotic/oxidant-resistant cells. Therefore, if *Syzygium* antioxidant components are present in concentrations and ratios consistent with pro-oxidant activity, the extract would be expected to induce apoptosis and therefore would have anticancer activity. If the levels of components are consistent with a reducing environment, antioxidant activity would occur, and the extract would not have anticancer activity. Conversely, if the protocol is repeated on a tumor with apoptotic-resistant/oxidant-resistant cells, the converse would apply and tumor progression would be observed.

3.6 ANTI-INFLAMMATORY ACTIVITY AND IMMUNOMODULATION

Inflammation is a complex response by our body to injury, which typically follows a variety of insults including burns, wounds, bites, stings, etc. It is characterized by a wide variety of symptoms including swelling, redness, and pain [103]. Though *Syzygium* species contain multiple active phytochemicals, many with antioxidant activity, it is likely that several of these may be required to address different aspects of the inflammatory process. The inflammatory processes require cellular release of several classes of molecules. Vasoactive substances (e.g., bradykinin, prostaglandins, and vasoactive amines) are required to dilate blood vessels, thereby opening junctions between cells to allow leukocytes to pass through capillaries. Any compound capable of blocking these vasoactive substances would potentially have a therapeutic effect as an anti-inflammatory agent. The phytosterols β-sitosterol and β-sitosterol glycoside are present in abundance in several *Syzygium* species [107]. Other phytosterols including sitost-4-en-3-one, betulinic acid, oleanic acid, friedelin, and 3-β-friedelinol are also present in some species [107]. Phytosterols (including β-sitosterol) stimulate smooth muscle cells to release prostacyclin (PGI_2) [132]. Conversely, β-sitosterol treatment blocks the release of PGI_2 and prostaglandin E_2 (PGE_2) from macrophages [11]. Thus, β-sitosterol treatment would be expected to affect vasodilation and therefore have a therapeutic effect on inflammation.

Numerous recent studies have reported anti-inflammatory properties of *S. cumini* extracts and essential oils. Various parts of the plant have been reported to inhibit inflammation, although the activity of leaf extracts and selected oils has been the most extensively examined. Several studies have reported significant anti-inflammatory activity of *S. cumini* leaf extracts using a rodent paw edema model [21, 74, 99]. While both these studies reported good anti-inflammatory activity of *S. cumini* leaf extracts, they differed substantially in their potency. Jain et al. [74] reported a significant reduction in carrageenan-induced paw swelling in adult Wistar rats at oral doses of 200 and 400 mg/kg. In contrast, Brito et al. [21] and Lima et al. [99] reported substantially higher potency in an adult Swiss mouse model, with a reduction of paw swelling evident at doses as low as 0.01 μg/kg. These studies used different rodent models, which may contribute to these different potencies. Furthermore, different extraction protocols were utilized to prepare the extracts. Jain et al. [74] tested methanol and ethyl acetate leaf extractions, whereas Brito et al. [21] and Lima et al. [99] screened infusions for bioactivity. As methanol, ethyl acetate, and water are likely to extract quite different phytochemical profiles, the different potencies are perhaps not surprising. Notably, flavonoids and hydrolysable tannins were present in relatively high abundance in the *S. cumini* leaf infusions, and the authors postulated that these contributed to the observed anti-inflammatory activity [99]. While the specific compounds were not determined in that study (only the class of compounds), flavonoids and tannins have well-established anti-inflammatory properties [33], and thus, their presence in this extract is consistent with the observed activity.

Essential oils produced from *S. cumini* leaves also have notable anti-inflammatory activity. A recent study evaluated the effects of the oils and their terpenoid-enriched fractions using a lipopolysaccharide (LPS)-induced pleurisy model and reported that eosinophil migration was significantly inhibited by the *S. cumini* leaf essential oil and its fractions by 63–76% [148]. The ability to inhibit eosinophil migration correlated with more volatile fractions. These higher volatility extracts were rich in α-pinene, β-pinene, limonene, ocimene isomers, β-caryophyllene, and humulene, all of which have been shown to have anti-inflammatory activity in the LPS-induced pleurisy model [58, 101, 128, 169]. The potency of the individual oil terpenoids α-pinene and β-caryophyllene and of an artificial mixture of these compounds (at ratios similar to those in the essential oil) has also been reported [148]. Interestingly, that study indicated that the synergy between these terpenoids is likely due to the addition of α-pinene, which significantly potentiated the anti-inflammatory activity of β-caryophyllene.

Similarly, *S. cumini* seed extracts have been reported to have anti-inflammatory activity in rat paw edema studies [92, 93]. All these studies reported similar efficacies in the range 200–250 mg/kg. Interestingly, these crude extracts had similar anti-inflammatory potency to that of the pure compound sodium diclofenac (100 mg/kg), which was used as a positive control [111]. A further study reported that *S. cumini* seed extract significantly inhibited neutrophil chemotaxis [54]. A different study reported that *S. cumini* bark extract also had potent anti-inflammatory activity and reduced swelling by 84% in a mouse paw edema model [117].

S. aromaticum extracts [87, 176] and essential oil [6] also inhibit inflammation in rodent model systems. The inhibition was of a similar potency to that of the *S. cumini* extracts, with 250 mg/kg concentration suppressing 52% of rat paw edema [176]. However, the study also reported an LD_{50} value of just over 500 mg/kg. The similarity of the effective and toxic doses would impact the usefulness of *S. aromaticum* extract for systemic uses. Further studies have demonstrated that an aqueous *S. aromaticum* flower bud extract inhibits systemic anaphylaxis and IgE-induced cutaneous reactions via an inhibition of histamine release [87]. The study also reported that these effects were related to the activation of adenylate cyclase, resulting in an increase in intracellular cAMP levels.

Other Asian *Syzygium* species also have potent anti-inflammatory activity. The *S. alternifolium* root extracts (500 mg/kg) induced rapid reduction in carrageenan-induced Wistar rat paw edema within 1 hour [80]. Furthermore, the effects of the extract were still evident 5 hours after the initial induction of inflammation. Carrageenan-induced edema is believed to be biphasic, with the early phase (1–2 hours) mediated by histamine, serotonin, and increased synthesis of prostaglandins in the affected tissue [199]. The later phase is sustained by prostaglandin release and mediated by bradykinin and leukotrienes produced by tissue macrophages [199]. The study also examined the in vitro enzyme activity of 5-lipoxygenase (5-LOX) activity and reported a significant reduction, similar to that of the standard drug Zileutin. Thus, *S. alternifolium* root extracts have potential in the prevention and treatment of inflammation.

Similarly, *S. samarangense* leaf extracts also have significant anti-inflammatory activity in a mouse paw edema test [114]. That study reported that 200 mg/kg concentration reduced paw edema by 65%, compared to 84% for a similar dose of diclofenac. A different study identified the *S. samarangense* chalcone aurentiancin as a possible anti-inflammatory mediator and investigated its effects on LPS-induced inflammation in mouse macrophages [86]. The study reported that aurentiancin significantly inhibited LPS-induced nitric oxide production via suppression of nitric oxide synthase expression [86]. Aurentiancin exposure also down-regulated mRNA levels of the proinflammatory cytokines TNF-α, interleukin-6, NF-κB, and p65. Furthermore, aurentiancin treatment attenuated the phosphorylation of MAPKs and p65 [86]. Thus, the authors concluded that aurentiancin mediates its anti-inflammatory activity via the inhibition of NF-κB activation. Similarly, *S. jambos* extracts and the pure compounds ursolic acid and myricitin also inhibit the release of the proinflammatory cytokines interleukin-8 and TNF-α by 74–99% [159].

Few studies have examined the anti-inflammatory potential of African or Australian *Syzygium* species. A recent study used an induced mouse paw edema study to examine the effects of leaf extracts of Western African species *S. guineense* on inflammation [73]. The study reported significant anti-inflammatory activity at relatively high extract concentrations (1000 mg/kg). However, lower concentrations (200 mg/kg) did not significantly affect inflammation. While authors of this chapter were unable to find reports of direct effects of other *Syzygium* spp. on inflammation, several studies have reported that the African species *S. cordatum* [39, 40] as well as the Australia species *S. australe* and *S. leuhmannii* [41] can inhibit some bacterial triggers of autoimmune inflammatory diseases and thus have potential in the prevention and treatment of these inflammatory diseases. However, none of these studies tested whether these extracts also have direct effects on inflammatory mechanisms and such studies are required. If anti-inflammatory activity is subsequently demonstrated, these species may be particularly useful in the treatment of these diseases via pluripotent mechanisms.

Much more research work is required to evaluate the anti-inflammatory activity of other *Syzygium* species. Previous and current studies have focused on relatively few Asian *Syzygium* species due to their use in traditional healing systems. However, due to the taxonomic relationships between species of this genus as well as their phytochemical similarities, it is likely that other species of this genus might also have anti-inflammatory potential. In particular, investigation of the anti-inflammatory potential of African and Australian species is urgently required.

3.7 ANTIDIABETIC ACTIVITY

The potential for *Syzygium* species in providing considerable beneficial effects on glucose metabolism and storage, and thus in diabetes treatment, has greatly expanded in recent years [13]. Much has been discovered on the effects of plant extracts, and, in some cases, constituents purified from extracts, in treating diabetes and diabetes-related ailments.

Early studies on *Syzygium cumini* indicated hypoglycemic properties of plant leaves consumed as an infusion [20, 196]. Due to the wide cultivation of the plant worldwide [65, 170, 185], interest in these properties have increased with extracts considered to act as a likely antidiabetic agent [67]. Despite this, the first research outcomes using the infusion or extract preparations were conflicting. Interestingly, evidence began to emerge that revealed a paradoxical lack of anti-hyperglycemic effect on animal subjects and no hypoglycemic effect on fasting blood glucose levels in clinical trials in infusions prepared from leaves of *S. cumini* [177–180]. Ethanolic extracts of the plant did, however, induce pancreatic insulin secretion in diabetic rats [153], but glucose levels were not altered in either diabetic or nondiabetic mice [124]. In contrast, in rats induced with diabetes using streptozotocin, 100 mg/kg doses of ethanolic *S. cumini* kernel extracts restored normal blood glucose and liver glycogen levels comparable to those of control animals and of diabetic animals treated with the hypoglycemic drug glibenclamide [142]. High fructose diets in rats lead to hyperinsulinemia, which can be prevented by daily oral administration of 400 mg aqueous *S. cumini* extracts [190]. Later studies using flavonoid-rich ethanolic extracts of *S. cumini* seeds showed a similar effect [158] as well as generated a gene expression profile that enhances glucose tolerance, which may explain a possible attenuation of the diabetic condition [161]. Collectively, these studies suggest that infusions prepared from the leaves possess little or no activity, whilst extracts from other parts of the plant such as the seeds or bark contain the active ingredient(s).

Shinde et al. [164] showed that ethanolic extracts of *S. cumini* seeds inhibited a key enzyme (α-glucosidase) involved in the metabolism of carbohydrates, leading to postprandial hyperglycemia. Animals treated with the extract showed a concomitant decrease in plasma glucose levels, suggesting an improvement in glucose tolerance in animals challenged with maltose [164]. These findings were supported by others who demonstrated the in vitro inhibition of this enzyme by either ethanolic or aqueous *S. cumini* extracts [110] or of α-amylase by *S. cumini* aqueous seed extracts [82]. In addition, the pathological changes in pancreatic islets and β-cells in type 2 diabetic rats were ameliorated by pre-treatment with 400 mg/kg saline suspension of *S. cumini* extracts [157]. Liver enzymes, which were increased in the alloxan-induced diabetic mice, were also significantly reduced by treatment with 250 mg/kg aqueous extracts of *S. cumini*, thus reducing serum glucose levels [160]. Together, these findings support both hypoglycemic and liver protective roles of *S. cumini* extracts in treating type 2 diabetes.

Advances in research methodologies allowed for more definitive tests to be performed using *S. cumini*. In particular, the phytochemical constituents of extracts from various parts of the plant have been identified to determine the component responsible for the observed antidiabetic properties. The apparent inhibition of enzymes responsible for the breakdown of complex carbohydrates into monomeric glucose may be due to the presence of alkaloids and glycosides in *S. cumini* seed extracts [13, 183]. Such extracts are also known to be rich in flavonoids, which may exert hypoglycemic [158] and pancreatic β-cell regenerative effects [189] in diabetic animals. Sharma et al. [161] partially purified a fraction from ethanolic *S. cumini* seed extracts, which was found to contain saturated fatty acid, Δ^5 lipid, and sterol. When administered to diabetic rabbits, the fraction increased plasma insulin, increased liver and muscle glycogen, and restored glycolysis and gluconeogenesis enzyme activities close to those of untreated animals [161]. Recently, the compound vitalboside A (VBA) was purified from a bioactive fraction obtained from *S. cumini* seeds [181] and found to stimulate insulin uptake, prevent the downregulation of insulin signaling, and promote glucose transport and uptake. While infusions prepared from *S. cumini* leaves had no effect on diabetes, the preparation of a hydroalcoholic (50% water, 50% ethanol) leaf extract was shown to contain high concentrations of phenols and the presence of myricetin glycosides [15], with the latter postulated to have antidiabetic properties [31, 100]. Betulinic acid and 3,5,7,4'-tetrahydroxy flavanone extracted from *S. cumini* seeds appear responsible for the inhibition of carbohydrate metabolism via α-amylase blockade [82].

Evidence of antidiabetic activity has arisen from studies involving other *Syzygium* species. *S. aromaticum* extracts are capable of downregulating the expression of genes encoding liver gluconeogenesis enzymes in vitro [133] and inhibit small intestine carbohydrate-metabolizing enzymes (such as α-amylase, sucrose, and α-glucosidase) with a concomitant decrease in in vivo blood glucose [1, 85, 123]. Thus, *S. aromaticum* extracts appear to act in a similar manner to insulin and have even been shown to increase cellular glycolysis, and thus glucose consumption, in mouse myocytes [187]. Indeed, methanolic extracts almost completely abolished glycation-induced protein crosslinking [129], a process known to be accelerated in diabetes [63]. *S. aromaticum* extracts are rich in free polyphenols and flavonoids [2] and the essential oils contain significant quantities of α-pinine, β-pinene, and 1,8-cineole [123], which may assist in alleviating type 2 diabetes-related oxidative stress [53].

S. aqeuem leaf extracts possess α-amylase and α-glucosidase inhibitory properties, which has been found to be linked to multiple flavonoid constituents contained within the ethanolic fractions [105]. In addition, these compounds prevent glucose uptake and enhance adipogenesis [106], thus acting as insulin-like agents. This may explain the reported use of this plant in a traditional setting for diabetes treatment [139]. *S. samarangense*, or pink wax apple, contains a myriad of phytochemicals including quercetin glycosides [166], flavonoids [143], and triterpenoids [171]. Vescalagin, a diastereoisomer of castalagin [192], was identified in young fruit of *S. samarangense* and found to possess hypoglycemic effects [163]; it was later demonstrated to be effective in ameliorating glucose homeostasis in rats fed high-fructose diets [71]. Ripened *S. densiflorum* fruits contain flavonoids, terpenoids, phenols, quinones, and other phytochemicals [90] within the ethanolic extracts, which appear to be capable of eliciting anti-hyperglycemic effects and restoring pancreatic β-cell capacity [89]. Blood glucose in diabetic rats was significantly reduced following the administration of *S. cordatum* leaf [118] or *S. alternifolium* seed [83, 140] extracts. The methanolic extract of *S. polyanthum* (Wight.) also reduced blood glucose levels in diabetic rats, which may be accounted for by the presence of tannins, flavonoids, glycosides and alkaloids within the extract [198]. Bark extracts (methanolic) of *S. caryophyllatum* also cause dose-dependent decrease in plasma glucose levels in diabetic Wistar rats [10], while this effect was also elicited by *S. calophyllifolium* bark extracts with an accompanying improvement in the hepatic, renal, and pancreatic cellular architecture [29].

3.8 CONCLUSIONS

This survey on the traditional medicinal and pharmacognosy literature has highlighted the therapeutic importance of this useful *Syzygium* species. Various *Syzygium* species have been widely used in diverse traditional medicine systems such as Ayurveda, Siddha, Western, and Southern and Central African medicinal systems as well as in various Australian Aboriginal medicinal systems. Despite this history of traditional usage, until recently, rigorous scientific research has been confined to the study of only a few species. of these, the Ayurvedic medicinal plants (especially *S. aromaticum* and *S. cumini*) have received the most attention. In these species, multiple therapeutic bioactivities have been documented. These species have been reported to have good antioxidant, anticancer, antidiabetic, antiseptic, and anti-inflammatory effects. In some cases, the active phytochemicals have been established, although for many medicinal properties, the active principles have been only partially characterized. Instead, often the partially purified compounds of a crude extract are itemized, but the active component(s) remain unidentified.

In other studies, the active compounds have not been characterized and instead only the classes of compounds in the crude mixture have been determined. Much work is still required to fully understand the phytochemistry of the genus *Syzygium*. Even for more extensive studied species (*S. aromaticum* and *S. cumini*), research studies are required to fully determine the extent of medicinal properties and phytochemistry. Furthermore, few of these previous studies have provided substantial mechanistic details to explain how the active principles achieve their medicinal effects.

The scope for future pharmacognostic studies is perhaps broader for many of the *Syzygium* species not traditionally used in Ayurveda due to the lack of even most basic studies into their therapeutic potential and phytochemistry. Some interesting bioactivity studies have begun to examine African and Australian *Syzygium* species in recent years. However, most of the tested species have only been tested for one or two medicinal properties, and these studies are usually driven purely by an ethnobotanical history of use. Given the taxonomic similarities between species of this genus, it is evident that more bioactivity studies are needed, and where a therapeutic bioactivity is detected, phytochemical and mechanistic studies are also necessary.

Australian *Syzygium* species have been particularly poorly studied despite an oral history of medicinal usage and the species diversity of this region. Because several recent studies have reported an extremely high antioxidant content of *S. australe* and *S. leuhmannii*, there has been a recent interest in these species. However, these studies have generally focused only on bioactivity screening or phytochemical screening, without combining these disciplines for a bioactivity-driven approach to characterize the active components and mechanisms. While *S. australe* and *S. leuhmannii* are receiving attention, the other Australian species remain largely unstudied, and substantial work is needed in this area.

Presently, the *Syzygium* species tested are largely only those with a history of use in traditional medicinal systems. Yet, the similarity between many species indicates that testing all species for bioactivities evident in the more established species is warranted. More investigation is required for a more complete understanding not only of the potential of plants from this genus but also of their phytochemistry. Such an approach will definitely lead to the establishment of useful nutraceuticals and may also result in the development of further pharmaceuticals.

3.9 SUMMARY

Plants of the genus *Syzygium* are widely used traditionally for medicinal purposes in tropical and subtropical regions. Many species are used for their antibacterial, antifungal, antiprotozoal, antiviral, antidiarrheal, antidiabetic, analgesic, antimalarial, antioxidant, anti-inflammatory, and anticancer activities. The antidiabetic and antimicrobial effects of *Syzygium cumini* and *Syzygium aromaticum* have been well documented. The analgesic effects of *S. aromaticum* are also widely known.

However, few studies have rigorously examined the therapeutic properties/mechanisms and phytochemistry of many other species. The therapeutic properties of Australian *Syzygium* species have been particularly poorly reported, despite this region having the richest diversity of *Syzygium* species. This may be because much of our understanding of the use of the Australian species has been irretrievably lost as ethnobotanical knowledge was passed between generations of the first Australians by oral tradition. Thus, in contrast to Asian and African traditional healing systems, there is now generally little understanding of the traditional uses of Australian *Syzygium* species. This has provided considerably fewer avenues for drug discovery.

The last decade has seen a large increase globally in the number of studies on the use of *Syzygium* species as therapeutic agents. Several species used in Ayurvedic medicine (*S. aromaticum* and *S. cumini*) in particular have received much attention. Recent reports have also highlighted the medicinal potential of species from Africa, Australia, and the Oceania region. This chapter summarizes recent research on the medicinal properties, phytochemistry, and therapeutic mechanisms of *Syzygium* species and also highlights future areas of research on the medicinal activities of *Syzygium* species.

KEYWORDS

- **anticancer**
- **antihyperglycemic**
- **antimicrobial**
- **antioxidant**
- **bark**
- **bioactivities**
- **blood glucose**
- **clove**
- **diabetes**
- **ethnobotany**
- **eugenol**
- **extracts**
- **flavonoids**
- **herbal**
- **inflammation**
- **infusions**
- **medicinal plants**
- **myricetin**
- **myrtaceae**
- **phytochemistry**
- **polyphenols**
- **quinones**
- **seeds**
- ***Syzygium aromaticum***
- ***Syzygium australe***
- ***Syzygium cordatum***
- ***Syzygium cumini***
- ***Syzygium leuhmannii***
- **traditional medicine**
- **triterpenoids**

- **vescalagin**
- **vitalboside A**
- **α-amylase**

REFERENCES

1. Adefegha, S. A., Oboh, G., Adefegha, O. M., Boligon, A. A., & Athayde, M. L., (2014). Antihyperglycemic, hypolipidemic, hepatoprotective and antioxidative effects of dietary clove (*Szyzgium aromaticum*) bud powder in a high-fat diet/streptozotocin-induced diabetes rat model. *Journal of the Science of Food and Agriculture, 94*(13), 2726–2737.
2. Ademiluyi, A. O., Akpambang, V. O. E., & Oboh, G., (2009). Polyphenol contents and antioxidant capacity of tropical clove bud (*Eugenia aromatic* Kuntze). *Rivista Italiana Delle Sostanze Grasse*, *86*, 131–137.
3. Aisha, A. F. A., Abu-Salah, K. M., Alrokayan, S. A., Siddiqui, M. J., Ismail, Z., & Majid, A. M. S. A., (2012). *Syzygium aromaticum* extracts as a good source of betulinic acid and potential anti-breast cancer. *Brazilian Journal of Pharmacognosy*, *22*(2), 335–343.
4. Amor, E. C., Villasenor, I. M., Antemano, R., Perveen, Z., Concepcion, G. P., & Choudhary, M. I., (2007). Cytotoxic C-methylated chalcones from *Syzygium samarangense*. *Pharmaceutical Biology, 45*(10), 777–783.
5. Aneja, K. R., & Joshi, R., (2010). Antimicrobial activity of *Syzygium aromaticum* and its bud oil against dental cares causing microorganisms. *Ethnobotanical Leaflets, 14*, 960–975.
6. Apparecido, N. D., Sartoretto, S. M., Schmidt, G., Caparroz Assef, S. M., Bersani-Amado, C. A., & Cuman, R. K. N., (2009). Anti-inflammatory and antinociceptive activities of eugenol essential oil in experimental animal models. *Brazilian Journal of Pharmacognosy*, *19*(1B), 212–217.
7. Aqil, F., Gupta, A., Munagala, R., Jeyabalan, J., & Kausar, H., (2012). Antioxidant and antiproliferative activities of anthocyanin/ellagitannin-enriched extracts from *Syzygium cumini* L. (Jamun, the Indian blaclberry). *Nutrition and Cancer, 64*(3), 428–438.
8. Arora, D. S., & Kaur, G. J., (2007). Antibacterial activity of some Indian medicinal plants. *Journal of Natural Medicine, 61*, 313–317.
9. Arun, R., Prakash, M. V. D., Abraham, S. K., & Premkumar, K., (2011). Role of *Syzygium cumini* seed extract in the chemoprevention of in vivo genomic damage and oxidative stress. *Journal of Ethnopharmacology, 134*, 329–333.
10. Attanayake, A. P., Jayatilaka, K. A. P. W., Pathirana, C., & Mudduwa, L. K. B., (2013). Study of antihyperglycaemic activity of medicinal plant extracts in alloxan induced diabetic rats. *Ancient Science of Life, 32*(4), 193–198.

11. Awad, A. B., Toczek, J., Carol, C. S., & Fink, S., (2004). Phytosterols decrease prostaglandin release in cultured P388D$_1$/MAB macrophages. *Prostaglandins, Leukotrienes and Essential Fatty Acids, 70*(6), 511–520.
12. Ayele, Y., Urga, K., & Engidawork, E., (2010). Evaluation of in vivo antihypertensive and in vitro vasodepressor activities of the leaf extract of *Syzygium guineense* (Willd) DC. *Phytotherapy Research, 24*(10), 1457–1462.
13. Ayyanar, M., & Subash-Babu, P., (2012). *Syzygium cumini* (L.) Skeels: A review of its phytochemical constituents and traditional uses. *Asian Pacific Journal of Tropical Biomedicine, 2*(3), 240–246.
14. Babu, T. M. C., Rammohan, A., Baki, V. B., Gunasekar, D., & Rajendra, W., (2016). Development of novel HER2 inhibitors against gastric cancer derived from flavonoid source of *Syzygium alternifolium* through molecular dynamics and pharmacophore-based screening. *Drug Design, Development and Therapy, 10*, 3611–3632.
15. Baldissera, G., Sperotto, N. D. M., Rosa, H. T., Henn, J. G., Peres, V. F., Moura, D. J., Roehrs, R., Denardin, E. L. G., Lago, P. D., Nunes, R. B., & Saffi, J., (2016). Effects of crude hydroalcoholic extract of *Syzygium cumini* (L.) leaves and continuous aerobic training in rats with diabetes induced by a high-fat diet and low doses of streptozotocin. *Journal of Ethnopharmacology, 10*, 56–66.
16. Banerjee, S., Panda, C. K., & Das, S., (2006). Clove (*Syzygium aromaticum* L.), a potential chemopreventative agent for lung cancer. *Carcinogenesis, 27*(8), 1645–1654.
17. Banerjee, A., Dasgupta, N., & De, B., (2005). In vitro study of antioxidant activity of *Syzygium cumini* fruit. *Food Chemistry, 90*, 727–733.
18. Barh, D., & Viswanathan, G., (2008). *Syzygium cumini* inhibits growth and induces apoptosis in cervical cancer cell lines: a primary study. *eCancer, 2*, 83.
19. Benecia, F., & Courrges, M. C., (2000). In vitro and in vivo activity of eugenol on human herpesvirus. *Phytotherapy Research, 14*, 495–500.
20. Brahmachari, H. D., & Augusti, K. T., (1962). Orally effective hypoglycaemic agents from plants. *The Journal of Pharmacy and Pharmacology, 14*, 254–255.
21. Brito, F. A., Lima, L. A., Ramos, M. F. S., Nakamura, M. J., Cavalher-Machado, S. C., Siani, A. C., Henriques, M. G. M. O., & Sampaio, A. L. F., (2007). Pharmacological study of anti-allergic activity of *Syzygium cumini* (L.) Skeels. *Brazilian Journal of Medicinal and Biological Research, 40*, 105–115.
22. Brown, N. S., & Bicknell, R., (2001). Hypoxia and oxidative stress in breast cancer. Oxidative stress: its effects on the growth, metastatic potentialand response to therapy of breast cancer. *Breast Cancer Research, 3*, 323–327.
23. Bryant, K., & Cock, I. E., (2016). Growth inhibitory properties of *Backhousia myrtifolia* Hook. & Harv. and *Syzygium anisatum* (Vickery) Craven & Biffen extracts against a panel of pathogenic bacteria. *Pharmacognosy Communications, 6*(4), 194–203.
24. Burt, S. A., & Reinders, R. D., (2003). Antibacterial activity of selected plant essential oils against *Escherichia coli* O157:H7. *Letters in Applied Microbiology, 36*, 162–167.
25. Cai, L., & Wu, C. D., (1996). Compounds from *Syzygium aromaticum* possessing growth inhibitory activity against oral pathogens. *Journal of Natural Products, 59*, 987–990.

26. Chikowe, G., Mpala, L., & Cock, I. E., (2013). Antibacterial activity of selected Australian *Syzygium* species. *Pharmacognosy Communications*, *3*(4), 77–83.
27. Chai, L. Y., Denning, D. W., & Warn, P., (2010). *Candida tropicalis* in human disease. *Critical Reviews in Microbiology*, *36*(4), 282–298.
28. Chami, F., Chami, N., Bennis, S., Bouchikhi, T., & Remmal, A., (2005). Oregano and clove essential oils induce surface alteration of *Saccharopmyces cerevisiae*. *Phytotherapy Research, 19*(5), 405–408.
29. Chandran, R., Parimelazhagan, T., Shanmugam, S., & Thankarajan, S., (2016). Antidiabetic activity of *Syzygium calophyllifolium* in streptozotocin-nicotinamide induced type-2 diabetic rats. *Biomedicine and Pharmacotherapy*, *82*, 547–554.
30. Chandrasekaran, M., & Venkatesalu, V., (*2004*). Antibacterial and antifungal activity of *Syzygium jambolanum* seeds. *Journal of Ethnopharmacology, 91*, 105–108.
31. Choi, H. N., Kang, M. J., Lee, S. J., & Kim, J. I., (2014). Ameliorative effect of myricetin on insulin resistance in mice fed a high-fat, high-sucrose diet. *Nutrition Research and Practice*, *8*(5), 544–549.
32. Choi, S., & Chung, M. H., (2003). A review on the relationship between *Aloe vera* components and their biological effects. *Seminars in Integrative Medicine, 1*(1), 53–62.
33. Cock, I. E., (2013). The phytochemistry and chemotherapeutic potential of *Tasmannia lanceolata* (Tasmanian pepper): A review. *Pharmacognosy Communications, 3*(4), 13–25.
34. Cock, I. E., (2012). Antimicrobial activity of *Syzygium australe* and *Syzygium leuhmannii* leaf methanolic extracts. *Pharmacognosy Communications, 2*(2), 71–77.
35. Cock, I. E., (2011). Problems of reproducibility and efficacy of bioassays using crude extracts, with reference to *Aloe vera*. *Pharmacognosy Communications, 1*(1), 52–62.
36. Cock, I. E., (2011). Medicinal and aromatic plants – Australia. In: *Ethnopharmacology section, Biological, Physiological and Health Sciences, Encyclopedia of Life Support Systems* (EOLSS), Developed under the Auspices of the UNESCO, EOLSS Publishers, Oxford , UK, (http://www.eolss.net).
37. Cock, I. E., (2008). Antibacterial activity of selected Australian plant extracts. *The Internet Journal of Microbiology, 4*, 2.
38. Cock, I. E., & Van Vuuren, S. F., (2015). South African food and medicinal plant extracts as potential antimicrobial food agents. *Journal of Food Science and Technology, 52*(11), 6879–6899.
39. Cock, I. E., & Van Vuuren, S. F., (2015). The potential of selected South African plants with anti-*Klebsiella* activity for the treatment and prevention of ankylosing spondylitis. *Inflammopharmacology, 23*, 21–35.
40. Cock, I. E., & Van Vuuren, S. F., (2014). Anti-*Proteus* activity of some South African medicinal plants: Their potential for the prevention of rheumatoid arthritis. *Inflammopharmacology, 22*, 23–36.
41. Cock, I. E., Winnett, V., Sirdaarta, J., Van Vuuren, S. F., & Matthews, B., (2015). The potential of selected Australian medicinal plants with anti-*Proteus* activity for the treatment and prevention of rheumatoid arthritis. *Pharmacognosy Magazine, 11*(42), S190–S208.
42. Cordier, W., Gulumian, M., Cromarty, A. D., Van Vuuren, S. F., & Steenkamp, V., (2013). Attenuation of oxidative stress in U937 cells by polyphenolic-rich bark frac-

tions of *Burkea africana* and *Syzygium cordatum*. *BMC Complementary and Alternative Medicine, 13*(1), 1.
43. Cordier, W., (2012). *Effects of Polyphenolic-Rich Bark Extracts of Burkea Africana and Syzygium Cordatum on Oxidative Stress*. Doctoral dissertation, University of Pretoria.
44. Courtney, R., Sirdaarta, J., Matthews, B., Van Vuuren, S. F., & Cock, I. E., (2015). Tannin components and inhibitory activity of Kakadu plum leaf extracts against microbial triggers of autoimmune inflammatory diseases. *Pharmacognosy Journal, 7*(1), 18–31.
45. Da Franca Rodrigues, K. A., Amorim, L. V., Dias, C. N., Moraes, D. F. C., Carneiro, S. M. P., & De Amorim Carvalho, F. A., (2015). *Syzygium cumini* (L.) Skeels essential oil and its major constituent α-pinene exhibit anti-*Leishmania* activity through immunomodulation in vitro. *Journal of Ethnopharmacology, 160*, 32–40.
46. Davis, W., Ze'ev Ronai, J. R., & Tew, K. D., (2001). Cellular thiols and reactive oxygen species in drug-induced apoptosis. *The Journal of Pharmacology and Experimental Therapeutics, 296*(1), 1–5.
47. Deans, S. G., Noble, R. C., Hiltunen, R., Wuryani, W., & Penzes, L. G., (1995). Antimicrobial and antioxidant properties of *Syzygium aromaticum* (L.) Merr. & Perry: impact upon bacteria, fungi and fatty acid levels in ageing mice. *Flavour and Fragrance Journal, 10*(5), 323–328.
48. Djipa, C. D., Delmée, M., & Quetin-Leclercq, J., (2000). Antimicrobial activity of bark extracts of *Syzygium jambos* (L.) Alston (Myrtaceae). *Journal of Ethnopharmacology, 71*, 307–313.
49. Djoukeng, J. D., Abou-Mansour, E., Tabacchi, R., Tapondjou, A. L., Bouda, H., & Lontsi, D., (2005). Antibacterial triterpenes from *Syzygium guineense* (Myrtaceae). *Journal of Ethnopharmacology, 101*(1), 283–286.
50. Duraipandiyan, V., Ayyanar, M., & Ignacimuthu, S., (2006). Antimicrobial activity of some ethnomedicinal plants used by Paliyar tribe from Tamil Nadu, India. *BMC Complimentary and Alternative Medicine, 6*(35), 1–7.
51. Duraipandiyan, V., Ignacimuthu, S., & Valanarasu, M., (2008). Antibacterial and antifungal activity of *Syzygium lineare* Wall. *International Journal of Integrative Biology, 3*(3), 159–162.
52. Dwivedi, V., Shrivastava, R., Hussain, S., Ganguly, C., & Bharaadwaj, M., (2011). Comparative anticancer potential of clove (*Syzygium aromaticum*) – An Indian spice against cancer cell lines of various anatomical origin. *Asian Pacific Journal of Cancer Prevention, 12*, 1989–1993.
53. El-Hosseiny, L. S., Alqurashy, N. N., & Sheweita, S. A., (2016). Oxidative stress alleviation by sage essential oil in co-amoxiclav induced hepatotoxicity in rats. *International Journal of Biomedical Science, 12*(2), 71–78.
54. Ezekiel, U., & Heuertz, R., (2015). Anti-inflammatory effect of *Syzygium cumini* on chemotaxis of human neutrophils. *International Journal of Pharmacognosy and Phytochemical Research, 7*(4), 714–717.
55. Faria, A. F., Marques, M. C., & Mercadante, A. Z., (2011). Identification of bioactive compounds from jambolão (*Syzygium cumini*) and antioxidant capacity evaluation in different pH conditions. *Food Chemistry, 126*, 1571–1578.

56. Fayemiwo, K. A., Adeleke, M. A., Okoro, O. P., Awojide, S. H., & Awoniyi, I. O., (2014). Larvicidal efficacies and chemical composition of essential oils of *Pinus sylvestris* and *Syzygium aromaticum* against mosquitoes. *Asian Pacific Journal of Tropical Biomedicine, 4*(1), 30–34.
57. Feres, M., Figueiredo, L. C., Barreto, I. M., Coelho, M. H., Araujo, M. W., & Cortelli, S. C., (2005). In vitro antimicrobial activity of plant extracts and propolis in saliva samples of healthy and periodontally-involved subjects. *Journal of the International Academy of Periodontology, 7*(3), 90–96.
58. Fernandes, E. S., Passos, G. F., Medeiros, R., Da Cunha, F. M., Ferreira, J., Campos, M. M., Pianowski, L. F., & Calixto, J. B., (2007). Anti-inflammatory effects of compounds alpha-humulene and (-)-*trans*-caryophyllene isolated from essential oil of *Cordia verbenacea*. *European Journal of Pharmacy, 569*(3), 228–236.
59. Fidel, P. L., Vazquez, J. A., & Sobel, J. D., (1999). *Candida glabrata*: Review of epidemiology, pathogenesis, & clinical disease with comparison to *C., albicans*. *Clinical Microbiology Reviews, 21*(1), 80–96.
60. Friedman, M., Henika, P. R., & Mandrell, R. E., (2002). Bactericidal activities of plant essential oils and some of their isolated constituents against *Campylobacter jejuni, Escherichia coli, Listeria monocytogenes*, & *Salmonella enterica*. *Journal of Food Protection, 65*(10), 1545–1560.
61. Gayosa, C. W., Lima, E. O., Oliveira, V. T., Pereira, F. O., Souza, E. L., Lima, I. O., & Navarro, D. F., (2005). Sensitivity of fungi isolated from onychomycosis to *Eugenia cariophyllata* essential oil and eugenol. *Fitoterapia, 76*(2), 247–249.
62. Gemechu, Z., (2016). Antimalarial activity of 80% methanol extract of the stem bark of *Syzygium guineense* (Wild.) DC (Myrtaceae) in mice infected with *Plasmodium berghei*. Doctoral dissertation, submitted to Addis Ababa University; pp. 76; [http://etd.aau.edu.et/bitstream/123456789/8365/1/Gemechu%20ZelekeThesis%20Paper%201.pdf].
63. Goh, S. Y., & Cooper, M. E., (2008). Clinical review: The role of advanced glycation end products in progression and complications of diabetes. *The Journal of Clinical Endocrinology and Metabolism, 93*(4), 1143–1152.
64. Gould, K. S., Thodey, K., Philpott, M., & Ferguson, L. R., (2006). Antioxidant activities of extracts from traditional Maori food plants. 1–4.
65. Goyal, M., (2015). Traditional plants used for the treatment of diabetes mellitus in Sursagar constituency, Jodhpur, Rajasthan: An ethnomedicinal survey. *Journal of Ethnopharmacology, 174*, 364–368.
66. Halliwell, B., (1996). Vitamin C: Antioxidant or pro-oxidant in vivo? *Free Radical Research, 25*, 439–454.
67. Helmstadter, A., (2008). *Syzygium cumini* (L.) Skeels (Myrtaceae) against diabetes-125 years of research. *Die Pharmazie, 63*(2), 91–101.
68. Hemaiswarya, S., & Doble, M., (2009). Synergistic interaction of eugenol with antibiotics against Gram negative bacteria. *Phytomedicine, 16*(11), 997–1005.
69. Hertog, M. G., Bueno-de-Mesquita, H. B., Fehily, A. M., Sweetnam, P. M., Elwood, P. C., & Kromhout, D., (1996). Fruit and vegetable consumption and cancer mortality in the caerphilly study. *Cancer Epidemiology, Biomarkers and Prevention, 5*, 673–677.

70. Holdsworth, D. K., (1991). Traditional medicinal plants of Rarotonga, Cook Islands, Part II., *International Journal of Pharmacognosy, 29*(1), 71–79.
71. Huang, D. W., Chang, W. C., Wu, J. S., Shih, R. W., & Shen, S. C., (2016). Vescalagin from Pink wax apple [*Syzygium samarangense* (Blume) Merrill and Perry] alleviates hepatic insulin resistance and ameliorates glycemic metabolism abnormality in rats fed a high-fructose diet. *Journal of Agricultural and Food Chemistry, 64*(5), 1122–1129.
72. Hussein, G., Miyashiro, H., Nakamura, N., Hattori, M., Kakiuchi, N., & Shimotohno, K., (2000). Inhibitory effects of Sudanese medicinal plant extracts on hepatitis C virus (HCV) protease. *Phytotherapy Research, 14*(7), 510–516.
73. Ior, L. D., Otimenyin, S. O., & Umar, M., (2012). Anti-inflammatory and analgesic activities of the ethanolic extract of the leaf of *Syzygium guineense* in rats and mice. *IOSR Journal of Pharmacy, 2*(4), 33–36.
74. Jain, A., Sharma, S., Goyal, M., Dubey, S., Jain, S., Sahu, J., Sharma, A., & Kaushik, A., (2010). Anti-inflammatory activity of *Syzygium cumini* leaves. *International Journal of Phytomedicine, 2*, 124–126.
75. Jamieson, N., Sirdaarta, J., & Cock, I. E., (2014). The anti-proliferative properties of Australian plants with high antioxidant capacities against cancer cell lines. *Pharmacognosy Communications, 4*(4), 71–82.
76. Jeyasankar, A., Raja, N., & Ignacimuthu, S., (2011). Insecticidal compound isolated from *Syzygium lineare* Wall. (Myrtaceae) against *Spodoptera litura* (Lepidoptera: Noctuidae). *Saudi Journal of Biological Sciences, 18*(4), 329–332.
77. Joel, L. S., (1995). The dual roles of nutrients as antioxidants and pro-oxidants: Their effects of tumor cell growth. In: *Proceedings of the Prooxidant Effects of Antioxidant Vitamins, Experimental Biology Meeting, Atlanta, GA.*, Edited by Herbert, V., American Society of Nutrition, Bethesda, MD, pp. 1221–1226.
78. Kamatou, G. P., Vermaak, I., & Viljoen, A. M., (2012). Eugenol – from the remote Maluku Islands to the international market place: a review of a remarkable and versatile molecule. *Molecules, 17*(6), 6953–6981.
79. Kamboj, V. P., (2000). Herbal medicine. *Current Science, 78*, 35–39.
80. Kandati, V., Govardhan, P., Reddy, C. S., Nath, A. R., & Reddy, R. R., (2012). *In-vitro* and *in-vivo* anti-inflammatory activity of *Syzygium alternifolium* (wt) Walp. *Journal of Medicinal Plants Research, 6*(36), 4995–5001.
81. Karalliedde, L., & Gawarammana, I., (2008). *Traditional Herbal Medicines. A Guide to Their Safe Use*. Hammersmith Press Ltd., London, UK.
82. Karthic, K., Kirthiram, K. S., Sadasivam, S., & Thayumanavan, B., (2008). Identification of alpha amylase inhibitors from *Syzygium cumini* Linn seeds. *Indian Journal of Experimental Biology, 46*(9), 677–680.
83. Kasetti, R. B., Rajasekhar, M. D., Kondeti, V. K., Fatima, S. S., Kumar, E. G., Swapna, S., Ramesh, B., & Rao, C. A., (2010). Antihyperglycemic and antihyperlipidemic activities of methanol:water (4:1) fraction isolated from aqueous extract of *Syzygium alternifolium* seeds in streptozotocin induced diabetic rats. *Food and Chemical Toxicology, 48*(4), 1078–1084.
84. Keskin, D., & Toroglu, S., (2011). Studies on antimicrobial activities of solvent extracts of different spices. *Journal of Environmental Biology, 32*, 251–256.
85. Khathi, A., Serumula, M. R., Myburg, R. B., Van Heerden, F. R., & Musabayane, C. T., (2013). Effects of *Syzygium aromaticum*-derived triterpenes on postprandial

blood glucose in streptozotocin-induced diabetic rats following carbohydrate challenge. *PLoS One*, *8*(11), e81632.

86. Kim, Y. J., Kim, H. C., Ko, H., Amor, E. C., & Lee, J. W., (2012). Inhibitory effects of aurentiacin from *Syzygium samarangense* on lipopolysaccharide-induced inflammatory response in mouse macrophages. *Food and Chemical Toxicology*, *50*, 1027–1035.
87. Kim. H. M., Lee, E. H., Hong, S. H., Song, H. J., Shin, M. K., Kim, S. H., & Shin, T. Y., (1998). Effects of *Syzygium aromaticum* extract on immediate hypersensitivity in rats. *Journal of Ethnopharmacology*, *60*, 125–131.
88. Konczak, I., Zabaras, D., Dunstan, M., & Aguas, M., (2010). Antioxidant capacity and hydrophilic phytochemicals in commercially grown Australian fruits. *Food Chemistry*, *123*, 1048–1054.
89. Krishnasamy, G., Muthusamy, K., Chellappan, D. R., & Subbiah, N., (2016). Antidiabetic, antihyperlipidaemic, & antioxidant activity of *Syzygium densiflorum* fruits in streptozotocin and nicotinamide-induced diabetic rats. *Pharmaceutical Biology*, *54*(9), 1716–1726.
90. Krishnasamy, G., & Muthusamy, K., (2015). In vitro evaluation of antioxidant and antidiabetic activities of *Syzygium densiflorum* fruits. *Asian Pacific Journal of Tropical Disease*, *5*(11), 912–917.
91. Kuiate, J. R., Mouokeu, S., Wabo, H. K., & Tane, P., (2007). Antidermatophytic triterpenoids from *Syzygium jambos* (L.) Alston (Myrtaceae). *Phytotherapy Research*, *21*(2), 149–152.
92. Kumar, A., Illavarasan, R., Jayachandran, T., Deecaraman, M., Kumar, R. M., Aravindan, P., Padmanabhan, N., & Krishan, M. R. V., (2008). Anti-inflammatory activity of *Syzygium cumini* seed. *African Journal of Biotechnology*, *7*(8), 941–943.
93. Kumar, K. E., Mastan, S. K., Reddy, K. R., Reddy, G. A., Raghunandan, N., & Chaitanya, G., (2008). *International Journal of Integrative Biology*, *4*(1), 55–61.
94. Larhsini, M., Oumoulid, L., Lazrek, H. B., Wataleb, S., Bousaid, M., Bekkouche, K., & Jana, M., (2001). Antibacterial activity of some Moroccan medicinal plants. *Phytotherapy Research*, *15*(3), 250–252.
95. Lassak, E. V., & McCarthy, T., (2011). *Australian Medicinal Plants*. New Holland Publishers, Sydney – Australia; pp. 231.
96. Lau, K. Y., & Rukayadi, Y., (2015). Screening of tropical medicinal plants for sporicidal activity. *International Food Research Journal*, *22*(1), 421–425.
97. Lee, K. G., & Shibamoto, T., (2001). Antioxidant property of aroma extract isolated from clove buds [*Syzygium aromaticum* (L.) Merr. et Perry]. *Food Chemistry*, *74*(4), 443–448.
98. Liebmann, J. E., Cook, J. A., Lipschultz, C., Teague, D., Fisher, J., & Mitchell, J. B., (1993). Cytotoxic studies of paclitaxel (Taxol) in human tumor cell lines. *British journal of Cancer*, *68*(6), 1104.
99. Lima, L. A., Siani, A. C., Brito, F. A., Sampaio, A. L. F., Das Graças Muller Oliveira Henriques, M., & Da Silva Reihl, C. A., (2007). Correlation of anti-inflammatory activity with phenolic content in the leaves of *Syzygium cumini* (L.) Skeels (Myrtaceae). *Quimica Nova*, *30*(4), 860–864.
100. Liu, I. M., Tzeng, T. F., Liou, S. S., & Lan, T. W., (2007). Myricetin, a naturally occurring flavonol, ameliorates insulin resistance induced by a high-fructose diet in rats. *Life Sciences*, *81*(21–22), 1479–1488.

101. Lorente, I., Ocete, M. A., Zarzuelo, A., Cabo, M. M., & Jimenez, J., (1989). Bioactivity of the essential oil of *Bupleurum fruticosum*. *Journal of Natural Products, 52*(2), 267–272.
102. Machado, M., Dinis, A. M., Salgueiro, L., Custódio, J. B. A., Cavaleiro, C., & Sousa, M. C., (2011). Anti-*Giardia* activity of *Syzygium aromaticum* essential oil and eugenol: Effects on growth, viability, adherence and untrastructure. *Experimental Parasitology, 127*, 732–739.
103. Macpherson, G., (2011). *Inflammation. Blacks Medical Dictionary*. A and C Black, London, United Kingdom; pp. 315.
104. Magina, M. D., Dalmarco, E. M., Wisniewski, A., Simionatto, E. L., Dalmarco, J. B., Pizzolatti, M. G., & Brighente, I. M., (2009). Chemical composition and antibacterial activity of essential oils of *Eugenia* species. *Journal of Natural Medicines, 63*(3), 345–350.
105. Manaharan, T., Appleton, D., Cheng, H. M., & Palanisamy, U. D., (2012). Flavonoids isolated from *Syzygium aqueum* leaf extract as potential antihyperglycaemic agents. *Food Chemistry, 132*(4), 1802–1807.
106. Manaharan, T., Ming, C. H., & Palanisamy, U. D., (2013). *Syzygium aqueum* leaf extract and its bioactive compounds enhances pre-adipocyte differentiation and 2-NBDG uptake in 3T3-L1 cells. *Food Chemistry, 136*(2), 354–363.
107. Manoharan, K. P., Song, F. J., Benny, T. K. H., & Yang, D., (2007). Spectral assignments and reference data. *Magnetic Resonance in Chemistry, 45*, 279–281.
108. Mathabe, M. C., Nikolova, R. V., Lall, N., & Nyazema, N. Z., (2006). Antibacterial activities of medicinal plants used for the treatment of siarrhoea in Limpopo Province, South Africa. *Journal of Ethnopharmacology, 105*, 286–293.
109. Mayaud, L., Carricajo, A., Zhiri, A., & Aubert, G., (2008). Comparison of bacteriostatic and bactericidal activity of 13 essential oils against strains with varying sensitivity to antibiotics. *Letters in Applied Microbiology, 47*(3), 167–173.
110. Meshram, G. A., Yadav, S. S., Shinde, D., Patil, B., & Singh, D., (2011). Antibacterial and effect of ethanolic extracts of *Syzgium cumini* seeds powder on glucoamylase in vitro. *Journal Pharmaceutical Science and Research, 3*, 1732–1741.
111. Modi, D. C., Patel, J. K., Shah, B. N., & Nayak, B. S., (2010). Antiinflammatory activity of seeds of *Syzygium cumini* Linn. *Journal Pharmaceutical Education and Research, 1*(1), 68–70.
112. Mohamed, A. A., Ali, S. I., & El-Baz, K. K., (2013). Antioxidant and antibacterial activities of crude extracts and essential oils of *Syzygium cumini* leaves. *PLoS One, 8*(4), e60269.
113. Mohanty, S., & Cock, I. E., (2010). Bioactivity of *Syzygium jambos* methanolic extracts: Antibacterial activity and toxicity. *Pharmacognosy Research, 2*(1), 4–9.
114. Mollika, S., Islam, N., Parvin, N., Kabir, A., Sayem, W., & Luthfunnesa, S. R., (2014). Evaluation of analgesic, anti-inflammatory and CNS activities of the methanolic extract of *Syzygium samarangense* leave. *Global Journal of Pharmacy, 8*, 39–46.
115. Moon, S. E., Kim, H. Y., & Cha, J. D., (2011). Synergistic effect between clove oil and its major compounds and antibiotics against oral bacteria. *Archives of Oral Biology, 56*(9), 907–916.

116. Moskaug, J. O., Carlsen, H., Myhrstad, M. C., & Blomhoff, R., (2005). Polyphenols and glutathione synthesis regulation. *American Journal of Clinical Nutrition, 81*, 277S–283S.
117. Muruganandan, S., Srinivasan, K., Chandra, S., Tandan, S. K., Lal, J., & Raviprakash, V., (2001). Anti-inflammatory activity of *Syzygium cumini* bark. *Fitoterapia, 72*(4), 369–375.
118. Musabayane, C. T., Mahlalela, N., Shode, F. O., & Ojewole, J. A., (2005). Effects of *Syzygium cordatum* (Hochst.) [Myrtaceae] leaf extract on plasma glucose and hepatic glycogen in streptozotocin-induced diabetic rats. *Journal of Ethnopharmacology, 97*(3), 485–90.
119. Myrtle, N., Anderson, G. L., Doyle, M. P., & Smith, M. A., (2006). Antimicrobial activity of clove (*Syzygium aromaticum*) oil in inhibiting *Listeria monocytogenes* on chicken frankfurters. *Food Control, 17*, 102–107.
120. Nazif, N. M., (2007). The anthocyanin components and cytotoxic activity of *Syzygium cumini* (L.) fruits growing in Egypt. *Natural Product Sciences, 13*(2), 135–139.
121. Netzel, M., Netzel, G., Tian, Q., Schwartz, S., & Konczak, I., (2007). Native Australian fruits – A novel source of antioxidants for food. *Innovative Food Science and Emerging Technologies, 8*, 339–346.
122. Newman, D. J., & Cragg, G. M., (2007). Natural products as sources of new drugs over the last 25 years. *Journal of Natural Products, 70*(3), 461–477.
123. Oboh, G., Akinbola, I. A., Ademosun, A. O., Sanni, D. M., Odubanjo, O. V., Olasehinde, T. A., & Oyeleye, S. I., (2015). Essential oil from clove bud (*Eugenia aromatica* Kuntze) inhibit key enzymes relevant to the management of type-2 diabetes and some pro-oxidant induced lipid peroxidation in rats pancreas in vitro. *Journal of Oleo Science, 64*(7), 775–782.
124. Oliveira, A. C., Endringer, D. C., Amorim, L. A., Das Gracas, L. B. M., & Coelho, M. M., (2005). Effect of the extracts and fractions of *Baccharis trimera* and *Syzygium cumini* on glycaemia of diabetic and non-diabetic mice. *Journal of Ethnopharmacology, 102*(3), 465–469.
125. Osman, H., Rahim, A. A., Isa, N. M., & Bakhir, N. M., (2009). Antioxidant activity and phenolic content of *Paederia foetida* and *Syzygium aqueum. Molecules, 14*(3), 970–978.
126. Park, M. J., Gwak, K. S., Yang, I., Choi, W. S., Jo, H. J., Chang, J. W., Jeung, E. B., & Choi, I. G., (2007). Antifungal activities of the essential oils of *Syzygium aromaticum* (L.) Mer. Et Perry and *Leptospermum petersonii* Bailey and their constituents against various dermatophytes. *Journal of Microbiology, 45*(5), 460–465.
127. Parmar, J., Sharma, P., Verma, P., Sharma, P., & Goyal, P. K., (2011). Modulation of DMBA- induced biochemical and histopathological changes by *Syzygium cumini* seed extract duruni skin carcinogenesis. *International Journal of Current Biomedical and Pharmaceutical Research, 1*(2), 24–30.
128. Passos, G. F., Fernandes, E. S., Cunha, F. M., Ferreira, J., Pianowski, L. F., Campos, M. M., & Calixto, J. B., (2007). Anti-inflammatory and anti-allergic properties of the essential oil and active compounds from *Cordia verbenacea. Journal of Ethnopharmacology, 110*(2), 323–333.

129. Perera, H. K., & Handuwalage, C. S., (2015). Analysis of glycation induced protein cross-linking inhibitory effects of some antidiabetic plants and spices. *BMC Complementary and Alternative Medicine, 15*, 175.
130. Pinto, E., Vale-Silva, L., Cavaleiro, C., & Salgueiro, L., (2009). Antifungal activity of the clove essential oil from *Syzygium aromaticum* on *Candida, Aspergillus* and dermatophyte species. *Journal of Medical Microbiology, 58*(11), 1454–1462.
131. Pfaller, M. A., Diekema, D. J., Gibbs, D. L., Newell, V. A., Nagy, E., Dobiasova, S., Rinaldi, R., & Veselov, A., (2008). Global Surveillance Group. *Candida krusei*, a multidrug-resistant opportunistic fungal pathogen: Geographic and temporal trends from the Artemis Disk antifungal surveillance program, 2001 to 2005. *Journal of Clinical Microbiology, 46*(2), 515–521.
132. Plantnet. *Syzygium oleosum* (Muell, F.,), (2016). Hyland, B. http://plantnet.rbgsyd.nsw.gov.au/cgibin/NSWfl.pl?page=nswfl&lvl=sp&name=Syzygium~oleosum.
133. Prasad, R. C., Herzog, B., Boone, B., Sims, L., & Waltner-Law, M., (2005). An extract of *Syzygium aromaticum* represses genes encoding hepatic gluconeogenic enzymes. *Journal of Ethnopharmacology, 96*(1–2), 295–301.
134. Potter, J. D., (1997). Cancer prevention: Epidemiology and experiment. *Cancer Letters, 114*, 7–9.
135. Prtabhakaran, S., Gothandam, K. M., & Sivashanmugam, K., (2011). Phytochemical and antimicrobial properties of *Syzygium cumini* an ethnomedicinal plant of Javadhu hills. *Research in Pharmacy, 1*(1), 22–32.
136. Pulikottil, S. J., & Nath, S., (2015). Potential of clove of *Syzygium aromaticum* in development of a therapeutic agent for periodontal disease. A review. *SADJ, 70*(3), 108–115.
137. Pundir, R. K., Jain, P., & Sharma, C., (2010). Antimicrobial activity of ethanolic extracts of *Syzygium aromaticum* and *Allium sativum* against food associated bacteria and fungi. *Ethnobotanical Leaflets, 3*, 11.
138. Rahman, A., Shahabuddin, M., Hadi, S. M., & Parish, J., (1990). Complexes involving quercetin, DNA and Cu(II). *Carcinogenesis, 11*, 2001–2003.
139. Rahmatullah, M., Azam, N. K., Khatun, Z., Seraj, S., Islam, F., Rahman, A., Jahan, S., & Aziz, S., (2012). Medicinal plants used for treatment of diabetes by the Marakh Sect of the Garo Tribe living in Mymensingh District, Bangladesh. *African Journal of Traditional, Complementary, & Alternative Medicines, 9*(3), 380–385.
140. Rao, B. K., & Rao, C. H., (2001). Hypoglycemic and antihyperglycemic activity of *Syzygium alternifolium* (Wt.) Walp. seed extracts in normal and diabetic rats. *Phytomedicine: International Journal of Phytotherapy and Phytopharmacology, 8*(2), 88–93.
141. Ratnam, K. V., & Raju, R. R. V., (2008). In vitro antimicrobial screening of the fruit extracts of two *Syzygium* species (Myrtaceae). *Advances in Biological Research, 2*(1–2), 17–20.
142. Ravi, K., Sivagnanam, K., & Subramanian, S., (2004). Anti-diabetic activity of *Eugenia jambolana* seed kernels on streptozotocin-induced diabetic rats. *Journal of Medicinal Food, 7*(2), 187–191.
143. Resurreccion-Magno, M. H., Villasenor, I. M., Harada, N., & Monde, K., (2005). Antihyperglycaemic flavonoids from *Syzygium samarangense* (Blume) Merr. and Perry. *Phytotherapy Research, 19*(3), 246–251.

144. Rice-Evans, C., Miller, N., & Paganga, G., (1997). Antioxidant properties of phenolic compound. *Trends in Plant Science, 2*(4), 152–159.
145. Ruebhart, D. R., Wickramasinghe, W., & Cock, I. E., (2009). Protective efficacy of the antioxidants vitamin E and Trolox against *Microcystis aeruginosa* and microcystin-LR in *Artemia franciscana* nauplii. *Journal of Toxicology and Environmental Health, Part A, 72*, 1567–1575.
146. Ruefli, A. A., Davis, J. E., Sutton, V. R., Trapani, J. A., Smyth, M. J., & Johnstone, R. W., (2001). Disecting the apoptopic mechanisms of chemotherapeutic drugs and lymphocytes to design effective anticancer therapies. *Drug Development Research, 52*, 549–557.
147. Saeed, S., & Tariq, P., (2008). In vitro antibacterial activity of clove against gram negative bacteria. *Pakistan Journal of Botany, 40*(5), 2157–2160.
148. Saini, A. C., Souza, M. C., Henriques, M. G. M. O., & Ramos, M. F. S., (2013). Anti-inflammatory activity of essential oils from *Syzygium cumini* and *Psidium guajava*. *Pharmaceutical Biology, 51*(7), 881–887.
149. Samie, A., & Mashau, F., (2013). Antifungal activity of fifteen Southern African medicinal plants against five *Fusarium* species. *Journal of Medicinal Plants Research, 7*(25), 1839–1848.
150. Samy, M. N., Sugimoto, S., Matsunami, K., Otsuka, H., & Kamel, M. S., (2014). One new flavonoid xyloside and one new natural triterpene rhamnoside from the leaves of *Syzygium grande*. *Phytochemistry Letters, 10*, 86–90.
151. Santoro, G. F., Cardoso, M. G., Guimarães, L. G. L., Mendonça, L. Z., & Soares, M. J., (2007). *Trypanosoma cruzi*: activity of essential oils from *Achillea millefolium* L., *Syzygium aromaticum* L., and *Ocimum basilicum* L., on epimastigotes and trypomastigotes. *Experimental Parasitology, 116*(3), 283–290.
152. Sautron, C., & Cock, I. E., (2014). Antimicrobial activity and toxicity of *Syzygium australe* and *Syzygium leuhmannii* fruit extracts. *Pharmacognosy Communications, 4*(1), 53–60.
153. Schossler, D. R. C., Mazzanti, C. M., Luz, S. C. A., Filappi, A., Prestes, D., Silveira, A. F., & Cecim, M., (2004). *Syzygium cumini* and the regeneration of insulin positive cells from the pancreatic duct. *Brazilian Journal of Veterinary Research and Animal Science, 41*, 236–239.
154. Setzer, M. C., Setzer, W. N., Jackes, B. R., Gentry, G. A., & Moriarity, D. M., (2001). The medicinal value of tropical rainforest plants from Paluma, North Queensland, Australia. *Pharmaceutical Biology, 39*(1), 67–78.
155. Shafi, P. M., Rosamma, M. K., Jamil, K., & Reddy, P. S., (2002). Antibacterial activity of *Syzygium cumini* and *Syzygium travancoricum* leaf essential oils. *Fitoterapia, 73*, 414–416.
156. Shahreen, S., Banik, J., Hafiz, A., Rahman, S., Zaman, A. T., Shoyeb, A., Chowdhury, M. H., & Rahmatullah, M., (2012). Antihyperglycemic activities of leaves of three edible fruit plants (*Averrhoa carambola, Ficus hispida* and *Syzygium samarangense*) of Bangladesh. *African Journal of Traditional, Complementary and Alternative Medicines, 9*(2), 287–291.
157. Sharma, A. K., Bharti, S., Kumar, R., Krishnamurthy, B., Bhatia, J., Kumari, S., & Arya, D. S., (2012). *Syzygium cumini* ameliorates insulin resistance and β-cell dysfunction via modulation of PPARγ, dyslipidemia, oxidative stress, & TNF-α in type 2 diabetic rats. *Journal of Pharmacological Sciences, 119*(3), 205–213.

158. Sharma, B., Balomajumder, C., & Roy, P., (2008). Hypoglycemic and hypolipidemic effects of flavonoid rich extract from *Eugenia jambolana* seeds on streptozotocin induced diabetic rats. *Food and Chemical Toxicology, 46*(7), 2376–2383.
159. Sharma, B., Siddiqui, M. S., Kumar, S. S., Ram, G., & Chaudhary, M., (2013). Liver protective effects of aqueous extract of *Syzygium cumini* in Swiss albino mice on alloxan induced diabetes mellitus. *Journal of Pharmacy Research, 6*(8), 853–858.
160. Sharma, R., Kishore, N., Hussein, A., & Lall, N., (2013). Antibacterial and anti-inflammatory effects of *Syzygium jambos* L. (Alston) and isolated compounds on *Acne vulgaris*. *BMC Complementary and Alternative Medicine, 13*, 292.
161. Sharma, S. B., Rajpoot, R., Nasir, A., Prabhu, K. M., & Murthy, P. S., (*2011*). Ameliorative effect of active principle isolated from seeds of *Eugenia jambolana* on carbohydrate metabolism in experimental diabetes. *Evidence-based Complementary and Alternative Medicine*, 789–871.
162. Sharma, B., Viswanath, G., Salunke, R., & Roy, P., (2008). Effects of flavonoid-rich extract from seeds of *Eugenia jambolana* (L.) on carbohydrate and lipid metabolism in diabetic mice. *Food Chemistry, 110*(3), 697–705.
163. Shen, S. C., & Chang, W. C., (2013). Hypotriglyceridemic and hypoglycemic effects of vescalagin from Pink wax apple [*Syzygium samarangense* (Blume) Merrill and Perry cv. Pink] in high-fructose diet-induced diabetic rats. *Food Chemistry, 136*(2), 858–863.
164. Shinde, J., Taldone, T., Barletta, M., Kunaparaju, N., Hu, B., Kumar, S., Placido, J., & Zito, W., (2008). α-Glucosidase inhibitory activity of *Syzygium cumini* (Linn.) Skeels seed kernel in vitro and in Goto-Kakizaki (GK) rats. *Carbohydrate Research, 343*(7), 1278–1281.
165. Sibandze, G. F., Van Zyl, R. L., & Van Vuuren, S. F., (2010). The anti-diarrhoel properties of *Breonadia salicina, Syzygium cordatum,* and *Ozoroa sphaerocarpa* when used in combination in Swazi traditional medicine. *Journal of Ethnopharmacology, 132*, 506–511.
166. Simirgiotis, M. J., Adachi, S., To, S., Yang, H., Reynertson, K. A., Basile, M. J., Gil, R. R., Weinstein, I. B., & Kennelly, E. J., (2008). Cytotoxic chalcones and antioxidants from the fruits of *Syzygium samarangense* (Wax Jambu). *Food Chemistry, 107*, 813–819.
167. Singh, S., Farhan, A. S., Ahmad, A., Khan, N. U., & Hadi, S. M., (2001). Oxidative DNA damage by capsaicin and dihydrocapsaicin in the presence of Cu(II). *Cancer Letters, 169*, 139–146.
168. Singh, S. P., & Mohan, L., (2014). Mosquito Repellent activity of *Syzygium aromaticum* (clove) against malaria vector, *Anopheles stephensi*. *Advances in Bioresearch, 5*(1), 112.
169. Souza, M. C., Siani, A. C., Ramos, M. F. S., Menezes-de-Lima, O., & Henriques, M. G. M. O., (2003). Evaluation of anti-inflammatory activity of essential oils from two Asteraceae species. *Die Pharmazie: An international Journal of Pharmaceutical Sciences, 58*(8), 582–586.
170. Srivastava, S., & Chandra, D., (2013). Pharmacological potentials of *Syzygium cumini*: A review. *Journal of the Science of Food and Agriculture, 93*(9), 2084–2093.
171. Srivastava R, Shaw, A. K., & Kulshreshtha, D. K., (1995). Triterpenoids and chalcone from *Syzygium samarangense*. *Phytochemistry, 38*(3), 687–689.

172. Ssegawa, P., & Kasenene, J. M., (2007). Plants for malaria treatment in southern Uganda: traditional use, preference and ecological viability. *Journal of Ethnobiology, 27*(1), 110–131.
173. Steenkamp, V., Fernandes, A. C., & Van Rensburg, C. E., (2007). Screening of Venda medicinal plants for antifungal activity against *Candida albicans. South African Journal of Botany, 73*(2), 256–258.
174. Syzygium Working Group, (2016). Syzygium (Myrtaceae): Monographing a taxonomic giant via 22 coordinated regional revisions. *Peer Journal Preprints*, https://peerj.com/preprints/1930.pdf.
175. Takechi, M., & Tanaka, Y., (1981). Purification and characterization of antiviral substance from the bud of *Syzygium aromatica. Planta Medica, 42*(05), 69–74.
176. Tanko, Y., Mohammed, A., Okasha, M. A., Umar, A. H., & Magaji, R. A., (2008). Anti-nociceptive and anti-inflammatory activities of ethanol extract of *Syzygium aromaticum* flower bud in Wistar rats and mice. *African Journal of Traditional Complementary and Alternative Medicine, 5*(2), 209–212.
177. Teixeira, C. C., Fuchs, F. D., Weinert, L. S., & Esteves, J., (2006). The efficacy of folk medicines in the management of type 2 diabetes mellitus: results of a randomized controlled trial of *Syzygium cumini* (L.) Skeels. *Journal of Clinical Pharmacy and Therapeutoics, 31*(1), 1–5.
178. Teixeira, C. C., Pinto, L. P., Kessler, F. H., Knijnik, L., Pinto, C. P., Gastaldo, G. J., & Fuchs, F. D., (1997). The effect of *Syzygium cumini* (L.) skeels on post-prandial blood glucose levels in non-diabetic rats and rats with streptozotocin-induced diabetes mellitus. *Journal of Ethnopharmacology, 56*(3), 209–213.
179. Teixeira, C. C., Rava, C. A., Mallman da Silva, P., Melchior, R., Argenta, R., Anselmi, F., Almeida, C. R., & Fuchs, F. D., (2000). Absence of antihyperglycemic effect of jambolan in experimental and clinical models. *Journal of Ethnopharmacology, 71*(1–2), 343–347.
180. Teixeira, C. C., Weinert, L. S., Barbosa, D. C., Ricken, C., Esteves, J. F., & Fuchs, F. D., (2004). *Syzygium cumini* (L.) Skeels in the treatment of type 2 diabetes: results of a randomized, double-blind, double-dummy, controlled trial. *Diabetes Care, 27*(12), 3019–3020.
181. Thiyagarajan, G., Muthukumaran, P., Sarath Kumar, B., Muthusamy, V. S., & Lakshmi, B. S., (2016). Selective inhibition of PTP1B by Vitalboside A from *Syzygium cumini* enhances insulin sensitivity and attenuates lipid accumulation via partial agonism to PPARγ: In vitro and *in silico* Investigation. *Chemical Biology and Drug Design, 88*(2), 302–312.
182. Tome, M. E., Baker, A. F., Powis, G., Payne, C. M., & Briehl, M. M., (2001). Catalase-overexpressing thymocytes are resistant to glucocorticoid-induced apoptosis and exhibit increased net tumor growth. *Cancer Research, 61*, 2766–2773.
183. Tripathi, A. K., & Kohli, S., (2014). Pharmacognostical standardization and antidiabetic activity of *Syzygium cumini* (Linn.) barks (Myrtaceae) on streptozotocin-induced diabetic rats. *Journal of Complementary and Integrative Medicine, 11*(2), 71–81.
184. Trofa, D., Gacser, A., & Nosanchuk, J. D., (2008). *Candida parasilosis*, an emerging fungal pathogen. *Clinical Microbial Reviews, 21*(4), 606–625.

185. Trojan-Rodrigues, M., Alves, T. L., Soares, G. L., & Ritter, M. R., (2012). Plants used as antidiabetics in popular medicine in Rio Grande do Sul. Southern Brazil. *Journal of Ethnopharmacology, 139*(1), 155–163.
186. Tsuda, T., Horio, F., Uchida, K., Aoki, H., & Osawa, T., (2003). Dietary cyaniding 3-O-b-D-glucoside-rich purple corn colour prevents obesity and ameliorates hyperglycemia in mice. *Journal of Nutrition, 133*, 2125–2130.
187. Tu, Z., Moss-Pierce, T., Ford, P., & Jiang, T. A., (2014). *Syzygium aromaticum* L. (Clove) extract regulates energy metabolism in myocytes. *Journal of Medicinal Food, 17*(9), 1003–1010.
188. Van Wyk, B. E., Van Outdshoorn, B., & Gericke, N., (2009). *Medicinal plants of South Africa*. Briza Publications, Pretoria, South Africa; pp. 415.
189. Vessal, M., Hemmati, M., & Vasei, M., (2003). Antidiabetic effects of quercetin in streptozocin-induced diabetic rats. *Comparative Biochemistry and Physiology Toxicology and Pharmacology, 135*(3), 357–364.
190. Vikrant, V., Grover, J. K., Tandon, N., Rathi, S. S., & Gupta, N., (2001). Treatment with extracts of *Momordica charantia* and *Eugenia jambolana* prevents hyperglycemia and hyperinsulinemia in fructose fed rats. *Journal of Ethnopharmacology, 76*(2), 139–143.
191. Vita, J. A., (2005). Polyphenols and cardiovascular disease: Effects on endothelial and platelet function. *American Journal of Clinical Nutrition, 81*, 292S–297S.
192. Vivas, N., Laguerre, M., Pianet de Boissel, I., Vivas de Gaulejac, N., & Nonier, M. F., (2004). Conformational interpretation of vescalagin and castalagin physicochemical properties. *Journal of Agricultural and Food Chemistry, 52*(7), 2073–2078.
193. Vuong, Q. V., Hirun, S., Chuen, T. L., Goldsmith, C. D., Bowyer, M. C., Chalmers, A. C., Phillips, P. A., & Scarlett, C. J., (2014). Physicochemical composition, antioxidant and anti-proliferative capacity of a lilly pilly (*Syzygium paniculatum*) extract. *Journal of Herbal Medicine, 4*(3), 134–140.
194. Walsh, G., (2003). *Biopharmaceuticals: Biochemistry and Biotechnology*. 3rd Ed., Wiley, Chinchester.
195. Warnke, P. H., Becker, S. T., Podschun, R., Sivananthan, S., Springer, I. N., Russo, P. A., Wiltfang, J., Fickenscher, H., & Sherry, E., (2009). The battle against multiresistant strains: renaissance of antimicrobial essential oils as a promising force to fight hospital-acquired infections. *Journal of Cranio-Maxillofacial Surgery, 37*(7), 392–397.
196. Wastl, H., Boericke, G. W., & Foster, W. C., (1947). Studies of effects of *Syzygium jambolanum* on alloxan-diabetic rats. *Archives Internationales de Pharmacodynamie et de Therapie, 75*(1), 33–50.
197. Webb, M. A., (2000). *Bush Sense. australian Essential Oils and Aromatic Compounds*. Griffin Press, Adelaide, Australia.
198. Widyawati, T., Yusoff, N. A., Asmawi, M. Z., & Ahmad, M., (2015). Antihyperglycemic effect of methanol extract of *Syzygium polyanthum* (Wight.) leaf in streptozotocin-induced diabetic rats. *Nutrients, 7*(9), 7764–7780.
199. Winter, C. A., Risely, E. A., & Nuss, W., (1962). Carrageenan induced edema in hind paw of rats as an assay for anti-inflammatory drugs. *Proceedings of the Society for Experimental Biology and Medicine, 111*, 544–547.

200. Wong, K. C., & Lai, F. Y., (1996). Volatile constituents from the fruits of four *Syzygium* species grown in Malaysia. *Flavour and Fragrance Journal, 11*(1), 61–66.
201. Wright, M. H., Matthews, B., Greene, A. C., & Cock, I. E., (2015). Growth inhibition of the zoonotic bacteria *Bacillus anthracis* by high antioxidant Australian plants: New leads for the prevention and treatment of anthrax. *Pharmacognosy Communications, 5*(3), 173–189.
202. Yadav, S. S., Meshram, G. A., Shinde, D., Patil, R. C., Manohar, S. M., & Upadhye, M. V., (2011). Antibacterial and anticancer activity of bioactive fraction of *Syzygium cumini* L., seeds. *HAYATI Journal of Biosciences, 18*(3), 118–122.
203. Youdim, K. A., Spencer, J. P., Schroeter, H., & Rice-Evans, C. A., (2002). Dietary flavonoids as potential neuroprotectans. *Biological Chemistry, 383*, 503–519.
204. Yukawa, T. A., Kurokawa, M., Sato, H., Yoshida, Y., Kageyama, S., Hasegawa, T., Namba, T., Imakita, M., Hozumi, T., & Shiraki, K., (1996). Prophylactic treatment of cytomegalovirus infection with traditional herbs. *Antiviral Research, 32*(2), 63–70.

PART II

MEDICINAL PLANTS/PLANT PRODUCTS AND HEALTH PROMOTION

CHAPTER 4

EFFECTS OF *TALINUM TRIANGULARE* ON HEPATIC ANTIOXIDANT GENE EXPRESSION PROFILE IN CARBON TETRACHLORIDE-INDUCED RAT LIVER INJURY

GBENGA ANTHONY ADEFOLAJU, BENEDICT ABIOLA FALANA, ADEOYE OLUWOLE OYEWOPO, and ANTHONY MWAKIKUNGA

CONTENTS

4.1 INTRODUCTION

Talinum triangulare, popularly known as waterleaf, is a fleshy, saponin-rich, perennial herb that is widely cultivated as an edible and medicinal vegetable in West Africa, Asia, and South America [4]. In folk medicine (in Taiwan province of China), the plant has been employed in the prevention and management of diabetes, cancer, stroke, obesity, and measles. *T. triangulare* has been utilized in the prevention of hepatopathies and cancer in traditional medicine [4]. The hepatoprotective effects of *T. triangulare* against oxidative liver damage are well documented [1, 4, 21, 25, 28]. However, the molecular mechanisms underlying *T. triangulare's* antioxidant activities have not been totally elucidated. Extracts from *T. triangulare* have been shown to demonstrate potent antioxidant activities in in vitro and in vivo studies [1, 4, 21, 25]. These antioxidant properties have been associated with high proportions of betalains [28], phenols, and flavonoids [21, 25] in the plant. Excessive generation of reactive oxygen species (ROS) leads to oxidative stress, a process that mediates cellular damage. It has been shown that oxygen free radicals are involved in the development of cancer, diabetes, cardiovascular diseases, atherosclerosis, ischemia/reperfusion injury, neurodegenerative diseases, inflammation, ageing [2, 14, 15, 29], and various hepatopathies [6]. The peroxidation of unsaturated fatty acids in biological membranes leads to disruption of membrane structure and function that is implicated in numerous pathological changes [2]. There are many known endogenous systems and processes that limit ROS and their resultant deleterious effects; however, this protection may not be sufficient at all times, and hence, there is a need for extrinsic antioxidants [11, 16].

A number of studies [1, 4, 25] have demonstrated the hepatoprotective effects of *T. triangulare* on experimentally induced liver damage at the biochemical level by investigating the levels of relevant serum enzymes.

The study in this chapter is the first to describe the mRNA expression mechanisms underlying the antioxidant hepatoprotective effects of *T. triangulare* by describing the gene expression profile of hepatic antioxidant genes: superoxide dismutase-1 (*Sod1*), heme oxygenase (decycling)-1 (*Hmox-1*), catalase (*Cat*), and glutathione peroxidase-1 (*Gpx1*). This chapter also describes the mRNA expression profile of key hepatic marker enzymes such as alkaline phosphatase (*Alpl*) and glutamic pyruvate transaminase (alanine aminotransferase) (*Gpt*).

4.2 MATERIALS AND METHODS

4.2.1 CHEMICALS, REAGENTS, AND PLANT MATERIAL

Formaldehyde, absolute alcohol, and xylene were purchased from Sigma Chemical Co. (St. Louis, MO, USA). DNase I, RNase-free kit, RNase-free water, O'Gene Ruler Low Range DNA Ladder, and Proteinase-K were purchased from Thermo Scientific (Pittsburgh, PA, USA). Trizol reagent, High Capacity cDNA Reverse Transcription Kit and Power SYBR® Green PCR Master Mix were purchased from Life technologies (California, USA). Oligos for qPCR were purchased from Integrated DNA Technologies, Inc. (Coralville, Iowa, USA). Agarose D-1 Low EEO-GQT was purchased from Conda Laboratories (Madrid, Spain). *T. triangulare* leaves were purchased from the general market (Ilorin, Nigeria) and were verified and authenticated by Mr. Adebayo at the herbarium of Department of Plant Biology, University of Ilorin, Nigeria. The leaves were washed in distilled water and prepared as described by Adefolaju et al. [1].

4.2.2 ANIMALS AND THEIR TREATMENT

The maintenance and care of experimental animals complied with guidelines for the humane use of laboratory animals by the National Institute of Health; and the approval was obtained from the research committee of College of Health Sciences, University of Ilorin. Nigeria. Twenty-five healthy male Wistar rats (weight 150–200 g) were obtained from and kept at the animal holdings unit of the Department of Anatomy, University of Ilorin, in well-ventilated cages under standard photoperiodic conditions and were given rat pellets (Bendels Feeds, Yoruba road, Ilorin) and water ad libitum. The animals were randomly divided into five groups of five rats each:

- Group A (Control) was kept on normal diet and received water orally and administration of olive oil (intraperitoneally, i.p) at 0.5 mL/kg of body weight (b.w.) once daily for 7 days.
- Group B received carbon tetrachloride ((CCl_4) @ 0.5 mL/kg of b.w., 20% CCl_4/olive oil i.p) [27], once daily for 7 days.

- Group C received 200 mg/kg of extract concentration orally and 0.5 mL/kg of b.w., 20% CCl_4/olive oil i.p concurrently for 7 days.
- Group D received 400 mg/kg of aqueous extract orally and 0.5 mL/kg of b.w., 20% CCl_4/olive oil i.p concurrently for 7 days.
- Group E received 600 mg/kg of aqueous extract orally and 0.5 mL/kg of b.w., 20% CCl_4/olive oil i.p concurrently for 7 days.

The animals were anaesthetized and sacrificed on the 8th day, and liver tissues were formalin fixed and paraffin embedded.

4.2.3 TOTAL RNA ISOLATION AND REAL-TIME qPCR

Extraction of total RNA from formalin-fixed, paraffin-embedded tissue was performed as described by Ma et al. [22]. Briefly, 20-µm sections were cut from the interior of paraffin blocks using a microtome. Tissue slices were placed into 1.5 mL RNase-free 1.5 mL microcentrifuge tubes and 1 mL 100% xylene was added. Tubes were incubated at 50°C for 3 min to melt the paraffin and then centrifuged for 1 min at the maximum speed to pellet the tissue; the xylene was discarded without disturbing the pellet, and the xylene wash was repeated. The pellet was washed twice with 1 mL 100% ethanol and air dried. Protease K digestion buffer (150 µL) 1x was added to each sample and incubated at 55°C for 3 h. Trizol (1 mL) was added to each sample and incubated at 15 to 30°C for at least 5 min to dissociate nucleoprotein complexes. Chloroform (0.2 mL) was added, and the tubes were vortexed vigorously for 15 s and incubated at 30°C for 3 min. The samples were centrifuged at 12,000 × *g* for 15 min at 4°C. The aqueous phase was then transferred to a fresh tube, and 10 µg of glycogen was added and mixed. The total RNA was precipitated by mixing with 0.6 mL isopropyl alcohol; the precipitate was placed at -20°C for 1 h and then centrifuged at 12,000 × *g* for 10 min at 4°C. The RNA pellet was washed with 100% ethanol, briefly air dried, and dissolved in RNase-free water. RNA concentration and purity were determined using the Nanodrop-1000 spectrophotometer. RNA integrity was checked by gel electrophoresis. According to the manufacturer's instructions, genomic DNA was removed from total

RNA using the DNase I, RNase-free kit. The DNAase I-treated RNA was again cleaned with the GeneJET RNA Purification kit, re-quantified, and stored at -80°C until use.

According to the manufacturer's instructions, cDNA was synthesized using the MultiScribe™ Reverse Transcriptase from 700 ng RNA. The reverse transcriptase reaction was carried out in a GeneAmp® PCR System 9600 Thermal Cycler for 10 min at 25°C, 120 min at 37°C, and the enzyme was then deactivated for 5 min at 85°C. The cDNA aliquots were then utilized in qPCR reactions for *Sod1, Ho-1, Cat, Gpx1*, *Alpl,* and *Alt* with *Sdha, Actb,* and *B2m* used as the endogenous reference genes.

PCR reactions were amplified for 40 cycles prior to which the AmpliTaq Gold® DNA polymerase was activated for 10 min at 95°C. Each cycle consisted of a denaturing step for 15 s at 95°C and annealing/extension step for 1 min at 60°C. PCR amplification was performed in a final volume of 20 μL using the Power SYBR® Green PCR Master Mix with the ABI 7500 real-time PCR machine. Primer sequences and PCR product sizes are indicated in Table 4.1.

To confirm the absence of nonspecific amplification, the PCR products were separated on 3% agarose gels and stained with ethidium bromide, and images were acquired with the BioRad Gel Doc® XR (Model 170-8170 Segrate, Milan, Italy). Melt curves were generated for each PCR product using the Applied Biosystems ABI 7500 software. The relative mRNA expression levels of target genes in each sample were calculated using the qbasePLUS software (Biogazelle, Zulte, Belgium).

4.2.4 STATISTICAL ANALYSIS

Statistical analysis was performed using JMP® (Version 10.0, SAS Institute Inc., Cary, NC, USA). Data were reported as mean + SEM. After verifying the normal distribution and the homogeneity of the variance using an F-test ($P < 0.05$), a one-way analysis of variance (where a significance level of $P < 0.05$ was set) was used to compare the results. All mean values were then compared using the Tukey-Kramer HSD.

TABLE 4.1 Oligonucleotide Sequences Used for qPCR

Gene	NCBI accession No	Sequences (5'-3' direction)		Product size (bp)
Target Genes				
Sod1	NM_017050.1	**F**	TTGGCCGTACTATGGTGGTC	121
		R	TGGGCAATCCCAATCACACC	
Hmox-1	NM_012580.2	**F**	TCTGCAGGGGAGAATCTTGC	135
		R	TTGGTGAGGGAAATGTGCCA	
Cat	NM_012520.2	**F**	GCTCCGCAATCCTACACCAT	104
		R	GTGGTCAGGACATCGGGTTT	
Gpx1	NM_030826.3	**F**	CCCGGGACTACACCGAAATG	147
		R	CGGGTCGGACATACTTGAGG	
Alpl	NM_013059.1	**F**	CTTCCCACCCATCTGGGCTC	100
		R	ATGAGTTGGTAAGGCAGGGTCC	
Gpt	NM_031039.1	**F**	ATTTCCCCACTGGACCCTTC	171
		R	TCGAACTGCATACTCCACCC	
Reference Genes				
Sdha	NM_130428.1	**F**	CCCATTCCAGTCCTTCCCAC	102
		R	CACAATCTGATCCTGGCCGT	
Actb	NM_031144.3	**F**	CTGTGTGGATTGGTGGCTCT	134
		R	AGCTCAGTAACAGTCCGCCT	
B2m	NM_012512.2	**F**	CACACGCAGTCTGAAAACCC	170
		R	AAAGGAGCTTTGGGGACACA	

4.3 RESULTS

4.3.1 EFFECTS OF T. TRIANGULARE ON THE ALPL AND GLUTAMIC PYRUVATE TRANSAMINASE (ALANINE AMINOTRANSFERASE-GPT) MRNA LEVELS IN THE LIVER OF WISTAR RATS

As shown in Figure 4.1, CCl_4 injection produced a significant upregulation of both *Alpl* and *Gpt* mRNA levels in rat liver compared with that

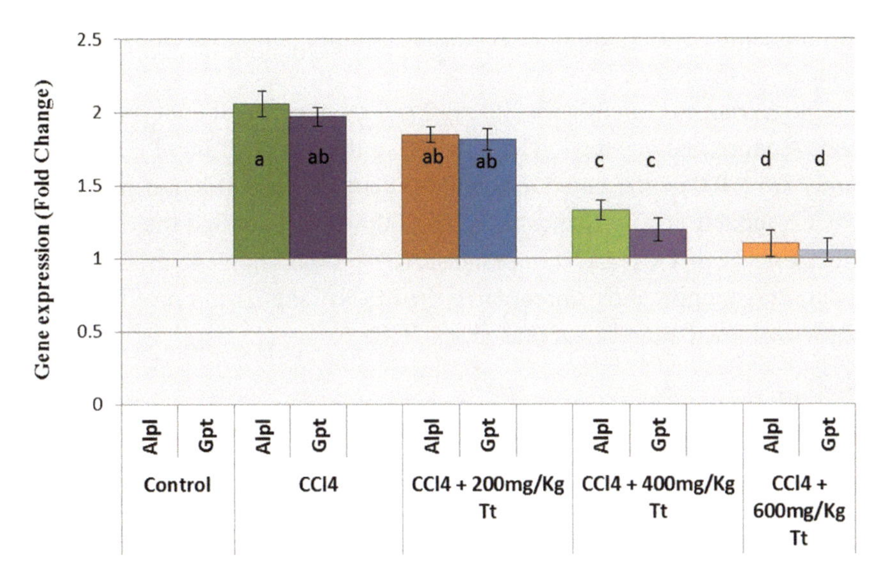

FIGURE 4.1 Effects of *Talinum triangulare* on Alpl and Gpt mRNA expression in CCl_4-induced liver injury in rats. Data (mean ± SEM) are represented as fold changes of gene expression relative to values from the control group (defined as 1) and representative of 3 independent experiments for RNA extraction. Groups not connected by the same letter are significantly ($P<0.05$) different.

in the control group ($P < 0.05$), reflecting the tissue damage in the liver. Administration of extracts of *T. triangulare* (@ 200, 400, and 600 mg/kg) reversed this elevation to normal. This hepatoprotective effects of *T. triangulare* was dose dependent, and high doses (400 and 600 mg/kg) exerted statistically significant effects comparable to the normal controls.

4.3.2 EFFECTS OF T. TRIANGULARE ON HEPATIC ANTIOXIDANT EXPRESSION PROFILE IN CCL_4-INDUCED LIVER INJURY IN WISTAR RATS

Figure 4.2 indicates that CCl_4 treatment significantly decreased the mRNA levels of Sod1, Hmox-1, Cat, and Gpx1 compared to normal control rats, thus indicating stronger oxidative stress and lipid peroxidation in rat liver tissues.

Concurrent treatment with 200 mg/kg *T. triangulare* did not protect the livers from oxidative stress. The 400 mg/kg dose of *T. triangulare* significantly ($P < 0.05$) prevented antioxidant gene down-regulation produced by CCl_4 induction. The highest dose of 600 mg/kg inhibited the oxidative stress effects of CCl_4 on antioxidant mRNA expression. At this dose, *T. triangulare* significantly upregulated the mRNA expression profiles of all genes examined even in the presence of CCl_4.

4.4 DISCUSSION

This study was designed to investigate the molecular mechanisms underlying the widely reported antioxidant/hepatoprotective effects of the

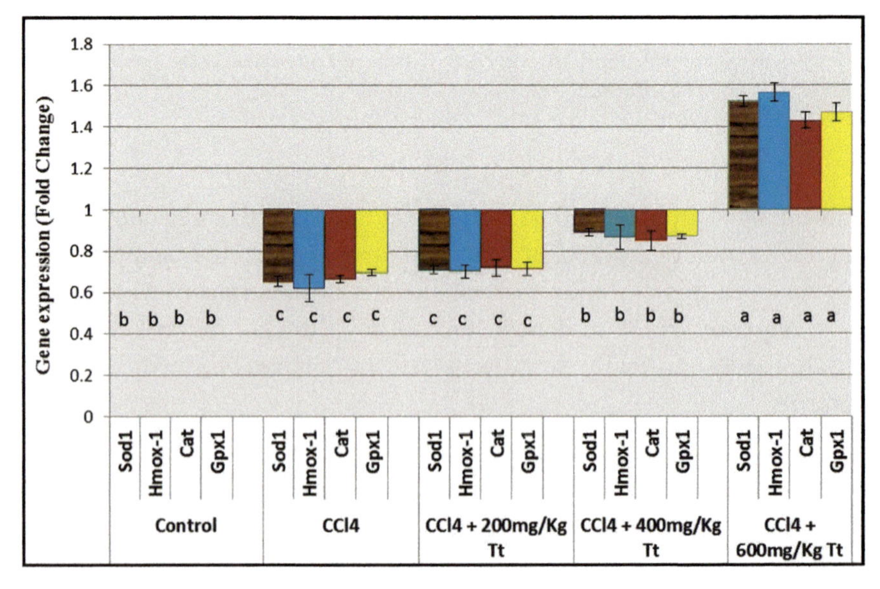

FIGURE 4.2 Effects of *Talinum triangulare* on Sod1, Hmox-1, Cat, and Gpx1 mRNA expression in CCl_4-induced liver injury in rats. Data (mean ± SEM) are represented as fold changes of gene expression relative to values from the control group (defined as 1) and representative of 3 independent experiments for RNA extraction. Groups not connected by the same letter are significantly ($P<0.05$) different.

extract of *T. triangulare* by studying the expression profile of key antioxidant genes *Sod1, Hmox-1, Cat*, and *Gpx1* as well as the gene expression profile of two important hepatic marker proteins. The higher doses of *T. triangulare* significantly upregulated the mRNA expression of all antioxidant genes investigated. This mRNA upregulation provides an insight into molecular mechanisms involved in the increased levels of antioxidant proteins in similar experimental models as reported by Liang et al. [21], who showed improved antioxidant *Sod-1* and glutathione (*Gsh*) enzyme levels as a result of *T. triangulare* treatment of CCl_4 -induced liver injury in mice.

The research study by authors of this chapter supports the molecular explanation for the increased enzyme levels, namely the upregulation of mRNA. The *Sod*-1 gene encodes the superoxide dismutase enzyme, which binds copper and zinc ions to destroy toxic, charged oxygen molecules also known as free superoxide radicals and converts them to molecular oxygen and hydrogen peroxide [5, 23, 24]. The *Hmox*-1 gene encodes the rate-limiting enzyme that degrades heme into biliverdin and is a powerful antioxidant enzyme [17, 31]. The *Cat* gene encodes catalase, which catalyzes the degradation of hydrogen peroxide (a deleterious byproduct of normal metabolism) to water and oxygen. It is a very potent antioxidant thus protecting cells from oxidative damage by reactive oxygen species (ROS) [8, 12]. Another important antioxidant gene is glutathione peroxidase, which encodes the *Gpx1* enzyme involved in the detoxification of hydrogen peroxide [7, 10, 20].

CCl_4-induced hepatotoxicity is one of the most commonly used experimental models in screening drugs/agents for their hepatoprotective activity [18, 21]. CCl_4 is degraded by cytochrome P450 2E1 enzymes to trichloromethyl ($\bullet CCl_3$) and proxy trichloromethyl ($\bullet OOCCl_3$) radicals that bind to macromolecules and produce peroxidative cellular membrane damage and hepatic cell death [26]. Hepatic injury is usually assessed by examining serum or tissue levels of hepatic marker genes/enzymes such as ALP and ALT. Previous studies [1, 3, 21] have shown that *T. triangulare* demonstrated hepatoprotective activities by significantly lowering serum/tissue levels of hepatic marker enzymes such as AST, GGT, ALP, ALT, and AST. The abundance of these proteins in hepatocytes makes them leak into extracellular and intravascular compartments because of hepatocellular degradation [19, 21]. The ALP gene encodes the ALP enzyme that is found in many cells including those lining the biliary ducts of the liver. ALP levels in plasma increase as a result of hepatocellular degradation [9]. GPT levels are also commonly clinically measured as a diagnostic evaluation of liver injury [13, 30].

4.5 CONCLUSIONS

The study was designed to evaluate the expression profile of key hepatic antioxidant and marker genes in Wistar rats fed with *Talinum triangulare* aqueous extract (TTAE) in a CCl_4 model of hepatic injury.

Twenty-five healthy male Wistar rats were randomly divided into five groups of five rats each. Group A served as control. Group B received distilled water orally and carbon tetrachloride (CCl_4) (0.5 mL/kg of b.w., 20% CCl4/olive oil i.p) once daily. Group C received *TTAE* 200 mg/kg, orally and 0.5 mL/kg of b.w., 20% CCl4/olive oil i.p concurrently. Group D received 400 mg/kg *TTAE* orally and 0.5 mL/kg of b.w., 20% CCl4/olive oil i.p concurrently. Group E received 600 mg/kg *TTAE* orally and 0.5 mL/kg of b.w., 20% CCl4/olive oil i.p concurrently. All administrations lasted for 7 days. The animals were sacrificed after 24 h, and RNA was isolated from formalin-fixed/paraffin-embedded liver tissues. Total RNA was reverse transcribed, and cDNA was amplified by real time quantitative PCR for the expression levels of antioxidant genes: superoxide dismutase 1 (*Sod1*), heme oxygenase (decycling) 1 (*Hmox-1*), catalase (*Cat*), glutathione peroxidase 1 (*Gpx1*), and the hepatic marker genes alkaline phosphatase (*Alpl*) and glutamic pyruvate transaminase (alanine aminotransferase) (*Gpt*).

A dose of 400-600 mg/kg of *TTAE* was the most effective in significantly ($P<0.05$) increasing hepatic antioxidant gene expression, while significantly ($P<0.05$) decreasing the mRNA levels of *Alpl* and *Gpt* markers of hepatic injury. The antioxidant/hepatoprotective effects of *T. triangulare* were dose-dependent and are a result of its actions on the expression profile of key hepatic markers and antioxidant genes.

4.6 SUMMARY

In this chapter, it was shown that CCl_4 led to the increased expression of *Alpl* and *Gpt* mRNA levels in the liver tissue, while decreasing the expression levels of the antioxidant genes *Sod1, Hmox-1, Cat,* and *Gpx1* all of which are indicative of hepatocellular injury. Concurrent treatment with *T. triangulare* significantly reversed these trends and downregulated the mRNA expression levels of *Alpl* and *Gpt*, indicating that the in vivo antioxidant hepatoprotective abilities of *T. triangulare* is as a result of its actions on the expression profile of key hepatic markers and antioxidant genes.

KEYWORDS

- **absolute alcohol**
- **ageing**
- **alkaline phosphatase (*alpl*)**
- **antioxidants**
- **asia**
- **atherosclerosis**
- **betalains**
- **cancer**
- **carbon tetrachloride**
- **cardiovascular diseases**
- **catalase (*cat*)**
- **deoxyribonucleic acid**
- **diabetes**
- **dnase**
- **flavonoids**
- **formaldehyde**
- **formalin**
- **gene expression**
- **glutamicpyruvate transaminase (*gpt*) (alanine aminotransferase)**

- **glutathione peroxidase 1 (*gpx1*) and hepatic marker genes**
- **heme oxygenase 1 (decycling) (*hmox-1*)**
- **hepatic injury**
- **inflammation**
- **intraperitoneal**
- **ischemia/reperfusion injury**
- **liver tissues**
- **measles**
- **medicinal vegetable**
- **neurodegenerative diseases**
- **obesity**
- **o'gene ruler low range DNA ladder**
- **olive oil**
- **oxidative stress**
- **oxygen free-radicals**
- **paraffin**
- **phenols**
- **polymerase chain reaction**
- **proteinase k**
- **reactive oxygen species**
- **reverse transcription**
- **ribonucleic acid**
- **Rnase free water**
- **South America**
- **stroke**
- **superoxide dismutase 1 (*sod1*)**
- ***Talinum triangulare* Trizol reagent**
- **West Africa**
- **wistar rats**
- **xylene**

REFERENCES

1. Adefolaju, G. A., Ajao, M. S., Olatunji, L. A., Enaibe, B. U., & Musa, M. G., (2008). Hepatoprotective effect of aqueous extract of water leaf (*Talinum Triangulare*) on carbon tetrachloride (CCL4) induced liver damage in wistar rats. *Internet Journal of Pathology*, *8*(1), 1.
2. Afonso, V., Champy, R., Mitrovic, D., Collin, P., & Lomri, A., (2007). Reactive oxygen species and superoxide dismutases: role in joint diseases. *Joint, Bone, Spine: Revue Du Rhumatisme*, *74*(4), 324–329.
3. Ajibade, A. J., Fakunle, P. B., Olayemi, O. T., Ehigie, L., & Abosede, O. Y., (2013). Ameliorative effects of aqueous water leaf extract (Talinum Triangulare) against carbon tetrachloride induced hepatotoxicity in adult wistar rats. *International Journal of Biological & Pharmaceutical Research*, *4*, 516–522.
4. Andarwulan, N., Batari, R., Sandrasari, D. A., Bolling, B., & Wijaya, H., (2010). Flavonoid content and antioxidant activity of vegetables from Indonesia. *Food Chemistry*, *121*(4), 1231–1235.
5. Barouki, R., (2006). Ageing free radicals and cellular stress. *Médecine Sciences: M/S*, *22*(3), 266–272.
6. Bruck, R., Aeed, H., Avni, Y., Shirin, H., Matas, Z., & Shahmurov, M., (2004). Melatonin inhibits nuclear factor Kappa B activation and oxidative stress and protects against thioacetamide induced liver damage in rats. *Journal of Hepatology*, *40*(1), 86–93.
7. Cao, C., Leng, Y., Huang, W., Liu, X., & Kufe, D., (2003). Glutathione peroxidase-1 is regulated by the c-Abl and Arg tyrosine kinases. *The Journal of Biological Chemistry*, *278*(41), 39609–39614.
8. Chelikani, P., Fita, I., & Loewen, P. C., (2004). Diversity of structures and properties among catalases. *Cellular and molecular life sciences: CMLS*, *61*(2), 192–208.
9. Coleman, J. E., (1992). Structure and mechanism of alkaline phosphatase. *Annual Review of Biophysics and Biomolecular Structure*, *21*, 441–483.
10. Esworthy, R. S., Ho, Y. S., & Chu, F. F., (1997). The Gpx1 gene encodes mitochondrial glutathione peroxidase in the mouse liver. *Archives of Biochemistry and Biophysics*, *340*(1), 59–63.
11. Fu, W., Chen, J., Cai, Y., Lei, Y., Chen, L., & Pei, L., (2010). Antioxidant, free radical scavenging, anti-inflammatory and hepatoprotective potential of the extract from *Parathelypteris Nipponica* (Franch. et Sav.) Ching. *Journal of Ethnopharmacology*, *130*(3), 521–528.
12. Gaetani, G. F., Ferraris, A. M., Rolfo, M., Mangerini, R., Arena, S., & Kirkman, H. N., (1996). Predominant role of catalase in the disposal of hydrogen peroxide within human erythrocytes. *Blood*, *87*(4), 1595–1599.
13. Ghouri, N., Preiss, D., & Sattar, N., (2010). Liver Enzymes, Nonalcoholic fatty liver disease, & incident cardiovascular disease: A narrative review and clinical perspective of prospective data. *Hepatology (Baltimore, Md.)*, *52*(3), 1156–1161.
14. Halliwell, B., (2006). Reactive species and antioxidants: Redox biology is a fundamental theme of aerobic life. *Plant Physiology*, *141*(2), 312–322.
15. Halliwell, B., & Gutteridge, J. M., (1990). Role of free radicals and catalytic metal ions in human disease: An overview. *Methods in Enzymology*, *186*, 1–85.

16. Huang, B., Ban, X., He, J., Zeng, H., Zhang, P., & Wang, Y., (2010). Hepatoprotective and antioxidant effects of the methanolic extract from *Halenia Elliptica*. *Journal of Ethnopharmacology*, *131*(2), 276–281.
17. Immenschuh, S., Baumgart-Vogt, E., & Mueller, S., (2010). Heme oxygenase-1 and iron in liver inflammation: A complex alliance. *Current Drug Targets*, *11*(12), 1541–1550.
18. Jia, X. Y., Zhang, Q. A., Zhang, Z. Q., Wang, Y., Yuan, J. F., Wang, H. Y., et al., (2011). Hepatoprotective effects of almond oil against carbon tetrachloride induced liver injury in rats. *Food Chemistry*, *125*(2), 673–678.
19. Kew, M. C., (2000). Serum aminotransferase concentration as evidence of hepatocellular damage. *The Lancet*, *355*(9204), 591–592.
20. Legault, J., Carrier, C., Petrov, P., Renard, P., Remacle, J., & Mirault, M. E., (2000). Mitochondrial GPx1 decreases induced but not basal oxidative damage to mtDNA in T47D cells. *Biochemical and Biophysical Research Communications*, *272*(2), 416–422.
21. Liang, D., Zhou, Q., Gong, W., Wang, Y., Nie, Z., & He, H., (2011). Studies on the antioxidant and hepatoprotective activities of polysaccharides from *Talinum Triangulare*. *Journal of Ethnopharmacology*, *136*(2), 316–321.
22. Ma, Z., Lui, W. O., Fire, A., & Dadras, S. S., (2009). Profiling and discovery of novel miRNAs from formalin-fixed, paraffin-embedded melanoma and nodal specimens. *The Journal of Molecular Diagnostics: JMD*, *11*(5), 420–429.
23. Milani, P., Gagliardi, S., Cova, E., & Cereda, C., (2011). SOD1 transcriptional and posttranscriptional regulation and its potential implications in ALS., *Neurology Research International*, *2011*, 458427.
24. Muller, F. L., Lustgarten, M. S., Jang, Y., Richardson, A., & Van Remmen, H., (2007). Trends in oxidative aging theories. *Free Radical Biology & Medicine*, *43*(4), 477–503.
25. Omoregie, E. S., & Osagie, A. U., (2012). Antioxidant properties of methanolic extracts of some Nigerian plants on nutritionally-stressed rats. *Nigerian Journal of Basic & Applied Science*, *20*, 7–20.
26. Ranawat, L., Bhatt, J., & Patel, J., (2010). Hepatoprotective activity of ethanolic extracts of bark of zanthoxylum armatum DC in CCl4 induced hepatic damage in rats. *Journal of Ethnopharmacology*, *127*(3), 777–780.
27. Shyu, M. H., Kao, T. C., & Yen, G. C., (2008). Hsian-Tsao (*Mesona Procumbens HemL.*) prevents against rat liver fibrosis induced by CCl_4 via inhibition of hepatic stellate cells activation. *Food and Chemical Toxicology: An International Journal Published for the British Industrial Biological Research Association*, *46*(12), 3707–3713.
28. Swarna, J., Lokeswari, T. S., Smita, M., & Ravindhran, R., (2013). Characterization and determination of in vitro antioxidant potential of betalains from *Talinum Triangulare* (Jacq.) Willd. *Food Chemistry*, *141*(4), 4382–4390.
29. Valko, M., Leibfritz, D., Moncol, J., Cronin, M. T. D., Mazur, M., & Telser, J., (2007). Free radicals and antioxidants in normal physiological functions and human disease. *The International Journal of Biochemistry & Cell Biology*, *39*(1), 44–84.
30. Wang, C. S., Chang, T. T., Yao, W. J., Wang, S. T., & Chou, P., (2012). Impact of increasing alanine aminotransferase levels within normal range on incident diabetes. *Journal of the Formosan Medical Association: Taiwan Yi Zhi*, *111*(4), 201–208.
31. Xu, Y. Q., Zhang, D., Jin, T., Cai, D. J., Wu, Q., & Lu, Y., (2012). Diurnal variation of hepatic antioxidant gene expression in mice. *PloS One*, *7*(8), e44237.

CHAPTER 5

REVIEW ON POTENTIAL OF SEEDS AND VALUE-ADDED PRODUCTS OF BAMBARA GROUNDNUT (*VIGNA SUBTERRANEA*): ANTIOXIDANT, ANTI-INFLAMMATORY, AND ANTI-OXIDATIVE STRESS

YVONNE YEUKAI MUREVANHEMA, VICTORIA ADAORA JIDEANI, and OLUWAFEMI OMONIYI OGUNTIBEJU

CONTENTS

5.1 INTRODUCTION

Socio-economic and health-related issues, including climate change and premature death associated with life-style degenerative diseases, have recently become a state of urgency. Erratic rainfall, attributed to climate change, is a threat to food security [27]. Bambara groundnut (*Vigna subterranea* (L.) Verdc) (BGN) is one of the drought-tolerating crops, and it is envisaged to be the food of the future.

Food intake, absorption, assimilation, biosynthesis, catabolism, and excretion are taken into consideration in nutrition. The phenolic phytochemicals present in BGN, with regard to nutrition in relation to heath, inflammation, and oxidative stress, need more attention. These interactions can either be one of the following four possibilities:

- Nutraceutical by providing health and medical benefits;
- Therapeutic for treatment of diseases;
- Medicinal by possessing similar properties as conventional pharmaceutical drugs; and
- Pharmacological by containing a natural moleculethat exerts a biological effect on the cell, tissue, and organs. In this review, the therapeutic potentials of BGN and other legumes are of interest.

By definition, therapy is a consequence of medical treatment of any kind, the results of which are judged to be desirable and beneficial for the treatment of a disease. In this regard, ethnobotany knowledge has shown the healing or curative effects of BGN. Recent studies on BGN have shown potential of securing food security owing to its drought resistance, and more studies are being conducted to determine the potentials of BGN. However, its economic status is still undermined; hence, it is a neglected and underutilized crop. Scientifically proven, the therapeutic potential of BGN and other legumes is envisaged to strengthen their economic status and break new grounds for alternatives to synthetic drugs for the management of lifestyle-related degenerative diseases including inflammation and oxidative stress.

Inflammation and oxidative stress have been implicated in the pathogenesis of a number of degenerative diseases including autoimmune diseases, metabolic syndrome, neurodegenerative diseases, cancers, and cardiovascular diseases (CVDs), which cause premature death of 16 million people every year as reported by World Health Organization (WHO) [96]. Acute inflammation is a part of a regulated defense system [56] against invading pathogens or injury [4]. Conversely, the nonregulated form accelerates chronic inflammation [56], which is responsible for pathogenesis of most degenerative diseases. Compelling evidence has demonstrated the connection between inflammation and oxidative stress [17] and their role in pathogenicity of many unrelated diseases as depicted in Figure 5.1, which is a representation of the possible relationship amongst nutrition (antioxidants), inflammation, and oxidative stress. Oxidative stress is due to an imbalance between the systemic manifestation of reactive oxygen species (ROS) and reactive nitrogen species (RNS) and the biological system's ability to readily detoxify the reactive intermediates or to repair the resulting damage. Overgeneration of RONs results in oxidative stress, which is a deleterious process that can lead to damage of important cell structures, including carbohydrates, lipids and membranes, and protein and deoxyribonucleic acid (DNA); thus, its association with cellular injury has been seen in many pathological conditions.

An antioxidant-rich diet has been suggested to address chronic inflammation and oxidative stress by providing redox active compounds that can neutralize free radicals and by enhancing gene expression of the endogenous antioxidants [62].

Neglected and underutilized species (NUS) such as BGN, being rich in antioxidant, is attributed to have the potential of treating diseases linked to inflammation and oxidative stress-related diseases. Based on indigenous knowledge, BGN have been used for medicinal purposes including treatment of inflammation. However, no scientific evidence has been documented to substantiate these therapeutic claims on BGN. Novel products

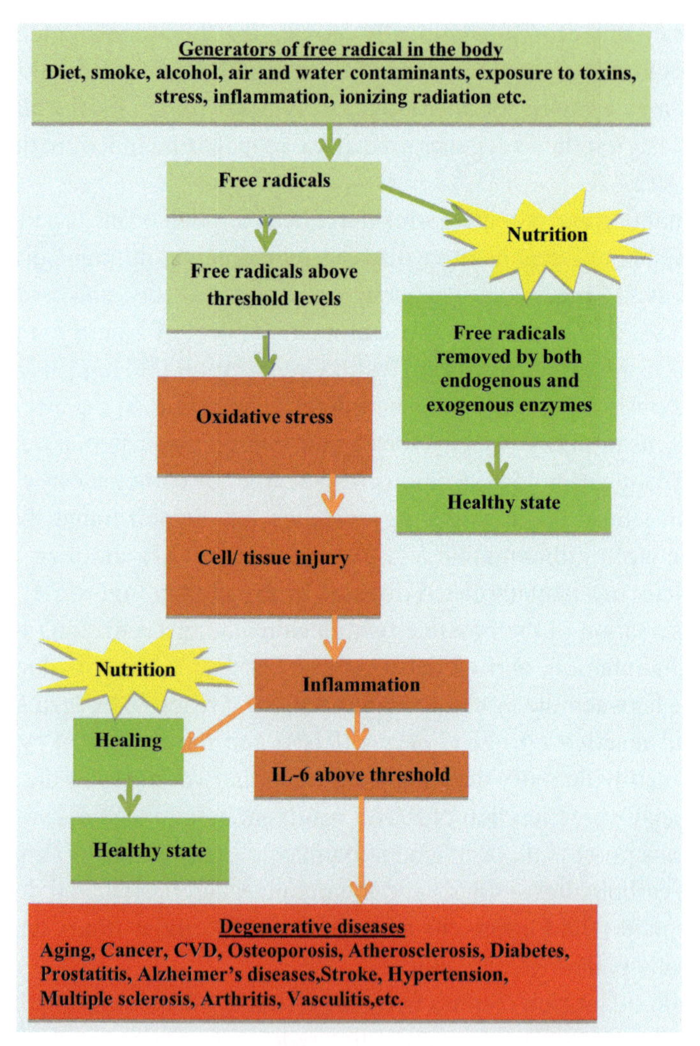

FIGURE 5.1 Relationship among nutrition, inflammation, and oxidative stress.

including milk and probiotic beverage have been developed from BGN with evidence of some antioxidant activity [57].

Therefore, the antioxidant efficacy of BGN seeds, milk, and the probiotic beverage and other legumes for modulatory effect on inflammation and oxidative stress will be highlighted in the review in this chapter. Scientifically and clinically, unlocking the value of BGN is gaining a high emphasis currently. Subsequently, outputs from this review will impact the stakeholders, health scientists, nutritionists, dieticians, pharmacologists, public health scientists, state agricultural departments, food manufacturers and processors, policymakers, and marketing and economic strategists, thus positively affecting the economic and healthcare status of BGN, which is regarded as the poor man's food or women's crop, thereby contributing to food security and helping to alleviate nutritional problems.

5.2 SEEDS AND VALUE-ADDED PRODUCTS OF BAMBARA GROUNDNUT

BGN, an indigenous African legume, which is well adapted to poor soil and dry climates, has been cultivated in Africa for years. It is grown in the African continent from Senegal to Kenya, and from the Sahara to South Africa and Madagascar [88]. Outside of Africa, BGN is cultivated in Brazil (where it is known as mandubid Angola), West Java, and southern Thailand. The BGN is also cultivated in other tropical locations such as Middle East, Syria, and Greece. Small-scale cultivation trials of BGN have been successful in Florida, United States [64].

BGN is known in South Africa as *jugoboon* in Afrikaans; *jugo* in Xhosa; *indlubu* in Zulu/Ndebele; *nduhu* in Venda; *ditloo* in Sesuthu; *nyimo* in Shona (Zimbabwe); and *okpaotuanya* in Nigeria. Figure 5.2 illustrates an uprooted BGN plant with a compact well-developed tap root with many short (up to 20 cm long) lateral stems on which leaves are borne, the red seeds, and milk from seeds [50].

BGN is a highly nutritious plant that plays a crucial role in local diets [35]. Apart from nutritional benefits, BGN is important for its therapeutic properties. However, vigorous elucidation and scientific investigations on medicinal claims are needed. BGN has higher content of the essential amino acid methionine than other legumes. Therefore, it has the potential to reduce malnutrition in many parts of Africa. Despite the growing importance of BGN, the use of Bambara in South Africa has been limited. BGN can be explored as a good source of fiber, calcium, iron, potassium, and methionine and for value-added products (such as BGN milk). The

FIGURE 5.2 Bambara groundnut (*Vigna subterranea*) plant with a tap root (top left); red seeds (top right), and value-added product such as milk (bottom).

probiotic beverage from BGN has potential to promote its consumption, production, and processing.

Value-added products are commodities that are processed in a way such that it changes its physical state or form to a high-quality end product, in order to meet the taste/preferences of consumers. Value-added products can open new markets and enhance the public appreciation of underutilized legumes. BGN is an underutilized legume, which has a potential for its therapeutic benefits associated with its consumption and scientifically proven nutritional value. Despite the therapeutic benefits and high protein content of BGN, its processing has been limited to boiling and roasting, thus discouraging its production and consumption due to complicated cooking phenomena.

BGN milk is a "vegetable milk" produced by hydrating BGN flour according to Murevanhema and Jideani [58]. This milk contains all beneficial phytochemicals as those in the seeds, especially the pigments on hulls, which include phenolics and flavonoids. The opaque appearance of the milk is due to its chemical composition, pigment components, certain minerals, and content of suspended particles such as protein. BGN milk produced from the red seeds has a pronounced chocolate brown color due to the pigment components, which are present in the seed coat and can easily leach into the aqueous phase during the hydration step impacting the characteristic testa color of the milk. Starch gelatinization during the heat processing step produces rich appearance of the milk.

BGN probiotic beverage acceptable to consumer has been produced [59]. The probiotics affect pH, appearance, and viscosity of the BGN milk through their growth and metabolism. Taste, appearance, and color were important parameters that were considered by the panelists in their preference. Preference for BGN probiotic beverage was high among the >40 age groups, demonstrating the potential of BGN milk, when more sucrose is added and commercialized. Fermenting BGN has been reported by Ademiluyi and Oboh [2], who indicated that it improves the nutritional quality and enhances the antioxidant properties, by increasing the polyphenolic constituents. Similar observations were made by Murevanhema [57], who reported on antioxidative activity exhibited by BGN probiotic beverage.

Research reports in literature concur that BGN has a number of potentials, including drought resistance and as functional, therapeutic, and complete food [7, 13, 17, 21, 46, 57, 58]. The list of benefits is still growing with the recent interest in BGN as highlighted by Jideani [36] in the *Journal of Institute of Food Technologists* and studies on BGN addressing food insecurity [33]. However, scientific evidence confirming the antioxidant, anti-inflammatory, and antioxidative stress potentials of BGN seeds/milk/probiotic beverage and other legumes is necessary to substantiate ethnobotanical medical claims obtained from indigenous knowledge on BGN.

5.3 TRADITIONAL THERAPEUTIC USES OF BGN

Ethnobotany refers to the scientific study of the relationship that exist between people and plants. Ethnobotanical data obtained from local communities has brought attention to the therapeutic effects of BGN exhibited in tradition medicine therapies. In Burkina Faso, the seeds are used to treat a number of human diseases. The Igbo tribe in Nigeria uses the seeds to treat venereal diseases, while the Luo tribe from Kenya uses them for treating diarrhea. Treatment of polymenorrhea, internal bruising, and speedy effects on the resorption of hematomas by BGN has been documented. The cream-colored testa is used in weaning formulae owing to the low condensed tannin content [42].

Owing to its high concentration of soluble fiber [21], BGN is expected to reduce the incidence of heart disease and prevent colon cancer. The black seeds are recognized for the treatment of impotence in Botswana. Nausea and vomiting are believed to be stopped when one chews and swallows immature fresh seeds. A similar remedy is used to treat morning sickness in pregnant women. Ethnopharmacological data of BGN reported by Brink et al. [13] included anti-inflammatory and healing of infectious wounds. Rural people of Cot d'Ivoire use seeds to treat anemia, hence its application helps peripartum women within one month of delivery. Further treatment applications include use as a hemostatic drink for the treatment of menorrhea and rectal bleeding,and treatment of gonorrhea and ulcers. BGN-containing phenolic compounds are heat labile, and they leach into the water during the boiling process, which may explain the medicinal effects observed after one drinks the cooked water as reported by Koneet al. [42].

Prasad and Kumar [75] reported on the ethnobotanical potential of 28 medicinal legumes, and they concluded that there is still a dependence on indigenous knowledge for healthcare in Western Ghats in India. The 28 medicinal legumes can cure asthma, dental disorder, diabetes, etc. BGN being a legume is reputed to possess similar medicinal potentials; and this could be a rewarding area of therapeutic research. Plants have the potential for the discovery of new drugs due to easy access and relatively low cost.

Etuk et al. [22] validated 29 plants to be potent medicinal plants. The bark of the willow tree (*Spiraeaulmaria, Salix* species) was traditionally used in number of cultures for the treatment of inflammation, and fever. Aspirin is used in the treatment of inflammation-associated diseases. Using modern technologies, aspirin was extracted as salicylic acid from a willow tree [78]. Nonetheless, misuse of the wrong species of medicinal plant – incorrect dosage and use of products contaminated with potential hazardous substances such as toxic metals and pathogenic microorganisms – cannot be understated. Reports of severe kidney failure while taking herbal preparations containing aristolochic acid have been reported [78]. Furthermore, it is invaluable to investigate the efficacy of plants with medicinal properties to identify active components and to advocate for the ban of poisonous plants and contaminants.

Other plant-based foods have been reported to have strong epidemiological link to reduce the occurrence of CVDs, neurodegenerative diseases, certain types of cancers, chronic inflammation, and oxidative stress-related complications [4, 38, 90]. Phenolic phytochemicals have been attributed to medicinal and therapeutic effects observed in legumes. Multifunctionality of antioxidants in BGN is anticipated to work in synergistic interaction; thus, the different antioxidants with complementary mechanism of action are hypothesized to be invaluable in the defense system. There is a need to understand how the bioactive compounds modify a multitude of processes related to the defense system/mechanism.

5.4 NUTRITIONAL LINK TO THERAPEUTIC EFFECTS

Hippocrates (460–370 BC) said "*let food be your medicine and medicine your food*." This statement is true today than ever based on recent food trends. Currently, there is a great interest in phytochemical research due to possibility of improved public health through diet, where preventative health can be promoted through the consumption of legumes such as BGN. However, synthetic drugs have been the mainstay for the treatment of degenerative diseases, regardless of their known side effects. Recio et al. [78] and Safavi et al. [80] summarized the effects associated with prolonged use of ant-inflammatory steroids, such as:

- suppressing steroid biogenesis in the adrenal cortex;
- hyperglycemia and diabetic complications (aggravation of ulcers, diminished resistance to infections); and
- gradual development of osteoporosis, joint destruction, skin atrophy, impaired wound healing, excessive hair growth, fat redistribution, accelerated atherosclerosis, and hypertension.

Consequently, paving way for alternative therapies is most important to identify plant substances that are capable of treating and managing degenerative diseases in a way that is homeostatic, modulatory, efficient, and well tolerated by the body [78]. Several authors [1, 18, 24, 52, 82, 89] have mentioned that antioxidants have the potential to retard and prevent oxidative stress injury to vital body organs. The search for plant antioxidants has been accelerated by the misuse of antibiotics and chemotherapeutic agents, leading to drug resistance and side effects.

In the early 1960s, phenolic compounds were considered to be metabolic waste stored in plant vacuoles; however, these studies have been challenged by most recent studies. Bhattacharya [9] contends that phenolic compounds perform diverse physiological functions in the plant such as providing color for the leaves; flowers; fruit; possess antimicrobial and antifungal properties; protection against UV damage; chelation of toxic heavy metals; protection from free radicals generated during the photosynthetic process attract or repeal other organisms. However, a more profound symbiotic relationship between plants and humans has developed whereby the plants have become a dietary staple for humans. In this process, humans have benefited from the phenolic compounds furnished by plant both nutritionally and therapeutically. As a result of the innate ability of plants to synthesize nonenzymatic antioxidants, antioxidative activity has been exhibited by most plants. Upon ingestion, phenolic compounds are extensively degraded to their various simpler forms, some of which still possess their radical scavenging ability.

A number of studies have offered credible evidence showing that the products of plant origin have the capacity to prevent and treat diseases such as decreasing blood glucose level; weight reduction; anticarcinogenic; anti-inflammatory; antimicrobial; estrogenic activities; antiaging, and antithrombotic activity in humans and animals [8, 67, 71]. These biological activities are explained by their electron donation, reduction power, and metal chelating capabilities, and their enormous structural variability [85].

Plants are source of bioactive compounds and secondary metabolites, including alkaloids, phenolics, tannins, flavonoids, steroids and terpernoids, which have shown to possess important toxicological, pharmacological, and ecological benefits [26, 34, 69]. The secondary metabolites in plants have been shown to reduce risk of chronic diseases and conditions, including inflammation and oxidative stress [1, 4, 5, 6, 89]. Bernal et al. [8] acknowledges that although these metabolites have low potency as bioactive compounds, if they are ingested regularly and in significant amounts as part of a diet, noticeable long-term physiological effects can be expected.

There is compelling evidence that the oxidation reaction of free radicals contributes to many health problems. Free radicals are highly reactive and unstable atoms. Due to the uncoupled electrons on the free radicals, they tend to be highly reactive with other molecules in the biological systems. Deleterious effects of free radicals are quickly observed on protein, lipid, DNA, and RNA, owing to their molecular size. Protein side chains and double bonds on unsaturated fatty acids are easily attacked by ROS. However, phenolic compounds supplemented diet may display an in vivo antioxidant activity, which is evident by the increase in the plasma antioxidant status [71]. In agreement with Pietta [71], extensive work has shown that the establishment of number of degenerative diseases is attenuated by polyphenols, owing to their role as antioxidants and modulators of cell signaling [25, 26, 39]. An adequate antioxidant status in the body can be maintained through intracellular antioxidant enzymes and dietary intake of antioxidants [45]. Thus, in theory, a patient should experience health benefits from consuming BGN milk and probiotic beverage.

BGN has the potential of improving the quality of nutrition and consequently improving the health of many rural and low income communities. Kanatt et al. [38] concur that legumes are natural source of bioactive phenolic compounds, with a wide range of antioxidative activity and promising health benefits. Therefore, the phenolic phytochemistry of BGN and other legumes needs to be investigated in detail so as to understand the biochemical basis of the medicinal effects in ethnobotany reports.

5.5 PHENOLIC PHYTOCHEMICAL STRUCTURE

Phytochemistry is the study of secondary metabolites of plants. Phenolic compounds, recognized for their antioxidant capacity, are secondary metabolites that are widely found in the plant, fungi, and bacteria kingdoms. Phenolic compounds comprise an extensive range of molecules, with more than 8000 compounds under study. There are numerous classes of phenolic compounds that have not yet been fully explored. Figure 5.3 depicts five classes of phenolic compounds: phenolic acids, flavonoids,

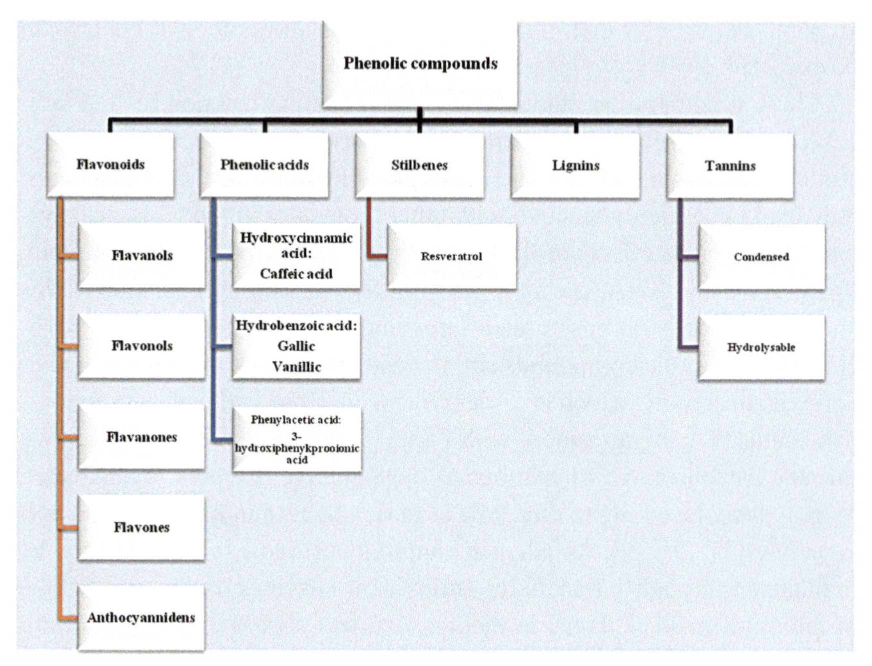

FIGURE 5.3 Classification of phenolic phytochemicals.

lignin, stilbenes, and tannins. The number of phenol rings determines the type of class. Phenolic compounds have a polyphenol structure consisting of several hydroxyl groups on aromatic rings [8, 32]. The amount of rings determines their antioxidative capacity. Phlorotannins have up to eight interconnected rings; therefore, they are more potent free radical scavengers than other types such as catechins, which have three to four rings.

Flavonoids have been reported to possess a stronger antioxidative capacity than nonflavonoids and bound forms of phenolic compounds, e.g., glycosides [2]. Flavonoids can act as reducing agents, hydrogen donors, metal chelators, and singlet oxygen quenchers due to their high redox potential. Flavonoids have been linked with reduction in the incidence of diseases such as cancer and heart diseases. Because preventive healthcare can be endorsed through diet, there is a scope for flavonoid research.

Bernal et al. [8] reported that the differences among bioactive and the other without noticeable bioactive phenolics are due to their metabolic origin. Remarkably, bioactive phenolic compounds are derived from two biosynthetic routes: shikimic acid and/or polyacetate pathways [8]. Investigation of phytochemicals produced from these pathways gave outstanding results in both in vitro and in vivo assays [40]. Thus, the occurrence of phenolic phytochemicals in plants is very variable. Consequently, highly sensitive equipment has to be used to identify the phenolic phytochemicals in BGN seeds, milk, and probiotic beverage. Bravo [12] acknowledged that the occurrence of phenolic phytochemicals as a mixture of phenolic compounds, which poses challenges in the study of their bioavailability, physiologic and nutritional effects. Furthermore, their analysis is a challenging task when compared to other pharmaceutical drugs [85]. The work revealed that phenolic acids are present in both free and bound forms, resulting in varying suitability to different extraction conditions. Hence, appropriate methods of extraction should be employed to effectively characterize phenolic phytochemicals in legumes. Furthermore, knowledge of the structure of the phenolic compounds may be crucial for understanding the mode of action in both in vitro and in vivo defense assessments.

5.6 PHENOLIC PHYTOCHEMICALS IN BGN

Antioxidants are redox active compounds that delay or prevent oxidative damage in a target molecule [28] by reacting nonenzymatically with a reactive oxidant (Figure 5.4). Lu et al. [45] acknowledges the neutralization of free radicals by accepting electrons from an antioxidant. Identifying and characterizing the antioxidants in BGN seeds, milk, and probiotic beverage will aid in the understanding and establishment of their role to attenuate inflammation and oxidative stress, and to consolidate the traditional therapeutic claims on BGN. Pioneers in BGN phenolic phytochemistry have already laid initial foundation with regard to antioxidant activity

Using a combination of column chromatography and preparative thin layer chromatography, Pale et al. [68] investigated the anthocyanin present in BGN. They detected five pigments consisting of three anthocyanins:

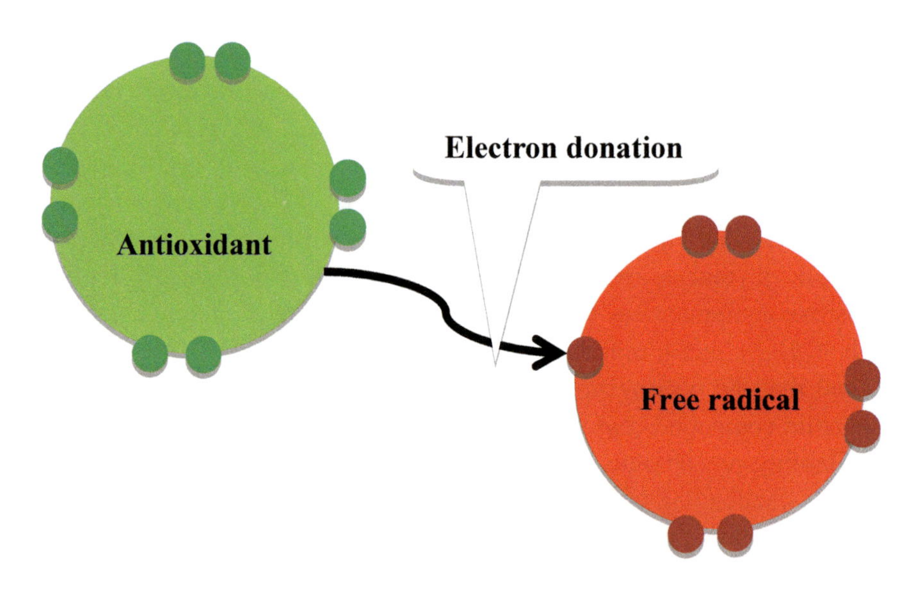

FIGURE 5.4 Reduction of a free radical by an antioxidant.

delphinidin 3-O-β glucoside; petunidin 3-O-β-glucoside; and malvidin 3-O-β-glucoside.

Anthocyanins are water-soluble phenolic compounds from the flavonoid class. These pigments are renowned for both pharmacological properties and strong biological functions such as anti-inflammatory and antioxidants activities. Their solubility in water may explain why therapeutic effects are observed when someone with hematomas drinks BGN flour dispersed in water. Nonetheless, further investigation on the presence of these pigments in BGN milk and probiotic beverage is needed to understand if they play any role in the attenuation of inflammation and oxidative stress. Reports on the anti-inflammatory activity of anthocyanin have been documented [78], further supporting the therapeutic reports on BGN and other legumes. Furthermore, the use of dark-colored seeds of BGN for producing milk and probiotic beverage can be explored.

Using a reverse-phase HPLC, Kaufman et al. [41] surveyed 80 different taxa of legumes for the anticancer metabolites genistein and daidzen and confirmed their presence by mass spectrometry. The results indicated that a number of legumes are excellent source of isoflavones, genistein, and daidzein. Many *Vigna* species from the study had these two isoflavones, genistein and daidzein. BGN seeds being *Vigna* species, its products, milk, and probiotic beverage are expected to contain genistein and daidzein that could explain the therapeutic effects according to indigenous data. The study by Kaufman et al. [41] mentioned that the isoflavones in question have an estrogenic effect, implying that they could cause some serious effects in women of childbearing age. However, this might not hold true considering that the ingested quantities are very minute. Moreover, the compounds are excreted from the body after a certain period of time. Detailed investigations are needed on the nutritional effects of main group of polyphenolic compounds, including their metabolism, effects on nutrient bioavailability, and antioxidant activities.

Onyilagha et al. [66] offers a convincing report after studying eleven species of *Vigna* for canavanine, proanthocyanidin, and flavonoids profiles, indicating that the prevalent flavonoid seems to be kaempferol among the *Vigna* species. However, kaempferol-3-O-glucoside-7-rhamnoside is restricted to BGN, thus questioning its relationship with other members of *Vigna*. As a polyphenol antioxidant, kaempferol has great health benefits and reduces the risk of cancer [19]. According to Calderón-Montaño et al. [16], kaempferol has been isolated in many plants commonly used in traditional medicine. This further confirms the therapeutic benefits associated with BGN. In addition, they reported on the positive association between the consumption of kaempferol-rich foods and the reduction of development of several disorders including cancer and CVDs. While preclinical studies have shown its anti-inflammatory capabilities, the antioxidative capacity of kaempferol has been recorded in both *in vitro* and in vivo assays.

Ademiluyi and Oboh [2] stated that fermented BGN predominantly contains carvocol, p-coumaric acid, vanillic acid, p-hydroxybenzoic acid, caffeic acid, ferulic acid, genistein, apigenin, shagaol, glycitein, kaempferol, luteolin, capsaicin, isorhamnetin, myriteina, and rosmarinic acid. However, it will be of interest to identify phenolic compounds in lactic acid bacteria (LAB) for fermented BGN milk. The metabolism of the LAB in milk is envisaged to have an effect on the predominant phenolic compounds in the BGN probiotic beverage.

Marathe et al. [47] observed variation in phenolic content in legumes that is attributed to genetic factors, degree of maturation, and environmental condition. Consequently, it is assumed that by using one variant of BGN, phenolic content variation would be minimized, thereby standardizing the modulatory efficacy of the BGN milk and probiotic beverage. Murevanhema and Jideani [58] documented similar findings to that of Marathe et al. [47] and Nyau [65]. All these studies highlighted the correlation between seed color and antioxidant activity and/or total phenolic content. The lighter seed hulls exhibited lower antioxidant activity and/or total phenolic content. Therefore, the milk and probiotic beverage produced from the red variant of BGN would be most appropriate for further studies. Seed size was reported by Marathe et al. [47] to influence the total phenolic content and the antioxidant activity.

Nyau [65] explored the phenolic profile of BGN using a high performance liquid chromatography–photo diode array–electron spray ionization–mass spectrometry (HPLC-PDA-ESI-MS). Phenolic acids and flavonoids were most prevalent in the BGN. Flavonoids have been associated with anti-inflammatory potential [4]. Due to many phenolic compounds in plants, sensitive equipment for their precise determination is needed. Mass spectrometry (MS) is sensitive and highly specific, and coupling it with chromatographic techniques is desirable [74]. MS converts analyte molecules to an ionized state and the analysis is based on the mass to charge ratio. The electron spray ionization (ESI) is another ion source capable of interfacing to liquid chromatography (LC). However, a softer ionization than MS is most suitable for polar molecules, thus making it ideal for the analysis of plant secondary metabolites.

Polyphenol content in the BGN seeds has been found to be 872.35 mg/100 g [51]. Soaking and cooking the seeds reduced the phenolic content to 647.67 mg/100 g and 413.79 mg/100 g, respectively. The authors attributed this decrease to leaching of polyphenols into the soaking water and possible interaction with proteins during cooking. Therefore, determining the phenolic content of seed, milk, and probiotic beverage and other legume products will give an insight into the effect of processing on the products. Furthermore, this can explain any discrepancies in the attenuation of inflammation and oxidative stress by raw seeds and value-added products.

Ujowundu et al. [90] reported the presence of phytochemical components (such as alkaloids, flavonoids), cyanogenic glycoside, oxalate, tannins, saponins, and phytate in BGN. These compounds have been previously shown to have health benefits. In this study, phytochemical values of BGN after gas flaring showed that oxalate was (0.38 ± 0.04%), saponin (0.24 ± 0.02%), vitamin E (3.18 ± 0.15 mg/ 100 g), vitamin C (1.17 ± 0.20 mg/100 g), vitamin A (26.05 ± 0.14mg/100 g), and niacin (2.10 ± 0.06 mg/100 g) [90]. Gas flaring significantly increased the concentrations of oxalate, saponin, alkaloids, and flavonoids; however, the vitamins were significantly reduced. Environmental factors may have an effect on nutrients and phytochemicals furnished by BGN, which may lead to impaired functions and metabolic processes. Therefore, obtaining germ plasma from a gene bank for analysis can limit the discrepancies brought about by the environmental factors; hence, one can draw conclusive results on the potential of BGN and other legumes on attenuating inflammation and oxidative stress.

It has been reported that the seed hulls usually contain higher amounts of phenolic compounds [47, 84], notably tannins and saponins that are antinutrient [43]. Thus, these impair the bioavailability of proteins. However, more recent studies contend that these compounds are invaluable exogenous antioxidants acting as reducing agents, metal chelators, and singlet oxygen quenchers. Consequently, they are capable of inhibiting oxidative stress damage. Tannins have been reported to promote wound healing and inflamed mucous membrane, conferring chemo-protective benefits. Infection and microbial inversion have been controlled by saponins, which are considered to be natural antibiotics. However, the processing of BGN milk and probiotic beverage and other legumes may have an effect on their amounts in the final product. Ujowundu et al. [90] reported on the decrease in tannins and saponins after soaking and cooking, due to leaching into the soaking and cooking water as well as possible complex formation with other components.

5.7 BGN ANTIOXIDANT DEFENSE NETWORK

Extensive work has been done on biological activities of phenolic compounds, including antiallergenic, antiviral, anti-inflammatory, and vasodilation [4, 8, 71]. Attention has been devoted to the antioxidant activity of the phenolic compounds, with regard to their ability to reduce free radical and to scavenge free radical [47], and their role in the pathway of oxidative stress mechanism of pathogenesis. However, BGN antioxidant defense network against inflammation and oxidative stress has not been documented. Lu et al. [45] attributes this ability to the free radical scavenging activity of antioxidants. The antioxidant mode of action is in four folds, namely:

- react with reactive radicals and destroy them;
- inhibit the expression of free radical-generating enzymes, e.g., nicotinamide adenine dinucleotide phosphate (NAD(P)H) oxidase and xanthine oxidase;
- enhance the activities and expression of antioxidant enzymes like superoxide dismutase (SOD); and
- catalyze (e.g., CAT) and prevent DNA damage, protein modification, and lipid oxidation.

Govindappa [26] concurs that antioxidants inhibit or setback the oxidation of other molecules by inhibiting the initiation and/or propagation of the chain reactions. Their antioxidant capacity is attributed to their high redox potential, owing to their configuration, phenolic rings, and total number of hydroxyl groups [61] that allow them to act as reducing agents, hydrogen donors, and singlet oxygen quenchers [32]. The antioxidants' multifunctional potential is attributed to the phenolic compound molecular skeleton, phenol ring, which acts while an electron is trapped to scavenge proxy, superoxide-anion, and hydroxyl radicals [95]. Therefore, it is of interest to investigate the multifunctional potential of BGN seeds, hulls, milk, and probiotic beverage and other legumes by using standard methods.

5.8 ASSESSMENT OF ANTIOXIDANT CAPACITY IN VITRO AND IN VIVO

The extraction of phenolic compounds is a very important step in their isolation and identification and their use. The extractability of the phenolic compounds is reported to be governed by the polarity of the solvent used;

extraction time and temperature; degree of polymerization of phenolics, and interaction with other food constituents. These factors should be taken into consideration when determining the total phenolic content of the BGN samples. Solvent extraction and extraction with supercritical fluid are commonly used techniques for the isolation of phenolic compounds [32]. Ethyl acetate, methanol, ethanol, chloroform, diethyl ether, hot water, sodium hydroxide, petroleum ether, acetone, and n-hexane were listed by Ignat et al. [32] as the organic solvents to extract polyphenols. Temperature, liquid-solid ratio, flow rate, and particle size influence the concentration of the desired component in the extract. Organic solvents to be used for the extraction of phenolic compounds should be chosen carefully. However, the technical advantages of using supercritical fluid extraction make it a better choice for the extraction of phenolic compounds from BGN. The use of large amounts of toxic solvents is avoided; the methods are rapid, selective, and automatable [11]. Moreover, degradation that can occur while using traditional extraction techniques is negligible due to the absence of light and air during the extraction. Supercritical carbon dioxide is mainly used because of its low toxicity, nonflammability, low cost, readily separable from solute, and compatibility with processed foods [32], making it ideal for the BGN milk and probiotic beverage and other legumes.

Determination of phenolic compounds is equally important due to the increasing demand for highly sensitive and selective analytical methods. Spectrophotometric methods have been used in the quantification of plant phenolics. Different principles are employed to determine the different structural groups. The Folin-Ciocalteau assay is mainly used for the determination of total phenolic content due to its simplicity and low cost. The spectrophotometry, however, only gives estimation and does not separate or give quantitative measurements of individual components, thereby making it slightly unfavorable [32]. Murevanhema [57] reported the total phenolic content of BGN milk and probiotic beverage using spectrophotometric assay. A preferred approach will be to use HPLC for the separation and identification of the individual phenolic compounds.

HPLC is the most preferred equipment for separation and quantification. HPLC can be coupled with different detectors such as diode array detector (DAD), photo diode array (PDA), mass spectrometry (MS), and UV depending on the phenolic compound of interest. Drawbacks in detection limit and sensitivity are the limitation of HPLC for analyzing complex matrix. However, this can be overcome by initial pre-concentration and purification of phenolic compounds from the complex matrix. Adsorption-desorption have been used in the purification of phenolic compounds, and the usage of highly efficient sorbents namely C18 and highly cross-linked styrene-divinyl-benzene is very popular.

Determination of the phenolic compound structure can be effectively achieved by using liquid chromatography-mass spectrometry (LC-MS) [14]. High speed counter current chromatography (HSCCC), supercritical fluid chromatography (SFC), paper chromatography (PC), thin liquid chromatography (TLC), gas chromatography (GC) [23], centrifugal partition chromatography (CPC), capillary electrophoresis (CE) [20], micellar-electrokinetic capillary chromatography (MECC) [25], and gel permeation chromatography (GPC) [76] are some of the identification and quantification methods that have been used for phenolic compounds. With the growing interest in phenolic compounds, these can be ideal for the identification and quantification of phenolic compounds in BGN seeds, milk, and probiotic beverage and other legumes.

There are 19 documented in vitro methods for assessing the antioxidant activities of plant samples. Figure 5.5 highlights some of the common methods. This review cannot cover all, but several illustrations will serve. Free radical scavenging activity of 2,2-diphenyl-1-picrylhydrazyl (DPPH); 2,2'-azino-bis(3-ethylbenzothiazoline-6-sulfonic acid) (ABTS), ferric ion reducing antioxidant power (FRAP), estimation of metal ion (Fe^{2+}), and chelating activity and oxygen radical absorbance capacity (ORAC) assays are some of the commonly used assays. Furthermore, they are grouped into two types: (a) Inhibition assays (e.g., ORAC and trolox equivalent antioxidant capacity (TEAC)) in which antioxidants and substrate compete for thermally generated peroxyl radicals and (2) Reduction assays (e.g., FRAP, ABTS, and DPPH) of a colored oxidant.

Most of these assays have a similar principle, which involves the ability of a biological sample to scavenge the radical or to reduce the redox-active compound that is monitored by a spectrophotometer. TEAC or vitamin C equivalent antioxidant capacity (VCEAC) is the commonly used standards to quantify antioxidant capacity. However, there are limitations ascribed to each of these methods. Therefore, multiple strategies can

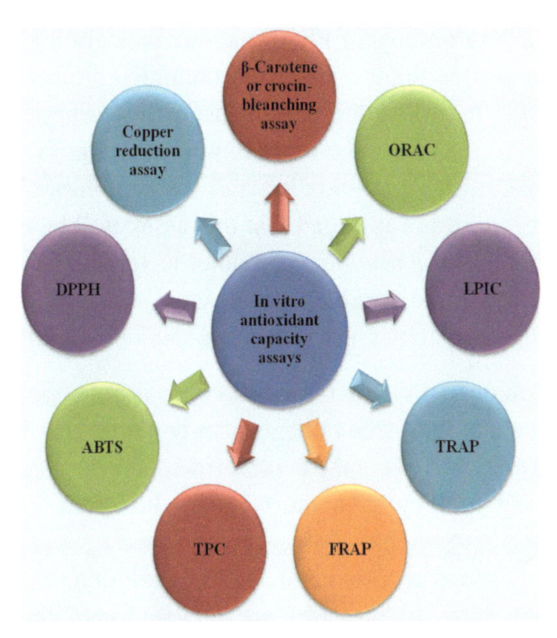

FIGURE 5.5 In vitro antioxidant capacity assays: ORAC (oxygen radical absorbance capacity); IOU (inhibited radical capacity); LPIC (lipid peroxidation inhibition capacity); TRAP (trapping antioxidant power assay); FRAP (ferric ion reducing antioxidant power assay); TPC (total phenolic content); ABTS (2,2'-azino-bis(3-ethylbenzothiazoline-6-sulfonic acid); and DPPH (2,2-diphenyl-1-picrylhydrazyl).

be adapted to confer antioxidant potential. Niki [62] summarized the factors that influence the actual free radical scavenging, such as: (1) Chemical reactivity for free radicals and stoichiometric number; (2) Fate of antioxidant-derived radical; (3) Concentration and mobility at the environment; and (4) Absorption, distribution, retention, and metabolism. Noteworthy is the interaction between antioxidants that may result in synergistic antioxidants, not additive.

In contrast, FRAP assays do not involve free radicals, but the reduction of ferric iron to ferrous iron is monitored. Metal chelating is an important mechanism of antioxidant activity. Ferrous ions have been noted to be most potent pro-oxidants among various kinds of transition metals. Fe^{2+} chelation is an essential antioxidative mechanism, which retards metal-catalyzed oxidation and thus protects against oxidative damage. In the system, Fe^{2+} ions are expected to participate in HO^- generating Fenton type reactions [47]. Shortcomings of FRAP include its inability to detect other small molecular weight thiols and sulfur-containing molecules like gallic.

Wang et al. [95] indicated ORAC to measure the peroxyl radical absorption capacity. It is regarded to be more biologically significant in extracts with multiple constituents co-existing and where complex reactions are involved. Therefore, it is envisaged that this assay will be most appropriate for the BGN probiotic beverage and other legumes. Pinchuk et al. [72] put forward three common problems associated with in vitro assays used to rank antioxidant capacity and to draw conclusion on humans:

- attention on quenching on radicals in vitro is not conclusive,because in vivo quenching of free radicals is only a fraction of the body network and enzymes plays the central role;
- capacity and potency of an antioxidant exhibited in vitro by various methods do not correlate with each other; and
- because in vitro assays are conducted in a solution or cell-free conditions, the findings are therefore not relevant to predict what happens in vivo, where a complex heterogeneous system exists.

These three observations are true. However, the second observation can be explained by varying mode of actions and antioxidant activity specific, retention, and biotransformation through enzymatic conjugation [62]. Therefore, a preferred practice would be to confirm potency using cell culture, animal models, and clinical trials on the attenuation of inflammation and oxidative stress using BGN seeds, milk, and probiotic beverage and other legumes.

Antioxidant capacity can be easily measured using commercial kits. There is also an added advantage of reproducible quantification despite frozen state of samples. However, there are shortcomings wherein the antioxidant activity in serum may not reflect that of cellular microdomains that are important to the pathogenesis of CVD. The antioxidant's mode of action in vivo can be through reaction with reactive radicals thus destroying them; or by inhibiting the expression of free radical-generating enzymes, e.g., NAD(P)H oxidase and xanthine oxidase; or by enhancing the activities and expression of antioxidant enzymes like SOD and CAT and prevent DNA damage, protein modification, and lipid oxidation. Niki [62] acknowledges that there are different types of antioxidants, which exhibit different functions in their role of defending the network in vivo. Similar to in vitro assessments, different in vivo assays are to be employed so as to draw irrefutable conclusions on the anti-inflammatory and antioxidative stress potentials of BGN seeds, milk, and probiotic beverage and other legumes.

Bioavailability of antioxidants is determined by their bioaccessibility, which is the quantity or fraction that is released from the food matrix in the gastrointestinal (GI) tract and becomes available for absorption. Therefore, it should be taken into consideration when assessing antioxidant capacity in vivo. For that reason, it is important to have an antioxidant profile and chemical structure of the BGN seeds, milk, and probiotic beverage. This will aid in understanding their biological and physiological properties. The antioxidants have to be absorbed, transported, distributed, and retained properly in the biological fluids, cells, and tissues for conclusive capacity assessments and healthcare claims to be confirmed for BGN seeds, milk, and probiotic beverage and other legumes. Simulated GI tract model assessments can also be employed to study bioavailability and bioaccessibility of phenolic compounds of BGN and other legumes.

Plasma, erythrocytes [6], urine [74], and cerebrospinal fluid [55] from both animal and humans have been assessed for the effect of antioxidant compounds. Studies have revealed some biomarkers that can be used reliably for inflammation and oxidative stress assays. Antioxidants have been used as oxidative biomarkers as depicted in Figure 5.6. Positive effects of antioxidants and antioxidant-rich diets to reduce inflammation and oxidative stress have been recorded [27, 48, 73]. However, the beneficial effects

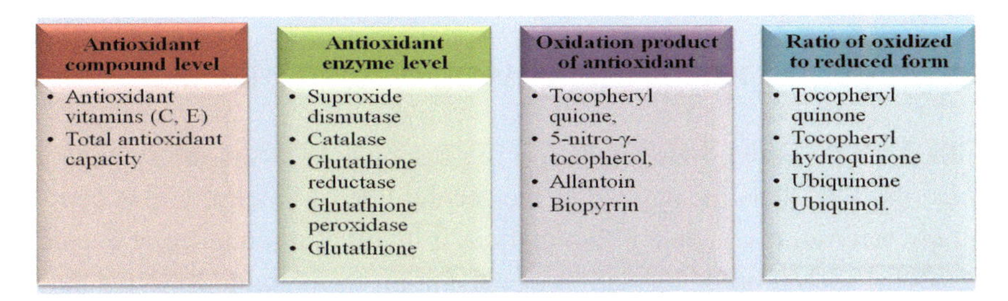

FIGURE 5.6 Antioxidants as biomarkers of oxidative stress.

are difficult to observe in normal health subjects with sufficient amounts of antioxidants.

FRAP and ORAC assays are commonly used to assess the total antioxidant capacity of biological samples. Free radical damage to a fluorescent probe, e.g., fluorescein, is monitored in the ORAC assay; the damage is caused by an oxidizing reagent resulting in a loss of fluorescent intensity over time. Antioxidative capacity of a compound is then determined by its correlation to the inhibition of oxidative damage [6]. The ability of a sample to reduce Fe^{3+} to Fe^{2+} is measured in FRAP assays, while ascorbic acid, uric acid, and α-tocopherols are monitored in the plasma. It is noteworthy that there is no correlation between antioxidative capacity and its reducing ability, owing to the varying antioxidant mode of action. Conversely, Kasote et al. [40] postulated on the relevance of plant antioxidants to therapeutic applications. Their argument is that antioxidant potential studies in vitro and in vivo are not always the same.

Physio-pharmacological process such as absorption, distribution, metabolism storage, and excretion are accountable for the relatively less therapeutic usefulness of antioxidants under in *vivo* conditions. Lu et al. [45] stated that too many factors in cell systems can affect the results compared to those obtained using cell-free systems. They further proposed that while antioxidants may not work cellularly and physiologically, they can work chemically.

As a result, direct extrapolation of in vitro findings to therapeutic usefulness can be viewed as a malpractice. Therefore, further investigation of antioxidant potentials of BGN and other legumes and their food products should be done in vitro and in vivo before conclusions can be drawn on their anti-inflammatory activities.

5.9 AN ETIOLOGY OF INFLAMMATORY ACTIVITIES

Inflammation is the body's non-specific response to tissue injury and/or infection [48]. It is recognized as the second line of defense and is commonly referred to as acute inflammation. It involves a sequence of events involving alteration of exudation of plasma proteins and leukocytes into the injured tissue or surrounding area, causing swelling, redness, pain, heat, and perhaps loss of function [48, 61, 78]. All these actions are necessary to deter the effects of injury or a dangerous agent in the body. Thus, it is a protective mechanism and an important basic concept in pathophysiology. Upon tissue injury, leukocytes accumulate resulting in an intensification of the response. In order to resolve the inflammation response, termination of pro-inflammatory signaling pathways and clearance of inflammatory cells need to be done, thus restoring of normal tissue function [78]. Su et al. [86] described cytokines as soluble factors generated by immune cells that are renowned for playing roles in differentiate maturation and activation of various cells. Their local microenvironment determines their proinflammation or anti-inflammation effects.

Upon occurrence of inflammation, different types of cells are activated, including monocytes, which locally differentiate into macrophages. Consequently, there is an activation of pro and anti-inflammatory mediators including cytokines such as tumor necrosis factor (TNF-α), interleukins (IL-1β, IL-6, IL-10), and chemokines (IL-8, monocyte chemoattractant protein-1, macrophage).

5.10 PROGRESSION TO CHRONIC INFLAMMATION

It may lead to detrimental health effect [39, 78] when the immune system is out of control for reasons such as autoimmune disorders, chronic irritation such as smoking, ROS, certain bacterial infections, or long-term abnormal immune response [91, 93] and/or when the cause of inflammation is not totally removed from the system. Consequently, the self-limiting process is unregulated resulting in excessive damage to host. Infiltration of mononuclear immune cells, tissue destruction, and attempts of healing will be observed in chronic inflammation; thus, the stimulant will persist, resulting in continuous need of monocytes [56, 92] and pathogenesis. ROS cause cell injury because they are not readily deactivated by antioxidants pathways, thus implying that they accumulate. ROS have been attributed to promote the production of proinflammatory cytokines [30].

Nuclear factor-kB (NF-kB) has been associated with the production of IL-6, TNF-α, and CRP cytokines; hence, it plays a role in amplifying inflammatory responses [30]. IL-6 has been shown to be an important stimulator from acute inflammation to chronic inflammation. Consequently, it is a possible marker for indicating the severity of degenerative diseases associated with inflammation. IL-6 functionality differs with its concentrations. It performs a protective role during the manifestation of certain inflammation responses. When the control limit is exceeded, the IL-6 level is relatively high, as a result of the activation of T cell differentiation and the induction of acute phase proteins. Thus, it counteracts the manifestation of certain anti-inflammatory response.

Pro-inflammatory cytokines such as IL-1β, IL-6, and TNF-α, have been recognized to be pivotal mediators in the inflammation process [44, 60]. Inducible enzymes (such as cyclooxygenase (COX)-2), include: vasoactive amines (histamine); adhesion molecules; and selections. Lipid-derived eicosanoids are activated during inflammation,and they play a critical role in controlling the inflammation process [61, 78]. Some of these inflammatory mediators have been reported to have a short half-life, which swiftly curb the inflammation response as soon as the stimulus is removed.

Metukuri et al. [54] studied the testicular expression of apoptotic mediators during acute inflammation by using a rat model. They reported that the induction of acute stress in the testis leads to mitochondrial dysfunction and activation of apoptotic pathways, while affecting sperm count, quality, and functionality.

5.11 BIOMARKERS OR INDICATORS OF INFLAMMATION

Biomarker is a characteristic that is objectively measured and evaluated as an indicator of normal biological process, pathological process, or pharmacological responses to a therapeutic intervention [55, 62, 87]. They are used for health examinations, diagnosis of pathogenic process at an early stage, prognosis, and individualization of therapy [87]. Moreover, biomarkers can be used for precise quantification of the magnitude of inflammation in in vivo models. Examples of inflammatory triggers, pathways, and conditions modulated by dietary anti-inflammatory agents are shown in Figure 5.7.

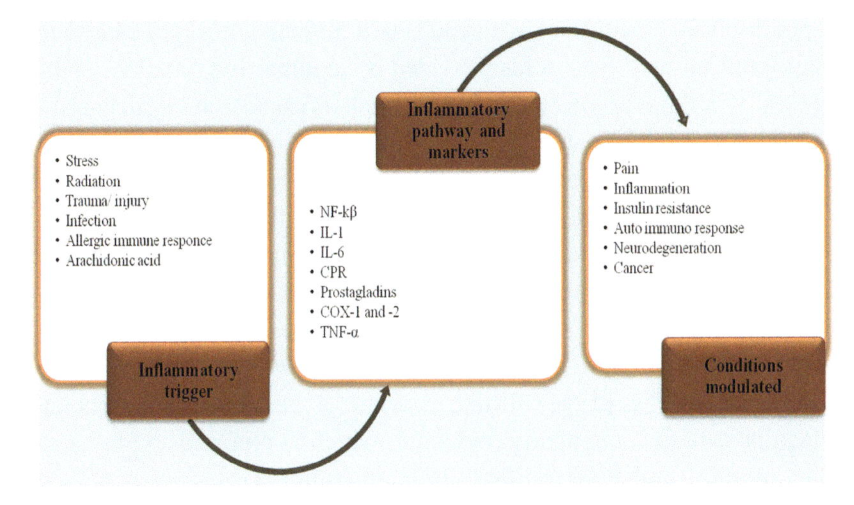

FIGURE 5.7 Examples of inflammatory triggers, pathways, and conditions modulated by dietary anti-inflammatory agents.

Cell adhesion molecules, pro-atherogenic molecules, and C-reactive protein (CRP) are important biomarkers [10]. These cytokines can be produced by virtually every nucleated cell; hence, their function as regulatory peptides or glycoproteins. There are other 40 possible biomarkers for inflammation; among them, IL-6, TNF-α, and CRP have been recognized as the most significant with many activities. IL-6 stimulates the final stages of B-cell maturation, causing B cells to differentiate into mature immunoglobulin-secreting plasma cells, thus regulating the immune and nervous systems. IL-6 is also involved in the metabolic control of the body and liver generation. CRP is produced by the liver in response to IL-6.

Recent studies have also shown CRP's involvement in the atherosclerotic process via leukocyte activation. Elevated levels of CRP in plasma is a cause for concern; hence, it is used as a biomarker. The lower level is considered to be <1.0 mg/ L, the mean value is 1.0 to 3.0 mg/L, and the highest level is ≥3.0 mg/ L.

The clinical study by Hermsdorff et al. [29] demonstrated the possibility of significant reduction in CRP due to consumption of legumes. Similar activity is expected for BGN supplemented diets. However, total dietary fiber (soluble and insoluble), resistant starch, high magnesium, L-arginine, and polyunsaturated fatty acids provided by legumes has an effect on biomarkers of inflammation. Determining their presence and concentration will be a better indication of the inhibition or anti-inflammatory potential of the BGN and other legumes.

TNF-α production may be influenced by either the characteristic of local and systemic or acute and chronic inflammation. Himmelfarb [30] hypothesized that TNF-α may contribute to increased oxidative stress. Compelling laboratory evidence has shown that the acceleration of atherosclerosis and vulnerable plaques can be clinically linked to the overexpression of TNF-α and MCP-1 [53]. Toll-like receptor 4 is a pathogen-associated molecular-pattern recognizing molecule, which is reported to play a role in the initiation and acceleration of inflammation [53].

According to the meta-analysis by Salehi-Abargouei et al. [81], detailed extensive studies will be required to determine the effect of dose-response of non-soy legume consumption on inflammation. The same recommendation can be drawn for BGN seeds, milk, and probiotic beverage.

5.12 MITIGATION OF PRO-INFLAMMATORY BIOMARKERS

Anti-inflammatory agents are substances that can treat and reduce inflammation or swelling. There are two types of anti-inflammatory agents: non-steroidal anti-inflammatory drugs (NSAIDs) and natural anti-inflammatory agents from medicinal plants. NSAIDs' mechanism of alleviating pain is by neutralizing the cyclooxygenase (COX) isoenzyme. Lu et al. [45] highlighted a similar mechanism of antioxidants, whereby these inhibit the free radical-generating enzymes. Zia-Ul-Haq et al. [97] were the first to report on the inhibition effect of a legume on COX-2. They suggested that methanolic extracts of black gram (*Vigna mungo* L.), green gram (*Vigna radiate* (L.) R. Witczek), soy bean (*Glycine max* (L.) Merr), and lentil (*Lens culinaris* medic.) had an effect on COX-2, which is responsible for PGE_2. Being a *Vigna* species, BGN is expected to deter the expression of COX-2. Furthermore, COX is accountable for inflammation because it synthesizes prostaglandins. Prostaglandins are lipid autacoids derived from arachidonic acid, and they may function in both promotion and resolution of inflammation [79]. Commonly used NSAIDs are aspirin; ibuprofen; diclofenac; naproxen; celecoxib; mefenamic acid; etoriooxub; and indomethacin. Most of these NSAIDs are generic medicines; thus, they are manufactured and distributed by number of companies under different trade names. However, some of these NSAIDs are limited by their mechanism of action, which address the symptoms of inflammation other than its root cause by free radicals [45]. Another drawback of NSAIDs is ascribed to health professional, who delays the commencement and intensification of treatment of disease [70].

Recently, there is a growing interest in the identification of anti-inflammatory agents in plants, thus encouraging the development of foods with therapeutic and medicinal benefits. Natural anti-inflammatory agents are commonly referred to as dietary supplements and herbal remedies [48]. Flavonoids, terpenes, quinones, catechins, and alkaloids are plant derivatives that are known to modulate the activation of pro-inflammatory signals [5].

The anti-inflammatory compounds impede the inflammatory pathway by counteracting the COX isoenzyme and/or by inhibiting nuclear factor-kB (NF-kB) inflammatory pathway. Therefore, understanding the mechanism of the phenolic phytochemical of BGN and other legumes in the attenuation of inflammation pathway is needed.

Hermsdorff et al. [29] reported on the reduction of TNF-α in plasma following the consumption of a legume-based hypocaloric diet. A legume-rich diet of 250 g/day has shown to reduce TNF-α.

Indigenous knowledge has pointed to some plants that have anti-inflammatory properties. However, this is yet to be proven conclusively for most of these plants. Notably successful anti-inflammatory drug from invested plants is still lacking, regardless of the numerous dietary supplements and herbal remedies developed. Ayepola et al. [6] using a malondialdehyde (MDA) biomarker demonstrated the resistance to oxidative stress in plasma of a diabetic rat model through dietary supplementation.

A number of studies have shown that the expression of pro-inflammatory cytokines can be attenuated or prevented via dietary antioxidant supplements. Using an animal model, Ayepola et al. [6] investigated the inflammatory markers including MCP-1, vascular endothelial growth factors (VEGF), and IL-1 in the serum of streptozotocin-induced diabetic rats using Bio-Plex Pro-magnetic Bead-based assay. They revealed that proinflammatory cytokines were significantly reduced through dietary supplementation. Attenuation of inflammation by BGN and other legumes is a possibility; thus, further investigation on the anti-inflammatory potentials of legumes is required.

White willow bark is regarded as one of the oldest remedies for pain and inflammation, and it is an inhibitor of COX-1 and COX-2; thus, it impedes the formation of inflammatory prostaglandins [48]. Pycnogenol from maritime pine bark is similar to the willow bark. It is considered to be one of the most potent antioxidants containing a wide range of active polyphenols, including catechin, taxifolin, procyanidins, and phenolic acids. Its mode of action is achieved by inhibiting TNF-α-induced NF-kB activation and adhesion molecule expression in the endothelium.

Curcumin from turmeric, conventionally used for coloring, has also been used as an anti-inflammatory and anticarcinogenic agent with antioxidative effects. Curcumin owes its pharmacological benefits to its regulation of certain molecular targets including proinflammatory cytokines, growth factors, adhesion molecules, and transcription factors [53]. Its anti-inflammatory role is achieved by suppressing NF-kB activator pathway and stemming its expression. Regulation of several enzymes and cytokines by inhibiting COX-1 and COX-2 has been observed in curcumin studies. Its demonstrated activity in the NF-kB, COX-1, and COX-2 pathway suggest that it can be used as an alternative to NSAIDs for the treatment of inflammation. In addition to the COX enzyme, NADH oxidase and monooxygenases' activation and expression needs to be modulated. Zia-Ul-Haq et al. [97] investigated the COX inhibition activities of some legumes. They reported the inhibition of COX by a legume. Based on this report, BGN seeds, milk, and probiotic beverage and other legumes are also expected to inhibit COX.

Lee et al. [44] demonstrated the anti-inflammatory effects of cowpea extracts and its bioactive compounds using lipopolysaccharide (LPS)-stimulated RAW 264.7 macrophage cells. Nitric oxide production was strongly inhibited by ethyl acetate and *n*-butanol fractions of cowpea. In conclusion, they attributed the anti-inflammatory activity of cowpea partially to the polyunsaturated fatty acids such as linolenic and linoleic acids. As BGN is a legume like cowpea with relatively similar phytochemical composition, these findings suggest the possibility of anti-inflammatory activities of BGN. Moreover, BGN seeds have been reported to have 4.5–7.4% of fat and the probiotic beverage had 8–9% of fat [57]; thus, it would be vital to investigate the fatty acid profile of BGN and other legumes.

Based on the antioxidant activities and phenolic phytochemistry profiling reported in the previous studies, BGN is a promising seed with a significant potential that requires further investigation to confirm most of the ethnobotany knowledge on anti-inflammatory and anti-oxidative imputed as a result of antioxidant capacity. Correlation between antioxidants capacity of legumes and successful pharmacological properties against inflammation and oxidative stress is needed.

5.13 OXIDATIVE STRESS: REGULATED AND UNREGULATED SYSTEMS

Oxidative stress is highly recognized for its involvement in pathogenesis of a number of degenerative diseases and other less well-defined variables that contribute to residual factors which in turn favor pathogenesis. Oxidative stress is described as the imbalance between the systemic manifestation of ROS and the biological system's ability to readily detoxify the reactive intermediates or to repair the resulting damage. According to Lu et al. [45], ROS attack macromolecules in a biological system. Either way, deleterious results are to be expected when the body system is somewhat exposed to any of the above-mentioned conditions.

When electrons are transferred from one atom to another during oxidation, it represents an essential part of aerobic life and metabolism [71]. Free radicals (superoxide (), peroxyl (ROO·) alkaoxyl (RO); hydroxyl (HO) and nitric oxide (NO)) are produced duringoxidation [45, 71]. Cell signaling, apoptosis, gene expression, and ion transportation are some of the important roles for which free radicals are recognized. Mitochondrial respiration chain, an uncontrolled arachidonic acid (AA) cascade, and NADPH oxidase are main sources of ROS and RNS [10]. Under normal circumstances, in the occurrence of excess free radicals in the system,they will be attacked by endogenous and exogenous antioxidants [40].

Similar to the case of plants, the human body has an antioxidant defense system to protect against free radicals. The defense system comprised endogenous and exogenous antioxidants. The endogenous system includes:

- enzymatic defense such as *Se*-glutathione; peroxidase; catalase; and SOD. Endogenous antioxidant or oxidative stress biomarkers are generally assessed invivo before and after oxidative stress. SOD, catalase, glutathione reductase and glutathione peroxidase are evaluated to determine enzymatic activity of endogenous antioxidants. Other researchers have also measured specific end products resulting from attacks on macromolecules by ROS. Furthermore, these enzymes metabolize superoxide, hydrogen peroxide, and lipid peroxides, thus arresting the generation of the toxic hydroxyl; and
- non-enzymatic defense such as: glutathione, histidine-peptides, the iron-binding proteins transferrin and ferritin, dihydrolipoic acid, reduced $C_oQ_{10;}$ melatonin, urate, and plasma protein thiols[83]. Radical-trapping capacity is accounted for by urate and plasma protein thiols[71].

Glutathione peroxidase plays an invaluable role of protecting hemoglobin, red cell enzymes and biological cell membranes against oxidative damage by increasing the level of reduced glutathione in the process of aerobic glycolysis [94]. Alteration of antioxidant defenses signifies oxidative stress. Ademiluyi et al. [3] reported reduction in glutathione and glutathione reductase activity in diabetic rats compared to that in the control and dietary supplemented group with fermented legumes.

Under pathophysiological conditions (such as inflammation, metabolism of foreign compounds), radiation leads to a load of free radicals. Free radicals are reactive and can attack molecules of nearby cells and cause damage. Free radicals initiate lipid peroxidation, leading to loss of membrane fluidity [40, 45]; stimulation of vascular smooth muscle cell proliferation and migration [30]; oxidation of nucleic acids, proteins including enzymes, thereby causing DNA damage [26]; and induction of cell dysfunction, necrosis, and apoptosis, thus inducing specific post-transitional modifications that alter the function of cellular proteins and signaling pathway. The resultant functional modification of proteins, either reversible or irreversible, ultimately leads to cell dysfunction [31]. This oxidative damage is implicated in aging and several degenerative diseases including cancer and heart diseases. Figure 5.8 shows the progression to pathogenesis brought about by oxidative stress.

Free radicals are of different biological, chemical, and physical properties, and are therefore involved in oxidative stress in varying capacities [62]. As a result, the byproducts are also different. Owing to the physiological, chemical, and physical properties of the byproducts, they can be used as biomarkers for the onset and progression of oxidative stress. Investigating the attenuation of oxidative stress biomarkers by fermented legumes has been done in the past. A similar study on BGN seeds, milk and probiotic is needed with a view of authenticating the medicinal claims of these byproducts. Furthermore, the output from the work will aid in explaining the involvement of specific biomarkers, which BGN can attenuate. The

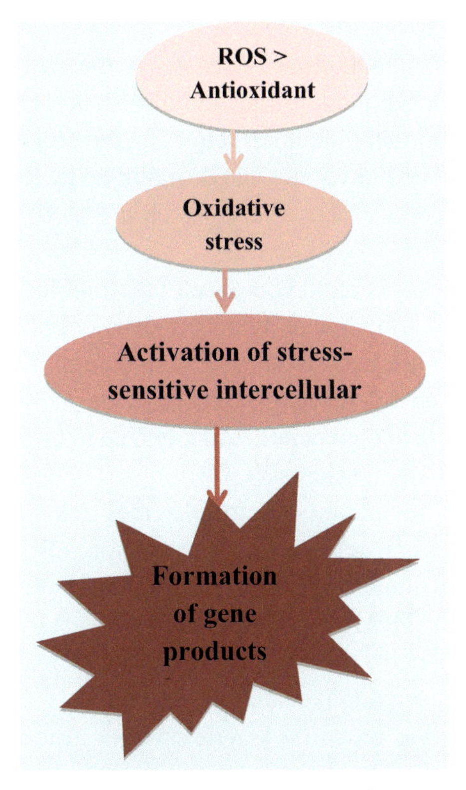

FIGURE 5.8 Pathogenicity of oxidative stress.

commonly studied biomarkers are highlighted and explained in the following sections.

5.14 OXIDATIVE STRESS: BIOMARKERS

Evidence of oxidative stress is verified by biomarkers [2, 6]. Byproducts of oxidative stress have certain characteristics so that these can be considered as biomarkers. Correlation of the biomarkers with the pathophysiological process of a disease and its utility as influenced by the ease and cost of measurement make it potential candidates as biomarkers. Moreover, the set criterion is based on; its sensitivity and specificity; and evidence for guiding management and improving patient outcome; chemical stability; and accurately quantifies reflection of specific oxidation pathway and concentration in biological samples correlated with the severity of diseases [31, 87].

Oxidative stress biomarkers are disease specific, i.e., those of CVD differ from those of cancer and reproduction disorders [1, 89]. ROS has been identified as an important biomarker reflecting the process of CVD. DNA, lipids, proteins, and carbohydrates can be modified by ROS in vivo; hence, these can be assessed as biomarkers, which are modified by interacting with ROS [31].

Lipids are prone to oxidation because of their molecular structure with many double bonds. Atherogenicity of low-density lipoprotein (LDL) has been shown to increase upon oxidative modification. Upon modification, the LDL is taken up by scavenger receptors on monocytes, which are turned into foam cells. Foam cells are recognized to be the earliest step for the development of atherosclerosis [30].

Impairment of membrane-bound enzymes and changes in fluidity and permeability of cell membranes are normally observed. Unstable substances that disintegrate into various aldehydes such as MDA, 4-hydrooxynonenal, TBARS, and conjugated dienes-hydroperoxides, are by-products of lipid peroxidation. Hence, they have been used extensively for the assessment of oxidative stress in *vivo* models. Lipid peroxidation has been characterized based on kinetic profiles. Antioxidants affect the kinetics of peroxidation by deterring the initial rate of lipid oxidation and the lag preceding propagation is longer. The antioxidants present in BGN and other legumes are expected to inhibit lipid peroxidation.

Isoprostanes (IsoPs) are unique byproducts of a non-enzymatic peroxidation of polyunsaturated fatty acids, for example arachidonic acid [10]. Their elevation is associated with renal dysfunction at varying stages. The IsoPs can be detected in serum and urine samples, and have been shown to be elevated in the presence of a range of CVD risk factors. However, the current method of quantification is impractical for large-scale screening (GC/MS); furthermore, it requires further validation using immunoassay kits.

MDA, 4-hydroxynonenal (HNE) isoprostanes, and acrolein are relatively stable end products of lipid peroxidation, making them ideal biomarkers for oxidative stress. However, lipid hydroperoxides and aldehydes can also be adsorbed from the diet and then excreted in urine.

Therefore, the measurements of urinary MDA and HNE will give ambiguous results; thus, they should not be used as an indicator of whole body lipid peroxidation unless diet is controlled [44]. In the pathogenesis of Alzheimer's disease (AD), lipid peroxidation has been implicated at great lengths. Under normal conditions, its end products, HNE-GSH adducts, are eliminated by the system's glutathione transferase (GST) and MRP-1. Thus, in subjects with oxidative stress, the elimination of HNE decreases, leading to the accumulation of HNE protein adducts in neuronal cells. Urinary isoprostane concentration, plasma lipid peroxide levels are some of the important biomarkers of oxidative stress.

MDA is one of the most known secondary products of lipid oxidation, and is frequently used as a biomarker for oxidative stress in *vivo*. MDA is easy to quantify by a spectrophotometer using TBARS assay. ELISA kits to detect MDA also have good performance. The progression of CVD and carotid atherosclerosis at 3 years can be predicted using MDA. However, TBARS assays are non-specific; thus, it can detect aldehydes other than MDA; moreover, sample preparation can influence the results [31].

Figure 5.9 lists the biomarkers associated with lipids, proteins, carbohydrates, and genetic matter advanced glycation end-products (AGEs) are formed during the oxidation of carbohydrates, whereas proteins are modified by reducing sugars in a non-enzymatic reaction called glycation

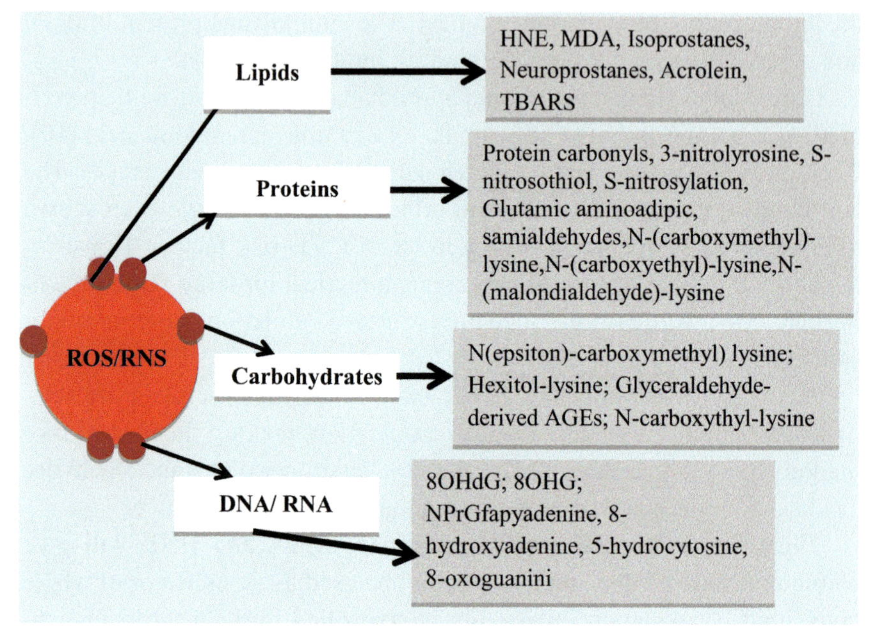

FIGURE 5.9 Oxidative stress biomarkers.

Protein glycation is associated with diseases, including diabetes mellitus, cardiac dysfunction, and neurodegenerative diseases. Hence, glycation is a potential useful biomarker for monitoring several diseases. Biomarkers of protein oxidation or nitration under a neurodegenerative disease can involve protein carbonyls and nitration, respectively [87]. Oxidation of protein will in most instances affect their functionality. Direct oxidation of certain amino acids side chains with products of lipid peroxidation has been implicated in the observed loss of functionality of protein [44].

Hydroxyl radicals produced in the vicinity of the nucleic acids DNA and RNA will modify the nucleic acid, because they are highly reactive and cannot diffuse from their site of formation [87]. Extensive studies have shown that guanine is the most reactive of the nucleic bases. Therefore, the oxidized bases 8-hydroxyguanosine (8-OHG) and 8-hydroxy-2'-deoxyguanosine (8-OHdG) are most available oxidized bases and are used as RNA and DNA oxidation biomarkers, respectively. Oxidative modification of mRNA causes reduction of protein levels and also induces translation errors in vivo with alteration of protein structure and functionality.

Oxidative stress can affect the kidney function [15, 30, 37]. Kidneys are involved in the vital roles in maintaining health by filtering waste materials from the blood and expel them from the body as urine. There are number of function tests that can be done to assess the kidney injury. Serum creatinine test examines the accumulation of creatinine in blood; under normal conditions, the kidneys would completely filter creatinine from the blood. Creatinine is a chemical waste product of muscle contraction. It is composed of creatine, a supplier of energy to muscles. Recommended creatinine levels in adults is 10.0 mg/dL. Thus, a high level of creatinine in the blood will suggest kidney dysfunction.

Similarly, uric acid is an indicator of kidney injury. Serum uric acid has been observed to be elevated in subjects with chronic kidney diseases [37]. Chronic kidney disease (CKD) has been extensively associated with increased oxidative stress, lipid peroxidation, advanced oxidation of protein products, and changes in glutathione content and increased acute phase inflammation. Studies have shown that patients with renal insufficiency had elevated levels of CRP and IL-6 compared to patients with normal kidney function. There is evidence to suggest that inflammation and oxidative stress are involved in the development of renal insufficiency. Urate is a salt derived from uric acid, and the plasma protein thiols accounts for the major damage contribution to the radical-trapping capacity of plasma.

While monitoring hepatic injury, function markers including aminotransferases and lactate hydrogenase are assessed. Lipid peroxidation markers, antioxidant enzymes, glutathione redox status, and cytokines levels are also used to assess the extent of liver injury. Hepatocytes make up a greater percentage of the liver; consequently, the liver actively participates in the metabolism of xenobiotics, thereby making it prone to toxic substances. ROS are generated in the detoxification of xenobiotics [4]. Notably, these RONS can either be deleterious or beneficial depending on their concentration. Their positive roles include cell signaling, defense against infectious agents, and induction of mitogenic responses, of course in low concentrations. However, at higher concentration, deleterious effects such as oxidative and nitrosative stress have been reported [4].

5.15 OXIDATIVE STRESS: MANAGEMENT

Glutathione (GSH) is a naturally occurring tripeptide derived from glutamic acid, cysteine, and glycine. GSH has both nucleophilic and reducing properties; thus, it is involved in protection against ROS, detoxification of xenobiotics, thiol-disulfide exchange, removal of hydroperoxides, amino acid transport across membranes, and maintenance of protein thiol (SH) groups. In metabolism involving glutathione peroxidase, glutathione S-transferase, and thioltransferase enzymes, GSH is required as a coenzyme. Owing to its strategic position between oxidants such as ROS and cellular reductants, glutathione is perfectly configured for signaling functions. Selenium-dependent glutathione peroxidase (GPx) is a central pillar of animal antioxidation metabolism, hence glutathione's has a role in animal protective mechanism [63]. Depletion of GSH may result in weakening the defense against oxidative stress, leading to damage to protein, DNA, RNA, and membrane lipids. Insufficient efficiency of the endogenous free-radical defense system and the existence of some physiopathological drivers will result in antioxidants derived from the diet being called upon to diminish the cumulative effects of ROS.

The health benefits have been attributed to the phenolic compounds' ability to counteract the oxidative stress status [61]. Consumption of legumes has demonstrated the capability of controlling the disturbances on the normal redox state within the human body. Oxidative stress affects the antioxidative enzymes, namely glutathione peroxidase, glutathione reductase, SOD; and other antioxidants such as glutathione [4]. Glutathione peroxidase enzyme is acknowledged to be responsible for the protection of erythrocyte against peroxides that are generated intracellularly or exogenously. Glutathione peroxidase plays an invaluable role of protecting hemoglobin, red cell enzymes, and biological cell membranes against oxidative damage by increasing the level of reduced glutathione in the process of aerobic glycolysis [94].

Aminotransferases are markers of clinical diagnosis of hepatic injury. However, dietary supplementation with antioxidant-rich legumes can reverse the injury. In the occurrence of hepatic injury, leakage of enzymes from the liver into the bloodstream has been observed. Ademiluyi and Oboh [2] demonstrated the attenuation of oxidative stress and protection of hepatic tissue damage by fermented BGN, soy, and locust bean condiments. In the study, the 14 days treatment of diabetic rats with fermented legumes reduced the elevated levels of the hepatic damage marker enzymes and MDA. These findings imply the possibility of managing oxidative stress through a BGN-rich diet. Therefore, consumption of tropical legumes could be a practical dietary approach of managing oxidative stress.

5.16 CONCLUSIONS

Inflammation and oxidative stress are unifying features in a number of degenerative diseases. Credible research has proved that nutrition is a possible avenue for addressing the pathogenesis and progression of the degenerative diseases. However, for this to occur, stakeholders need to team up in providing sustainable nutrition security. Nutraceutical, therapeutic, medicinal, and pharmacological efficacy of BGN and other legumes against inflammation and oxidative stress needs to be scientifically established through animal and human studies. Because BGN and other legumes are rich in phytochemicals, there might be countless possibilities of application in the medical arena.

5.17 SUMMARY

Lifestyle-related diseases associated with inflammation and oxidative stress claim 16 million lives every year. Synthetic drugs are the mainstay of the treatment and management of these diseases regardless of their known side effects. There is a quest of finding alternative medication that has led to extensive research on medicinal plants with anti-inflammatory and antioxidative stress activities. Bambara groundnut (*Vigna subterranea* (L.) Verdc) (BGN) is such a plant with anti-inflammatory and antioxidative potential, which has not been tapped scientifically. Phytochemistry studies on BGN have shown antioxidative activities; therefore, it is of interest to assess the potentials of BQN against inflammation and oxidative stress. Moreover, owing to BGN's versatility, value-added products can increase its consumption, thus improving its commercial status and leading to its wide production and improvement of its economic status.

Furthermore, value-added products such as milk and probiotic beverage have the potential to address Africa's health problems such as hunger, malnutrition, degenerative diseases, and gender inequality, especially in the rural communities. The review in this chapter aims to highlight the antioxidant, anti-inflammatory, and anti-oxidative stress potentials of BGN seeds, milk, and probiotic beverage with a view of making significant impacts on education, health, and economic costs through various stakeholders associated with BGN research, production, and processing.

KEYWORDS

- **alanine amino transferase**
- **aspartate amino transferase**
- **bambaragroundnut**
- **catalase**
- **conjugated diene**
- **ferric reducing antioxidant power**
- **gamma-glutamyltransferase**
- **glutathioneperoxidase**
- **glutathione-s-transferase**
- **inflammation**
- **interleukin-1 beta**
- **lactate dehydrogenase**
- **lipids peroxidation**
- **malondialdehyde**
- **monocyte chemotactic protein-1**
- **oxidative stress;**
- **oxidised glutathione**
- **probiotic beverage**
- **reactive nitrogen species**

- **reactive oxygen species**
- **reduced glutathione**
- **superoxide dismutase**
- **tumor necrosis factor-alpha**
- **value-addition**

REFERENCES

1. Aboua, G., Manirafasha, C., Mosito, R. B., Van der Linde, M., & Du Plessis, S. S., (2014). Potentials of phytotherapeutic treatment of erectile dysfunction. In: *Antioxidants – Antidiabetic Agents and Human Health*, Oguntibenju, O. O., (ed.), *In. Tech.*, Croatia, pp. 279–296.
2. Ademiluyi, A. O., & Oboh, G., (2011). Antioxidant properties of condiment produced from fermented bambara groundnut. *Journal of Food Biochemistry*, *35*, 1145–1160.
3. Ademiluyi, A. O., Oboh, G., Boligon A. A., & Athayde, M. L., (2012). Dietary supplementation with fermented legumes modulated hyperglycemia and acetylcholinesterase activities in Streptozotocin-induced diabetes. *Pathophysiology*, *22*, 195–201.
4. Ajuwon, O. R., Katengua-Thamahane, E., Van Rooyen, J., Oguntibeju, O. O., & Marnewick, J. L., (2013). Protective effects of rooibos (*Aspalathus linearis*) and/or red palm oil (*Elaeisguineensis*) supplementation on tert – butyl hydroperoxide-induced oxidative hepatotoxicity in Wister rats. *Evidence-Based Complementary and Alternative Medicine,* 1–19.
5. Ajuwon, O. R., Oguntibeju, O. O., & Marnewick, J. L., (2014). Amelioration of lipolysaccharide-induced liver injury by aqueous rooibos (*Aspalathus linearis*) extract via inhibition of pro-inflammatory cytokines and oxidative stress. *BMC Complementary and Alternative Medicine, 14*(392), 1–12.
6. Ayepola, O. R., Brooks, N. L., & Oguntibeju, O. O., (2014). Kolaviron improved resistance to oxidative stress and inflammation in the blood (erythrocyte, serum, & plasma) of streptozotocin-induced diabetic rats. *Scientific World Journal*, *24*, 1–8.
7. Berk, S., Tepe, B., & Arslan, S., (2011). Screening of the antioxidant, antimicrobial and DNA damage protection potentials of the aqueous extract of *Inula oculus-christi*. *African Journal of Pharmacy and Pharmacology, 5*(14), 1695–1702.
8. Bernal, J., Mendiola, J. A., Ibáñez, E., & Cifuentes, A., (2011). Advanced analysis of nutraceuticals. *Journal of Pharmaceutical and Biomedical Analysis*, *55*(4), 758–774.
9. Bhattacharya, A., Sood, P., & Citovsky, V., (2010). The roles of plant phenolics in defense and communication during Agrobacterium and Rhizobium infection. *Molecular Plant Pathology*, *11*(5), 705–719.
10. Blake, G. J., & Ridker, P. M., (2002). Inflammatory bio-markers and cardiovascular risk prediction. *Journal of Internal Medicine*, *252*, 283–294.

11. Bleve, M., Ciurlia, L., Erroi, E., Lionetto, G., Longoc, L., & Rescioa, L., (2008). An innovative method for the purification of anthocyanins from grape skin extracts by using liquid and sib-critical carbon dioxide. *Separation and Purification Technology, 64*, 192–197.
12. Bravo, L., (1998). Polyphenols: Chemistry, dietary sources, metabolism, & nutritional significance. *Nutrition Reviews, 56*(11), 317–333.
13. Brink, M., Ramolemana, G. M., & Sibuga, K. P., (2015). *Vignasubterranea* (L.) Verdc. In*: Plant Resources of Tropical Africa/ Resources végétales l'Afriquetropicalede*, Brink, M., Belay, G. (ed.), Wageningen, Netherlands: PROTA4U, 2006.
14. Bureau, S., Renard, C. M. G. C., Reich, M., Ginies, C., & Audergon, J. M., (2009). Change in anthocyanin concentrations in red apricot fruits during ripening. *Food Science and Technology, 42*, 372–377.
15. Cachofeiro, V., Goicochea, M., De Vinuesa, S. G., Oubina, P., Lahera, V., & Luno, J., (2008). Oxidative stress and inflammation, a link between chronic kidney disease and cardiovascular disease. *Kidney International, 74*(111), S4–S9.
16. Calderón-Montaño, J. M., Burgos-Moron, E., Perez-Guerrero, C., & Lopez-Lazaro, M., (2011). A review on the dietary flavonoid kaempferol. *Mini Reviews in Medicinal Chemistry, 11*(4), 298–344.
17. Calhau, C., & Santos, A., (2009). Oxidative stress in the metabolic Syndrme. In: *Soares, R., Costa, C. (Eds.), Oxidative Stress, Inflammation and Antiogenesis in the Metabolic Syndrome*, Springer Science and Business Media. pp. 33–64.
18. Ceriello, A., Testa, R., & Genovese, S., (2016). Clinical implication of oxidative stress and potential role of natural antioxidants in diabetic vascular complications. *Nutrition, Metabolism and Cardiovascular Diseases, 26*, 285–292.
19. Chen, A. Y., & Chen, Y. C., (2013). A review of dietary flavonoid, kaempferol on human health and cancer chemoprevention. *Food Chemistry, 138*(4), 2099–2107.
20. Da Costa, C. T., Horton, D., & Margolis, S. A., (2000). Analysis of anthocyanins in foods by liquid chromatography, liquid chromatography-mass spectrometry and capillary electrophoresis. *Journal of Chromatography A., 881*, 403–410.
21. Diedericks, C. F., & Jideani, V. A., (2015). Physicochemical and functional properties of insoluble dietary fiber isolated from Bambara groundnut (*Vignasubterranea* (L.) Verdc). *Journal of Food Science, 80*(9), 1933–1944.
22. Etuk, E. U., Bello, S. O., Isezuo, S. A., & Mohammed, B. J., (2010). Ethnobotanical survey of medicinal plants used for the treatment of diabetes mellitus in the North Western region of Nigeria. *Asian Journal of Experimental Biological & Sciences, 1*, 55–59.
23. Farag, M. A., Porzel, A., & Wessjohann, L. A., (2012). Comparative metabolite profiling and fingerprinting of medicinal licorice roots using a multiplex approach of GC-MS, LC-MS and 1-D NMR techniques. *Phytochemistry, 76*, 60–72.
24. Feng, Z., Hao, W., Lin, X., Fan, D., & Zhou, J., (2014). Antitumor activity of total flavonoids from *Tetrastigmahemsleyanum* Diels etGilg is associated with the inhibition of regulatory T cells in mice. *Onco Targets and Therapy, 7*, 947–956.
25. Fernandez-Orozco, R., Frias, J., Munoz, R., Zielinski, H., Piskula, M. K., Kozlowska, H., & Vidal-Valverde, C., (2008). Effect of fermentation conditions on the antioxidant compounds and antioxidant capacity of *Lupinusangustifolius* cv. *zapaton. European Food Research and Technology, 227*, 979–988.

26. Govindappa, M., (2011). Antimicrobial, antioxidant and in vitro anti-inflammatory activity and phytochemical screening of *Crotalaria pallidaiton. African Journal of Pharmacy and Pharmacology*, *5*(21), 2359–2371.
27. Granderson, A. A., (2014). Making sense of climate change risks and responses at the community level: A cultural-political lens. *Climate Risk Management*, *3*, 55–64.
28. Halliwell, B., (2007). Biochemistry of Oxidative stress. *Biochemistry Society Transactions*, *35*(3), 1147–1150.
29. Hermsdorff, H. H. M., Zulet M. A., Abete, I., & Martinez, A. J., (2010). A legume based hypocaloric diet reduces proinfammatory status and improves metabolic features in overweight/ obese subjects. *European Journal of Nutrition*, *50*, 61–69.
30. Himmelfarb, J., (2004). Linking oxidative stress and inflammation in kidney disease: Which is the Chicken and Which is the Egg ? *Seminars in Dialysis*, *17*(6), 449–454.
31. Ho, E., Galougahi, K. K., Liu, C., Bhindi, R., & Figtree, G. A., (2013). Biological markers of oxidative stress: Application to cardiovascular research and practice. *Redox Biology*, *1*(1), 483–491.
32. Ignat, I., Volf, I., & Popa, V. I., (2011). A critical review of methods for characterization of polyphenolic compounds in fruits and vegetables. *Food Chemistry*, *126*(4), 1821–1835.
33. International Foundation of Science (IFS), (2016). *Value chain for neglected crops in Africa*. http://www.ifs.se/ifs-news/value-chains-of-neglected-crops-in-africa.html.
34. Jideani, A. I. O., Silungwe, H., Takalani, T., Anyisa, T. A., Udeh, H., & Omolola, A., (2014). Antioxidant-rich natural grain products and human health. In: Oguntibeju, O. O. (Ed.), *Antioxidants, Antidiabetic Agents and Human Health, In. Tech.*, Croatia, pp. 209–238.
35. Jideani, V. A., (2016). *Utilizing Bambara Groundnut in Value-Added Products*. Institute of Food Technologist. http://www.ift.org/food-technology/past-issues/2016/january/features/utilizing-bambara-groundnut-in-value-added-products.aspx.
36. Jideani, V. A., & Mpotokwane, S. M., (2009). Modeling of water absorption of Botswana bambara varieties using Peleg`s equation. *Journal of Food Engineering*, *92*, 182–188.
37. Johnson, R. J., Nakagawa, T., Jalal, D., Sanchez-Lozada, L. G., Kang, D., & Ritz, E., (2013). Uric acid and chronic kidney disease: which is chasing which? *Nephrol Dial Transplant*, *28*, 2221–2228.
38. Kanatt, S. W., Arjun, K., & Sharma, A., (2011). Antioxidant and antimicrobial activity of legumes hulls. *Food Research International*, *44*, 3182–3187.
39. Kara, M., Uslu, S., Demirci, F., Temel, H. E., & Baydemir, C., (2014). Supplemental carvacrol can reduce the severity of inflammation by Influencing the production of mediators of inflammation. *Inflammation*, *38*(3), 1020–1027.
40. Kasote, D. M., Katyare, S., Hegde, M., & Bae, H., (2015). Significance of antioxidant potential of plants and its relevance to therapeutic applications. *International Journal of Biological Science*, *11*, 982–991.
41. Kaufman, P. B., Duke, J. A., Brielmann, H., Boik, J., & Hoyt, J. E., (1997). A comparative study survey of leguminous plants as sources of the isoflavones, genistein and daidzein: implications for human nutrition and health. *Journal of Alternative and Complementary Medicine*, *3*(1), 7–12.

42. Kone, M., Paice A. G., & Toure, Y., (2011). Bambara groundnut [*Vigna subterranean* (L.) Verdc. (Fabaceae)] Usage in human health. In: Preedy, V. R., Watson, R. R., Patel, V. B., (eds.), *Nuts and Seeds in Health and Diseases Preventation*, London, pp. 189–196.
43. Kumar, R., (1991). Anti-nutritional factors: Legume trees and other fodder trees as protein sources for livestock. In: *Proceedings of the FAO, Expert Consultation held at the Malaysia*, Agriculture Research and Development Institution, Kuala Lumpur, Malaysia, 14–18.
44. Lee, S. M., Lee, T. H., Cui, E., Baek, N., Hong, S. G., Chung, I., & Kim, J., (2011). Anti-inflammatory effects of cowpea (*Vignasinensis* K.) seed extracts and its bioactive compounds. *Journal of Korean Society for Applied Biological Chemistry, 54*(5), 710–717.
45. Lu, J., Lin, P. H., Yao, Q., & Chen, C., (2010). Chemical and molecular mechanism of antioxidants: experimental approaches and model systems. *Journal of Cellular and Molecular Medicine*, *14*(4), 840–860.
46. Maphosa, Y., & Jideani, V. A., (2016). Physicochemical characteristics of Bambara groundnut dietary fibers extracted using wet milling. *South African Journal of Science, 112*(1), doi.org/10.17159/sajs.2016/20150126.
47. Marathe, S. A., Rajalakshmi, V., Sahayog, N., & Sharna, A., (2011). Comparative study on antioxidant activity of different varieties of commonly consumed legumes in India. *Food and Chemical Toxicology*, *49*(9), 2005–2012.
48. Maroon, C., Bost, J. W., & Maroon, A., (*2009*). Natural anti-inflammatory agents for pain relief. *Surgical Neurology International, 1*(80), 1–17.
49. Marxen, K., Vanselow, K. H., Lippermeier, S., Hitze, R., Ruse, A., & Hanse, U., (2007). Determination of DPPH radical oxidation caused by methanolic extracts of some microalgal species by linear regression analysis of spectrophotometric measurements. *Sensors*, *7*(10), 2080–2095.
50. Massawe, F. J., Dickinson, M., Roberts, J. A., & Azam-Ali, S. N., (2002). Genetic diversity in bambara groundnut (*Vigna subterranean* (L.) Verdc). *Genome*, *45*, 1175–1180.
51. Mazahib, A. M., Nuha, M. O., Salawa, I. S., & Babiker, E. E., (2013). Some nutritional attributes of bambara groundnut as influenced by domestic processing. *International Food Research Journal*, *20*(3), 1165–1172.
52. Mehde, A. A., Mehdi, W., Zainulabdeen, J. A., & Abdulbari, A. S., (2014). Correlation of inhibin and several antioxidants in children with acute lymphoblastic leukemia. *Asian Pacific Journal of Cancer Prevention*, *15*(12), 4843–4846.
53. Meng, Z., Yan, C., Deng, Q., Gao, D., & Niu, X., (2013). Curcumin inhibits LPS-induced inflammation in rat vascular smooth muscle cells in vitro via ROS-relative TLR4-MAPK/NF-κB pathways. *Acta Pharmacologica Sinica*, *34*(7), 901–911.
54. Metukuri, M. R., Reddy, C. M. T., Reddy P. R. K., & Reddanna, P., (2010). Bacterial LPS mediated acute inflammation-induced spermatogenic failure in rats: Role of stress response proteins and mitochondrial dysfunction. *Inflammation*, *33*(4), 235–243.
55. Miller, E., Morel, A., Saso, L., & Saluk, J., (2014). Isoprostanes and neuroprostanes as biomarkers of oxidative stress in neurodegenerative Diseases. *Oxidative Medicine and Cellular Longetivity*, doi.org/10.1155/2014/572491.

56. Murakami, M., & Hirano, T., (2012). The molecular mechanisms of chronic inflammation development. *Frontiers in Immunology*, *3*(323), 1–2.
57. Murevanhema, Y. Y., (2013). Evaluation of Bambara groundnut (*Vignasubterrnea* (L.) Verdc.) milk fermented with lactic acid bacteria as a probiotic beverage. *M., Tech. Thesis, Faculty of Applied Sciences*, Cape Peninsula University of Technology, South Africa.
58. Murevanhema, Y. Y., & Jideani, V. A., (2015). Production and characterization of milk produced from Bambara groundnut (*Vignasubterranea*) varieties. *Journal of Food Processing and Preservation*, *39*, 1485–1498.
59. Murevanhema, Y. Y., & Jideani, V. A., (2013). Potential of Bambara groundnut (*Vignasubterranea* (L.) Verdc) milk as a probiotic beverage: A review. *Critical Reviews in Food Science and Nutrition*, *53*(9), 954–967.
60. Nedrebø, T., Reed, R. K., Jonsson, R., Berg, A., & Wiig, H., (2004). Differential cytokine response in interstitial fluid in skin and serum during experimental inflammation in rats. *Journal of Physiology*, *1*, 193–202.
61. Negrão, M. R., Keating, E., Faria, A., Azevedo, I., & Martins, M. J., (2006). Acute effect of tea, wine, beer, & polyphenols on ecto-alkaline phosphatase activity in human vascular smooth muscle cells. *Journal of Agricultural and Food Chemistry*, *54*(14), 4982–4988.
62. Niki, E., (2010). Assessment of antioxidant capacity in *vitro* and in *vivo*. *Free Radical Biology and Medicine*, *49*, 503–515.
63. Noctor, G., Mhamdi, A., Chaouch, S., Han, Y., Neukermans, J., Marquez-Garcia, B., & Foyer, C. H., (2012). Glutathione in plants: An integrated overview. *Plant, Cell and Environment*, *35*(2), 454–484.
64. NRC (National Research Council), (1996). *Lost Crops of Africa: Grains, volume 1.* Board of Science and Technology for International Development, National Academy Press, Washington, DC, pp. 59–76.
65. Nyau, V., (2013). Nutraceutical antioxidant potential and polyphenolic profiles of the Zambian market classes of Bambara groundnuts (Vignasubteranea (L.) Verdc) and common beans (Phaseolus vulgaris L.). *PhD Thesis*, Faculty of Science, University of Cape Town.
66. Onyilagha, J. C., Islam, S., & Ntamatungiro, S., (2009). Comparative phytochemistry of eleven species of *Vigna (Fabaceae)*. *Biochemical Systematics and Ecology*, *37*, 16–19.
67. Oyagbemi, A. A., Salihu, M., Oguntibeju, O. O., Esterhuyse, A. J., & Farombi, E. O., (2014). Some selected medicinal plants with antidiabetic potentials. In: Oguntibeju, O. O. (ed.), *Antioxidant – Antidiabetic Agents and Human Health, In. Tech.*, Croatia, pp. 209–238.
68. Pale, E., Nacro, M., Vanhaelen, M., & Vanhaelen-Fastre, R., (1997). Anthocyanins from Bambara groundnut (*Vignasubterranea*). *Agriculture and Food Science*, *45*, 3359–3361.
69. Pansti, W. G., Bester, D. J., Esterhuyse, A. J., & Aboua, G., (2014). Dietary antioxidant properties of vegetable oils and nuts: The race against cardiovascular disease progression. In: Oguntibeju, O. (ed.), *Antioxidant – Antidiabetic Agents and Human Health, In. Tech.*, Croatia, pp. 209–238.

70. Philis Tsimikas, A., (2009). Type-2 diabetes: Limitations of current therapies. *Consultant*, *49*, S5–S10.
71. Pietta, P. G., (2000). Flavonoids as antioxidants. *Journal of Natural Products*, *63*(7), 1035–1042.
72. Pinchuk, I., Shoval, H., Dotan, Y., & Lichtenbery, Y., (2012). Evaluation of antioxidants: Scope, limitation and relevance of assays. *Chemistry and Physics of lipids*, *165*, 638–647.
73. Pisoschi, A. M., & Pop, A., (2015). The role of antioxidants in the chemistry of oxidative stress: A review. *European Journal of Medicinal Chemistry*, *97*, 55–74.
74. Pitt, J. J., (2009). Principles and applications of liquid chromatography-mass spectrometry in clinical biochemistry. *Clinical biochemist. Reviews, 30*(1), 19–34.
75. Prasad, A. G. D., & Kumar, K., (2013). Ethno-botanical potential of medicinal legumes in the Westren Ghats of Karnataka. *Indo American Journal of Pharmaceutical Research*, *3*(1), 1300–1307.
76. Price, K. R., Griffiths, N. M., & Curl, C. L., (1985). Fenwick. Undesirable sensory properties of the dried pea (*Pisumsativum*): The role of saponins. *Food Chemistry*, 17, 105–115.
77. Prior, R. L., Wu, X., & Schaich, K., (2005). Standardized methods for the determination of antioxidant capacity and phenolics in foods and dietary supplements. *Journal of Agricultural and Food Chemistry*, *53*(10), 4290–4302.
78. Recio, M. C., Andújar, I., & Ríos, J. L., (2012). Anti-inflammatory agents from plants : progress and potential. *Current Medicinal Chemistry*, *19*, 2088–2103.
79. Ricciotti, E., & Fitzgerald, G. A., (2011). *ATVB in Focus Prostaglandins and Inflammation*, 986–1001. http://doi.org/10.1161/ATVBAHA.110.207449.
80. Safavi, M., Foroumadi, A., & Abdollahi, M., (2013). The importance of synthetic drugs for type-2 diabetes drug discovery. *Expert Opinion on Drug Discovery*, *8*(11), 1339–1363.
81. Salehi-Abargouei, A., Saraf-Bank, S., Bellissimo, N., & Azadbakht, L., (2015). Effect of non-soy legume consumption on C-reactive protein: A systematic review and meta-analysis. *Nutrition*, *31*, 631–639.
82. Sasmal, S., Majumdar, S., Gupta, M., Mukherjee, A., & Mukherjee, P. K., (2012). Pharmacognostical, phytochemical and pharmacological evaluation for the antipyretic effect of the seeds of *Saracaasoca* Roxb. *Asian Pacific Journal of Tropical Biomedicine*, 782–786.
83. Sivakumar, P. M., Prabhakar, P. K., & Doble, M., (2011). Synthesis, antioxidant evaluation and qual=ntitative structure-activity relationship studies of chalcones. *Medical Chemistry Research*, *20*, 482–492.
84. Slavin, J. L., (2000). Mechanism for the impact of whole grain foods on cancer risk. *Journal of American College of Nutrition*, *19*, S300–S307.
85. Steinmann, D., & Ganzera, M., (2011). Recent advances on HPLC/MS in medicinal plant analysis. *Journal of Pharmaceutical and Biomedical Analysis, 55*(4), 744–757.
86. Su, D. L., Lu, Z. M., Shen, M. N., Li, X., & Sun, L. Y., (2012). Roles of pro- and anti-inflammatory cytokines in the pathogenesis of SLE., *Journal of Biomedicine and Biotechnology*, http://dx.doi.org/10.1155/2012/347141.

87. Sultana, R., Cenini, G., & Butterfield, D. A., (2013). *Biomarkers of Oxidative Stress in Neurodegenerative Diseases. Basis of Oxidative Stress*. John Wiley & Sons, Inc. Hoboken, New Jersey, pp. 359–376.
88. Swanevelder, J. C., (1998). *Bambara food for Africa: Bombara Groundnut (Vignasubterranea)*. National Department of Agriculture. ARC-Grain Crops Institute, South Africa, p. 74.
89. Turner, T. T., & Lysiak, J. J., (2008). Oxidative stress: a common factor in testicular dysfunction. *Journal of andrology*, *29*(5), 488–498.
90. Ujowundu, C. O., Nwaoga, L. A., Ujowunda, F. N., & Belonwu, D. C., (2013). Effect of gas flaring on the phytochemical and nutritional composition of Treculiaafricana and Vigna subterranean. *British Biotechnolgy Journal*, *3*(3), 293–304.
91. Valente, C., Aboua, G., & Du Plessis, S. S., (2014). Garlic and its effect on health with special reference to the reproductive system. In: Oguntibenju, O. O. (ed.), *Antioxidants – Antidiabetic Agents and Human Health, In. Tech.*, Croatia, pp. 259–278.
92. Valko, M., Leibfritz, D., Moncol, J., Cronin, T. D., Mazur, M., & Telser, J., (2007). Free radicals and antioxidants in normal physiological functions and human disease. *International Journal of Biochemistry & Cell Biology*, *39*(1), 44–84.
93. Van Meter, K. C., & Hubert, R. J., (2014). *Gould's Pathophysiology for the Health Professions*. Elsevier Saunder, Missouri, pp. 361–389.
94. Waggiallah, H., & Alzohairy, M., (2011). The effect of oxidative stress on human red cells glutation peroxidase, glutathione reductase levels, & prevalence of anaemia among diabetics. *North American Journal of Medical Sciences*, *3*(7), 344–347.
95. Wang, S., Melnyk, J. P., Tsao, R., & Marcone, M. F., (2011). How natural dietary antioxidants in fruits, vegetables and legumes promote vascular health. *Food Research International*, *44*(1), 14–22.
96. World Health Organization (WHO), (2015). Lifestyle diseases kill 16 million premature. Lexis Nexis News, AFP I-Net Bridge Pvt. Ltd.
97. Zia-Ul-Haq, M., Landa, P., Kutil, Z., Qayum, M., & Ahmad, S., (2013). Evaluation of anti-inflammatory activities of selected legumes from Pakistan: In vitro inhibition of Cyclooxygenase-2. *Parkistan Journal of Pharmceutical Sciences*, *26*(1), 185–187.

CHAPTER 6

ANTIOXIDANT AND ANTIMICROBIAL ACTIVITY OF *OPUNTIA AURANTIACA* LINDL

WILFRED MBENG OTANG and ANTHONY JIDE AFOLAYAN

CONTENTS

6.1 INTRODUCTION

Opuntia aurantiaca (*Cactaceae*) is an inconspicuous, perennial succulent shrublet, which seldom exceeds 0.5 m in height in open pasture but can reach up to 2 m in closed vegetation. The plant grows naturally in Argentina, Paraguay, and Uruguay. The plant consists of sausage-like, fleshy segments or joints (also known as cladodes) and is a serious invasive weed on natural grasslands in Australia and South Africa [3, 4], where it was introduced as an ornamental species and spread rapidly via dispersal of vegetative parts.

In Mexican folk medicine, the pulp and juice of *O. aurantiaca* (tiger-pear, jointed cactus, or jointed prickly-pea) are used for the treatment of wounds and inflammation of the digestive and urinary tracts [13]. In China, *Opuntia* is used for moxibustion, where it is burnt on the skin for the treatment of skin diseases. *Opuntia* has also been used to treat burns by American Indians [10] and South Koreans. In Sicily, a decoction of flower of *Opuntia* is taken as a diuretic, while the cladodes are known for their anti-inflammatory potential against whooping cough, arthritis, and edema [11]. In South Africa, whites and blacks use the pounded joints of *Opuntia vulgaris* for tumors, boils, and open sores. The Xhosas in South Africa rub the juice of the joints into curt warts, claiming that they disappear after 2–3 days [10].

Vitamin E, alpha-tocopherol, and glutathione are skin antioxidants, which fight against the deleterious effects of reactive oxygen species (ROS). Among bacterial pathogens, *Staphylococcus aureus* and *Streptococcus pyogenes* are the most prevalent causative agents of common skin diseases such as abscesses, carbuncle, erysipelas cellulites, folliculitis, furuncle, and impetigo [5]. Ringworm and athlete's foot are the most common cutaneous fungal infections caused by *Trichophyton rubrum*, while *Candida albicans* is the main culprit responsible for gastrointestinal, vaginal, and oral candidiasis [5].

This chapter explores the efficacy of the extracts of *O. aurantiaca* against selected pathogens that cause human skin disorders and evaluate the antioxidant capacity for validation of folk uses of the plant that is used by the Xhosas of the Amathole District, Eastern Cape, South Africa.

6.2 MATERIALS AND METHODS

6.2.1 EXTRACTION PROCEDURE

O. aurantiaca leaves were sterilized with ethanol (70%), and distilled water was used to rinse them. Ground samples of the leaves in separate conical flasks containing ethanol and acetone were shaken in an orbital shaker for 24 hours. Filtration was done with Whatman No. 1 filter paper in a Buchner funnel. The filtrates were dried under reduced pressure in a rotavapor at 40°C. The extracts were re-constituted to yield a 20 mg/mL stock solution in their respective solvents.

6.2.2. MICROORGANISMS FOR THIS STUDY

Six gram-negative bacteria (*Escherichia coli, Salmonella typhimurium, Pseudomonas aeruginosa, Shigella flexneri, Shigella sonnei*, and *Klebsiella pneumonia*) and five gram-positive bacteria (*S. pyogenes, Klebsiella pneumoniae, S. aureus, Bacillus cereus, Enterococcus faecalis*, and *Bacillus subtilis*) were used for the antibacterial assays. *Candida glabrata* and *Candida krusei* were used for the antifungal assays. The choice of these microbes was tied to their significance as pathogenic microbes that cause cutaneous disorders in humans. American Type Culture Collection (ATCC) strains of all microbes were obtained from Total Lab (South Africa).

6.2.3 SUSCEPTIBILITY TESTS

Sabouraud dextrose agar (SDA) and nutrient agar were prepared according to the instructions by the manufacturer and transferred into disposable sterile petri plates. The plates were labeled, inoculated with 0.5 Mcfarland solutions (100 μL) of the respective microbe, and 50 mg/mL *O. aurantiaca* extract was loaded into 6-mm diameter agar wells. Positive controls (25–50 μg/well) for fungi and bacteria were nystatin and gentamicin, respectively. Zones of inhibition were measured after incubating at 37°C for 24 hours for bacteria and 72 hours for fungi. The minimum inhibitory concentration (MIC) was determined using the method described by Otang et al. [8].

6.2.4 ANTIOXIDANT ANALYSES

6.2.4.1 DPPH activity

DPPH scavenging activity was determined according to the method by Erukainure et al. [2]. Concentrations of 0.025 to 0.5 μg/mL of the plant extract were added to the same volume of 100 μM DPPH dissolved in methanol and kept for 30 minutes in the dark at room temperature. Standard controls were rutin and vitamin C. The experiment was run in triplicates. The absorbance at 518 nm was measured at and converted into antioxidant activity according to the formula:

$$\% \text{ DPPH Scavenging Activity} = [(A_0 - A_1)/A_0] \times 100 \quad (1)$$

where A_0 = absorbance of control; A_1 = absorbance of standard or extract; IC_{50} is the sample concentration needed to scavenge 50% DPPH radicals.

The IC_{50} values were calculated by linear regression of plots, where the horizontal axis represented the concentration of plant extracts and the vertical axis represented the percent scavenging activity.

6.2.4.2 Reducing Power

Different concentrations (0.025–0.05 μg/mL) of each extract in distilled water were mixed with 2.5 mL phosphate buffer (0.2 M, pH 6.6) and 2.5 mL potassium ferricyanide (1% w/v), incubated for 20 minutes at 50°C. Then, 2.5 mL of trichloroacetic acid (10% w/v) was added into the mixture and centrifuged (3000 rpm) for 10 minutes. The supernatant (2.5 ml) was mixed with 2.5 mL distilled water and 0.5 mL of $FeCl_3$ (0.1% w/v). Positive controls were rutin and vitamin C. The absorbance was recorded at 700 nm.

6.2.4.3 Scavenging of Hydrogen Peroxide

Hydrogen peroxide scavenging activity was determined according to the method of Erukainure et al. [2]. Hydrogen peroxide solution (40 mM) was prepared in phosphate buffer (pH 7.4). Different extract concentrations (0.025–0.05 μg/mL) in distilled water were added to 0.6 mL, 40 mM hydrogen peroxide solution. Absorbance at 230 nm was determined after 10 minutes against a blank solution of phosphate buffer alone. The hydrogen peroxide scavenging activity was determined using Eq. (1).

6.3 RESULTS AND DISCUSSION

6.3.1 YIELD OF EXTRACTION

The yield of the ethanolic solvent (9.3 g/100 g dry plant material) was higher than that of acetone (7.5 g/100 g dry plant material).

6.3.2 ANTIMICROBIAL ACTIVITY

The antimicrobial activity of the plant extracts and controls are summarized in Table 6.1. The maximum antibacterial activity (inhibition zone diameter > 19 mm) was obtained with the ethanol extract of *O. aurantiaca*

TABLE 6.1 Antibacterial Activity of *O. aurantiaca* Extracts

Microorganisms	Ethanol		Acetone		Positive control	
	ZI	MIC	ZI	MIC	ZI	MIC
Gram negative bacteria						
Escherichia coli	Na	Na	15±1.6	2.5	25±0.7	<0.01
Pseudomonas aeruginosa	15±1.5	2.5	15±2.1	2.5	24±2.1	0.01
Salmonella typhimurium	15±1.0	2.5	15±2.1	2.5	25±1.0	0.01
Shigellaflexneri	NA	NA	10±1.0	>5	24±1.1	<0.02
Shigellasonnei	20±2.1	0.02	15±1.1	2.5	24±2.6	<0.01
Gram positive bacteria						
Bacillus cereus	15±1.0	2.5	15±1.1	2.5	24±1.6	<0.01
Bacillus subtilis	Na	Na	15±1.0	2.5	22±1.2	0.02
Enterococcus faecalis	Na	Na	20±1.8*	0.01	24±1.4	<0.1
Klebsiella pneumoniae	Na	Na	15±1.1	2.5	24±2.2	<0.01
Staphylococcus aureus	16±1.0	2.5	NA	NA	24±0.6	<0.1
Streptococcus pyogenes	20±1.0	0.02	NA	NA	23±2.1	<0.01
Fungi						
Candida glabrata	20±1.7	<0.02	20±1. 1	0.02	20±1.7	0.02
Candida krusei	NA	NA	NA	NA	21±1.2	0.02

NA = Not active; * = Not significantly different (P < 0.05) from positive control.

The acetone extract of *O. aurantiaca* showed moderate antibacterial (inhibition zone diameter between 10 and 15 mm) potency against all gram-negative bacteria and no activity against *S. aureus.* The ethanol extract was not inhibitory against *E. coli, S. flexneri, B. subtilis*, and *K. pneumoniae.* Although both extracts showed strong antifungal activity against *C. glabrata,* yet none was inhibitory against *C. krusei.*

6.3.3 ANTIOXIDANT ASSAYS

The concentration necessary to obtain 50% scavenging activity (IC_{50}) is presented in Table 6.2. The lower was the IC_{50} value, the larger was the scavenging activity [12].

6.3.3.1 DPPH Activity

The results of the DPPH (Figure 6.1) assay revealed that the scavenging activity of the acetone extract was greater than that of the ethanol extract and did not differ significantly from those of rutin and vitamin C.

TABLE 6.2 The IC_{50} Values *O. aurantiaca* Extracts

	Reducing power	**DPPH**	**ABTS**
Acetone	0.21	0.12	0.10
BHT	ND	**	0.05
Ethanol	0.05	0.15	0.04
Gallic acid	ND	**	0.16
Rutin	0.06	0.11	ND
Vitamin C	0.12	0.10	ND

ND = Not determined; ** = No data.

against *S. sonnei* and *S. pyogenes* and the acetone extract against *E. faecalis.* The ethanol and acetone extracts showed strong antifungal activity against *C. glabrata* with zones of inhibition of 20 ± 1.1 mm and 20 ± 1.7 mm, respectively

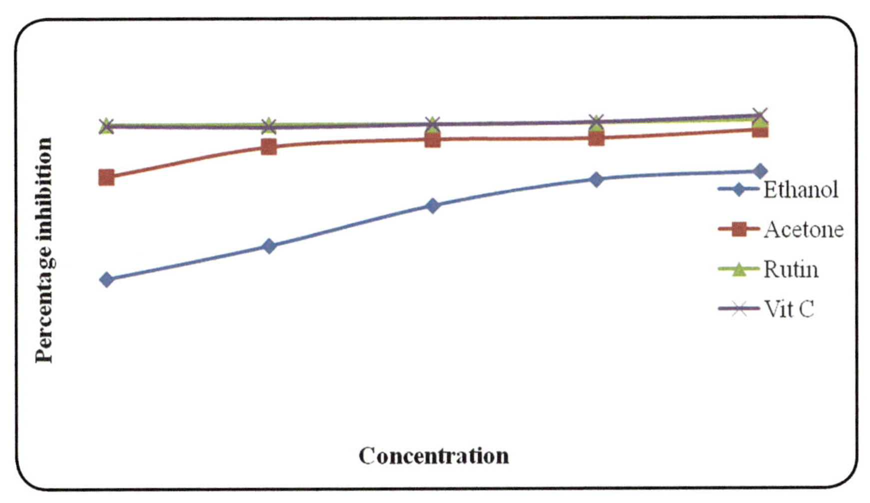

FIGURE 6.1 DPPH activity of *O. aurantiaca* extracts.

6.3.3.2 ABTS Scavenging Activity

The ABTS scavenging activity of the ethanol extract of *O. aurantiaca* was higher than that of the acetone extract, BHT, and gallic acid at all concentrations tested; and that of the acetone extract was greater than that of Gallic acid for extract concentrations higher than 0.05 mg/mL (Figure 6.2).

6.3.3.3 Reducing Power Assay

Reducing ability of the ethanol extract of *O. aurantiaca* was higher than that of rutin at concentrations lower than 0.15 mg/mL; and that of the acetone extract was lower than both controls at all concentrations (Figure 6.3).

In skin and soft tissue infections, *S. pyogenes is* one of the commonest bacterial agents that cause infections in wounds, carbuncles, impetigo furuncles, erysipelas, and abscesses. Many enterococcus species have been identified; however, only *Enterococcus faecium* and *E. faecalis* account for the greater share of human infections [6]. In the past, *C. glabrata* was assumed to be a nonpathogenic organism in humans, especially in the mucosal tissues. However, with the recent increased use of

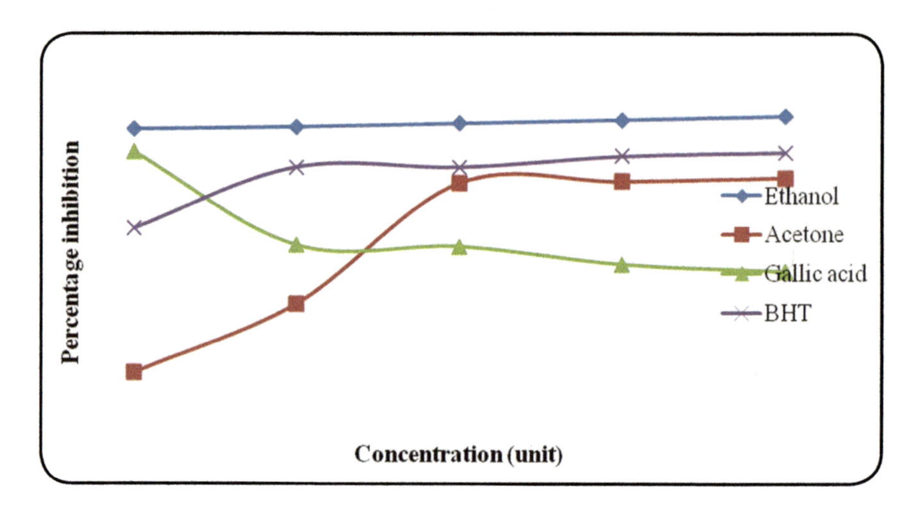

FIGURE 6.2 ABTS activity of *O. aurantiaca* extracts.

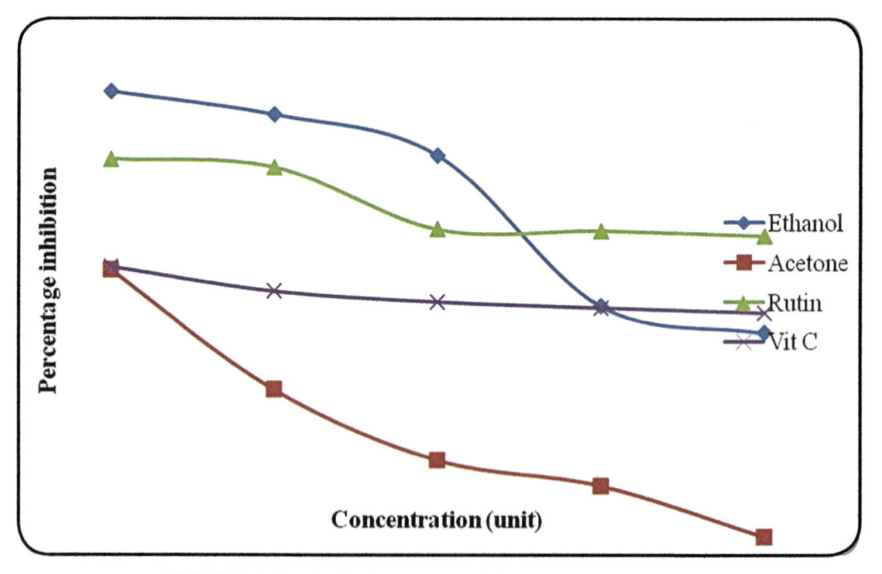

FIGURE 6.3 Reducing power of *O. aurantiaca* extracts.

immunosuppressive agents, both systemic and mucosal infections caused by *C. glabrata* have increased significantly, especially in AIDS patients [5]. *Opuntia* contains a range of phytochemicals such as gallic acid, vanillic acid, and catechins in variable quantities that may have biological activity [9]. The high sensitivity shown by *S. sonnei*, *S. pyogenes*, *E. faecalis*, *and C. glabrata* toward both extracts of *O. aurantiaca* may partially validate the local use of *O. aurantiaca* in traditional medicine against skin ailments.

S. aureus, *E. coli*, and *C. krusei* showed low sensitivity to the extracts of *O. aurantiaca*. Staphylococci infect wounds and are mainly responsible for abscesses, furuncles, erysipelas, carbuncles, and impetigo [9]. *E. coli,* which inhabits the intestines, may cause infections in wounds and sepsis. In one study, *E. coli, C. albicans*, and *S. aureus* were reported as the most common skin pathogens [7]. The fungus *C. krusei,* may occur in mucous membranes and on the skin without causing any infection if present at low frequency. However, they may become pathogenic and cause candidiasis and impetigo if they overgrow the normal flora, especially in diabetics and in individuals with suppressed immune systems [7].

As free radicals are definitely implicated in the pathogenesis of skin diseases, the antioxidant capacities of *O. aurantiaca were* also studied. ROS are usually produced in low quantities in aerobic organisms. When these species are released uncontrolled, they alter apoptotic pathways [1], and this may result in cutaneous malignancy. Although living cells have many antioxidant mechanisms that inactivate ROS to maintain homeostasis, yet unfortunately, these homeostatic mechanisms are sometimes overburdened, and the resultant increase in cutaneous ROS may accelerate the development of cutaneous disorders.

Hence, the availability of plant-derived antioxidants is gaining much importance since past few decades. In this chapter, the antioxidant capacity of *O. aurantiaca* was investigated using DPPH, reducing power, and ABTS radical scavenging assays. The DPPH scavenging activity of the acetone extract was significant, with IC_{50} value close to the scavenging activity of rutin and the ABTS; further, reducing power of the ethanol extract was comparable to those of the standard controls (BHT and rutin). The ability of these extracts to scavenge DPPH and ABTS free radicals suggest that they might donate electrons that combine with radicals, making them stable and thus terminating the radical chain reactions. The profound antioxidant activity of *O. aurantiaca* extracts is most likely due to polyphenolic compounds that are known to be present in the *Opuntia* genus. Although the antioxidant activity of *O. aurantiaca* has not been reported in literature, yet other species within the *Opuntia* genus such as *Opuntia ficus-indica* var. saboten extract exhibited concentration-dependent scavenging activity, superoxide anions, DPPH radicals, and hydroxyl radicals on different assay systems [9].

6.4 CONCLUSIONS

This study was undertaken to investigate the efficacy of *O. aurantiaca* extracts against selected pathogens responsible for human cutaneous diseases and to evaluate the antioxidant capacity for validation of folk uses of this plant. The most potent antibacterial activity was recorded with the ethanol extract of *O. aurantiaca* against *Shigella sonnei* and *Streptococcus pyogenes* and by the acetone extract against *Enterococcus faecalis.* Although both extracts showed strong antifungal activity against *Candida glabrata,* yet none was inhibitory against *Candida krusei.* The acetone extract of *O. aurantiaca* showed significant DPPH radical scavenging activity with IC_{50} value close to that of rutin, and the ABTS radical scavenging activity and reducing power of the ethanol extract were comparable to those of the standard controls (BHT and rutin). The high sensitivity shown by *S. sonnei, S. pyogenes, Enterococcus faecalis, and Candida glabrata* toward the extracts of *O. aurantiaca* may partially justify the use of the plant in traditional medicine against skin ailments.

6.5 SUMMARY

In view of the importance of *O. aurantiaca* in ethnopharmacology, the study in this chapter was undertaken to investigate the efficacy of the plant extracts against selected pathogens that cause human skin disorders and evaluation of the antioxidant capability for validation of folk uses of the plant. The agar well diffusion and microdilution techniques were used to test the antimicrobial activity of the plant against selected bacteria and fungi. The antioxidant activity of the plant was evaluated through DPPH (1,1-diphenyl-2-picrylhydrazyl), NO (nitric oxide), ABTS (2,2'-azino-bis 3-ethylbenzothiazoline-6-sulphonic acid), and reducing power assays.

The ethanol extract showed that the most potent antibacterial activity was against *Shigella sonnei* and *Streptococcus pyogenes,* followed by the acetone extract against *Enterococcus faecalis*. Although the two extracts showed strong antifungal activity against *Candida glabrata,* yet none was inhibitory against *Candida krusei*. The acetone extract of *O. aurantiaca* showed potent DPPH scavenging activity that was close to that of Rutin, and the ABTS radical scavenging activity and reducing power of the ethanol extract were not significantly different (P = 0.12) from those of the standard controls (Rutin and BHT). The high sensitivity shown by *Shigella sonnei*, *Streptococcus pyogenes*, *Enterococcus faecalis,* and *Candida glabrata* toward both extracts of *O. aurantiaca* may partially validate the local use of *Opuntia* in traditional medicine against skin ailments.

KEYWORDS

- **abscesses**
- **alpha tocopherol**
- **Amathole district**
- **antimicrobial activity**
- **antioxidant**
- **athlete's foot**
- **boils**
- *Candida albicans*
- **carbuncle**
- **eastern Cape**
- **efficacy**
- **erysipelas cellulites**
- **ethnopharmacology**
- **folk**
- **folliculitis**
- **furuncle**
- **gastrointestinal candidiasis**

- **glutathione reactive oxygen species**
- **impetigo**
- **open sores**
- ***Opuntia aurantiaca***
- **oral candidiasis**
- **pathogens**
- **ringworm**
- **skin diseases**
- **skin disorders**
- **South Africa**
- ***Staphylococcus aureus***
- ***Streptococcus pyogenes***
- ***Trichophyton rubrum***
- **tumors**
- **vaginal candidiasis**
- **vitamin E**
- **warts**
- **wounds**
- ***Xhosas***

REFERENCES

1. Bickers, D. R., & Athar, M., (2006). Oxidative stress in the pathogenesis of skin disease. *Journal of Investigative Dermatology, 126*(12), 2565–2575.
2. Erukainure, O. L., Oke, O. V., Ajiboye, A. J., & Okafor, O. Y., (2011). Nutritional qualities and phytochemical constituents of *Clerodendrum volubile*, a tropical non-conventional vegetable. *International Food Research Journal, 18*(4), 1393–1399.
3. Lee, J. C., Kim, H. R., Kim, J., & Jang, Y. S., (2002). Antioxidant property of an ethanol extract of the stem of *Opuntia ficus-indica* var. saboten. *Journal of Agricultural and Food Chemistry, 50*(22), 6490–6649.
4. Marchante, H., Freitas, H., & Hoffmann, J. H., (2011). The potential role of seed-banks in the recovery of dune ecosystems after removal of invasive plant species. *Applied Vegetation, Science, 14,* 107–119.

5. Nesy, E. A., & Mathew, L., (2014). Studies on Antimicrobial and antioxidant efficacy of *Thevetia neriifolia, Juss Leaf* extracts against Human Skin Pathogens. *International Journal of Pharmaceutical Sciences and Drug Research*, *6*(2), 164–168.
6. Obeidat, M., (2011). Antimicrobial activity of some medicinal plants against multidrug resistant skin pathogens. *Journal of Medicinal Plants Research*, *5*(16), 3856–3860.
7. Ördögh, L., Galgóczy, L., Krisch, J., Papp, T., & Vágvölgyi, C., (2010). Antioxidant and antimicrobial activities of fruit juices and pomace extracts against acne-inducing bacteria. *Acta Biologica Szegediensis*, *54*(1), 45–49.
8. Otang, W. M., Grierson, D. S., & Ndip, R. N. (2012). Antifungal activity of *Arctotis arctotoides* (L.f.) O., Hoffm. and *Gasteria bicolor* Haw. against opportunistic fungi associated with human immunodeficiency virus/acquired immunodeficiency syndrome. *Pharmacognosy Magazine*, *8*(30), 135–140.
9. Saleem, M., Kim, H. J., Han, C. K., Jin, C., & Lee, Y. S. (2006). Secondary metabolites from *Opuntia ficus-indica* var. saboten, *Phytochemistry*, *67*(13), 1390–1394.
10. Stintzing, F. C., & Carle, R., (2005). Cactus stems (*Opuntia* spp.): A review on their chemistry, technology, & uses. *Molecular Nutrition and Food Research*, *49*(2), 175–194.
11. Tesoriere, L., Fazzari, M., Allegra, M., & Livrea, M. A., (2005). Biothiols, taurine, & lipid-soluble antioxidants in the edible pulp of Sicilian cactus pear (*Opuntia ficus-indica*) fruits and changes of bioactive juice components upon industrial processing, *Journal of Agricultural and Food Chemistry*, *53*(20), 7851–7855.
12. Tung, Y. T., Wub, H. J., Hsieh, C., Ping-Sheng Chen, P. S., & Chang, S. T., (2009). Free radical scavenging phytochemicals of hot water extracts of *Acacia confusa* leaves detected by an on-line screening method. *Food Chemistry*, *115*(3), 1019–1024.
13. Vigueras, A. L., & Portillo, I., (2001). Uses of *Opuntia* species and the potential impact of *Cactoblastis cactorum* (lepidoptera: pyralidae) in Mexico. *The Florida Entomology*, *84*(4), 493–498.

CHAPTER 7

DIETARY INTERVENTION OF UTAZI (*GONGRENEMA LATIFOLIUM*) SUPPLEMENTED DIET USING WISTAR MALE RAT ANIMAL BRAIN MODEL

ESTHER EMEM NWANNA, GANIYU OBOH, and
OLUKEMI ABIMBOLA OKEDIRAN

CONTENTS

7.1 INTRODUCTION

Incidence rates of neurodegenerative diseases such as Alzheimer's disease (AD) and Parkinson's disease (PD) increase exponentially with age. According to World Health Organization (WHO), neurodegenerative diseases will become the world's second leading cause of death by the middle of the century, overtaking cancer [22]. According to the Global Burden of Disease Study, dementia and other neurodegenerative disorders will be the eighth cause of disease burden for developed regions by 2020 [27]. Oxidative stress is critical to the pathologies associated with brain damage and cognitive abilities [21]. Although multiple factors are involved in the development of neurodegenerative diseases, dysregulation in the inflammatory network and oxidative imbalance are key components in the pathogenesis of diseases such as Alzheimer's disease (AD), Parkinson's disease, brain tumors, and multiple sclerosis [14, 20]. Studies have shown low and moderate concentration of these neurotransmitters in patients with AD; therefore, inhibition of acetylcholinesterase (AChE), butyryl cholinesterase (BChE), and monoamine oxidase (MAO) activities have been accepted as an effective management strategy against AD [30, 35].

Neuroinhibitors (such as tacrine, donepezil, and rivastigmine) are commonly used synthetic drugs for the treatment of AD. However, these drugs are limited in use due to their adverse side effects [38]. The prevention of neurodegenerative condition has been one of the primary goals of present research using dietary regime [10] for early intervention, because they are pharmacologically safe, cost-effective, and available with minimal side effects. Hence, the focus is on plant phytochemicals as natural sources with psychotropic effects [7, 35]. In sub-Sahara Africa including Nigeria, most people with mental or behavioral disorder rely on traditional healing practices and medicinal plants for treatment of these conditions [10]. In addition, the traditional uses are dependent on ancestral experience [1].

Utazi (*Gongronema latifolium*) is a green leafy vegetable, which forms a major constituent of local diets in Nigeria. It is not only desired for its nutritional benefits but also for its medicinal properties as reported in folklore. Its leaf is consumed as a spice and in preparation of soups and stews [11]. The hypolipidermic, hyperglycemic, anti-inflammatory, and antioxidant properties of this vegetable have been reported by several investigators [23, 33]. A recent study on in vitro activities has shown some enzymes linked to neurodegeneration [30].

This chapter discusses the dietary intervention of Utazi supplemented diet using Wistar male rat animal model. Oxidative stress was induced in the rat brain using cyclophosphamide (CPH) cytoxan case No. (50-18-0). Cyclophosphamide, which is an anticancer drug, is known as an alkylating agent with alkylating properties that cause the disruption of nucleic acid function and the inhibition of DNA synthesis [8, 9]. This induced nucleic acid damage may lead to DNA mutations that can result in cytotoxicity, carcinogenicity, teratogenicity, and reproductive toxicity following chronic exposure to CPH [45].

7.2 MATERIALS AND METHODS

7.2.1 SAMPLE COLLECTION AND PREPARATION

Fresh leaves of Utazi (Figure 7.1) were harvested from local farm in Uyo, Nigeria, during the rainy season in June 2016. The samples were identified and authenticated at Forest Research Institute of Nigeria (FRIN), Ibadan, Nigeria. The leaves were plucked from its trunk, rinsed under running tap water, and dried for 7 days at room temperature (under shade) to constant weight. The dried leaves were pulverized using the method described by Mukhtar et al. [26] and kept in an airtight container for further use. The powder sample was analyzed using gas chromatography coupled with flame ionization detector (GC-FID), in Central Laboratory of Federal University of Technology, Akure, Ondo State.

FIGURE 7.1 Utazi vegetable.

7.2.2 CHEMICALS AND REAGENTS

Chemicals and reagents (such as gallic acid, quercetin, Folin–Ciocalteau's reagent, semicarbazide, benzylamine, acetylthiocholine, and butyryl thiocholine iodide) for this study were procured from Sigma-Aldrich, Inc. (St Louis, MO); dinitrophenyl hydrazine (DNPH) from ACROS Organics (New Jersey, USA), and methanol and acetic acid from BDH Chemicals Ltd. (Poole, England). Cyclophosphamide (case no. 50-18-2), donapezil hydrochloride E-2020, and all other kits used for bioassay were ordered from Randox Laboratories Ltd. Co., Crumlin, Antrim, UK.

7.2.3 EXPERIMENTAL DESIGN

Feed formulation and bioassay were carried out according to the modified method by Oboh et al. [32] and that of Srinivasan et al. [40], respectively. Male Wistar rats weighing 180–200 g were purchased from the Department of Veterinary Medicine, University of Ibadan, Nigeria. The rats were acclimatized for a period of 2 weeks. The animals were kept in wire-mesh cages under a controlled light cycle (12 h light/12 h dark) and placed on commercially available feed and water *ad libitum* during the period of acclimatization. A total of 42 rats were used for this study. The animals were allocated with specific dietary regimens consisting of 7 groups: control group I, negative control (induced with CPH); group II, positive control treated with 100 mg/kg body weight orally donepezil; group III, fed with only the basal diet (44.4% skimmed milk, 41.6% corn flour, 10% oil, and 4% mineral/vitamins premix); and the remaining groups IV to VII were fed with 42.4% skimmed milk, 41.6% corn flour, 10% oil, and 4% mineral/vitamin premix with 2% and 4% Utazi inclusion for 3 weeks. Mineral and vitamin premix contained 3200 I.U. vitamin A, 600 I.U vitamin D3, 2.8 mg vitamin E, 0.6 mg vitamin K3, 0.8 mg vitamin B1, 1 mg vitamin B2, 6 mg niacin, 2.2 mg pantothenic acid, 0.8 mg vitamin B6, 0.004 mg vitamin B12, 0.2 mg folic acid, 0.1 mg biotin H2, 7 mg Na chloride, 0.08 mg cobalt, 1.2 mg copper, 0.4 mg iodine, 8.4 mg iron, 16 mg manganese, 0.08 mg selenium, 12.4 mg zinc, and 0.5 mg antioxidant.

It should be noted that 2% and 4% Utazi inclusions were on equal-weight basis. The protein content of Utazi leaves was taken into consideration from the values obtained from the proximate analysis, which was used in the formulation of the protein content of diets. The feed was baked at 70°C for 30 mins to make the meal appealing to the rats; and 20 g of baked feed was administered to each rat in their compartmental cage on a daily basis.

7.2.4 ANIMAL ETHICS

All the animals received human care according to the criteria outlined in the Guide for the Care and the Use of Laboratory Animals prepared by the National Academy Science and published by the National Institute of Health (USA). The ethic regulations were followed in accordance with national and institutional guidelines for the protection of animals' welfare during experiments. The experiment was carried out at the Functional Food, Nutraceuticals and Phytomedicine Laboratory, Department of Biochemistry, Federal University of Technology, Akure, Ondo State, Nigeria.

7.2.5 PRETREATMENT STUDY

In the pre-treatment study, the rats (group III) were given suspension of donapezil (100 mg/kg body weight daily), while the treated groups (IV-VII) were given Utazi 2% and 4% supplemented diet. The experiment lasted for 21 consecutive days after which the administration of CPH was dissolved in distilled water (70 mg/kg b.w., i.p.) and was given to groups II to V on 24 h before termination, while groups I, VI, and VII were not induced with CPH. The animals were decapitated under mild diethyl ether anesthesia, and the brain tissue was rapidly isolated, placed on ice, and weighed. This tissue was subsequently homogenized in cold saline (1:10 w/v) solution with about 10 up-and-down strokes at approximately 1200 rpm in a Teflon glass homogenizer. The homogenate was centrifuged for 10 mins at 3000 *g* to yield a clear supernatant fraction [2]

7.2.6 ESTIMATION OF ANTIOXIDANT PARAMETERS

7.2.6.1 Determination of Protein Content

Protein content in brain tissue homogenate supernatant was determined as described previously by Lowry et al. [18]. The 0.01 mL of distilled water, protein standard (5.85 g/dL bovine serum albumin), and tissue homogenates was pipetted into sample tubes. Thereafter, 0.5 mL of biuret reagent (100 mmol/L NAOH, 16 mmol/l Na-K-ATPase tartarate, 15 mmol/L potassium iodide, and 6 mmol/L CuSO4·5H2O in 400 mL distilled water) was added and allowed to incubate for 30 minutes at room temperature. The absorbance of standard was read at 546 nm against the blank in a spectrophotometer. The total protein concentration was subsequently calculated against standard.

7.2.6.2 Determination of Glutathione Peroxidase Activity

Activity of glutathione peroxidase was determined by the method by Klafki et al. [17]. A total of 0.2 mL of 0.4M phosphate buffer (pH 7.0) was mixed with 0.1 mL of 10 mM sodium azide together with 2 mL plasma homogenate, 0.2 mL of 10 mM glutathione and 0.1 mL of 0.2 mL hydrogen peroxide. The mixture was incubated for 10 min followed by addition of 0.4 mL of 10% TCA to terminate the reaction. The mixture was centrifuged at 3200×*g* for 2 min, and the supernatant was assayed for glutathione content using Ellman's reagent (19.8 mg of DTNB in 100 mL of 0.1% sodium nitrate). The activity of *GPx* was expressed as mg of GSH consumed/min/g of protein.

7.2.6.3 Determination of Superoxide Dismutase Activity

The superoxide dismutase (SOD) was determined by the method by Misra et al. [25], but with a slight modification using a microplate reader at 490 nm. Enzyme activity was expressed as the amount of protein (μg) required to produce a 50% inhibition of auto-oxidation of 6-hydroxydopamine.

7.2.6.4 Determination of Vitamin C Content

Vitamin C content of the plasma tissue was determined using the method by Ngounou et al. [29]. Briefly, 75 μL DNPH (2-dinitrophenyl hydrazine, 230 mg thiourea, and 270 mg $CuSO_4 \cdot 5H_2O$ in 100 mL of 5M H_2SO_4) was added to 500 μL reaction mixture (300 μL of the tissue and control with 100 μL of 13.3% trichloroacetic acid (TCA), and water). The reaction mixture was subsequently incubated for 3 h at 37°C, and 0.5 mL of 65% H_2SO4 (v/v) was then added to the medium, and the absorbance was measured at 520 nm to find the vitamin C content of plasma.

7.2.7 LIPID PEROXIDATION AND THIOBARBITURIC ACID REACTION ASSAY

The 100 μL of the supernatant fraction of the brain tissue was mixed with a reaction mixture containing 30 μL of 0.1M Tris-HCl buffer (pH 7.4). The volume was increased to 300 μL by adding distilled water before incubation at 37°C for 60 min. The color reaction was developed by adding 300 μL of 8.1% sodium dodecyl sulfate to the reaction mixture, followed by 600 μL of acetic acid/ HCl (pH 3.4) mixture and 600 μL of 0.8% TBA. This mixture was incubated at 100°C for 60 min. The production of thiobarbituric acid reactive species (TBARS) was measured with UV-visible spectrophotometer (Jenway 6305) at 532 nm, and the absorbance was compared with that of a standard curve using malondialdehyde (MDA). Lipid peroxidation was assessed by measuring the formation of TBARS, as described by Ohkawa et al. [35]. TBARS were quantified using an extinction coefficient of 1.56 × 105 L mol^{-1} cm^{-1} and expressed as mmol of TBARS per mg of protein.

7.2.8 ENZYME ASSAYS

7.2.8.1 Cholinesterase Activity Assay

The acetyl cholinesterase (AChE) activity was measured in a reaction mixture containing 0.1M phosphate buffer (pH 8.0), 100 μL of a solution of 5,5′-dithio-bis(2-nitrobenzoic) acid (DTNB), 3.3 mM in 0.1 M phosphate buffered solution (pH 7.0) containing $NaHCO_3$ (6 mM), brain tissue (100 μL), and 500 μL of phosphate buffer (pH 8.0). After incubation for 20 min at 25°C, 100 μL of 0.05 mM acetylthiocholine iodide was added as the substrate. AChE activity was determined by monitoring changes in the absorbance at 412 nm for 3 min. Further, 100 μL of butyryl thiocholine iodide was used as a substrate to assay for BChE activity, while all other reagents and conditions remained the same [37]. The percentage inhibitory effect of extracts on AChE and BChE activities was subsequently calculated as:

$$\text{Percentage inhibitory effect} = [100 \times (\text{Absorbance of reference} - \text{Absorbance of sample})] / (\text{Absorbance of reference}) \quad (1)$$

where absorbance of reference is the absorbance without sample (brain tissue).

7.2.8.2 MAO Activity Assay

MAO activity was measured according to the procedure by Turski et al. [41], with slight modification. In brief, the mixture contained 0.025 M phosphate buffer (pH 7.0), 0.0125 M semicarbazide, 10 mM benzylamine, and 75 μL of 100 μL of supernatant. After 30 min, acetic acid was added and boiled for 3 min in water bath followed by centrifugation. The resultant supernatant (1 mL) was mixed with equal volume of 2,4-DNPH, and 1.25 mL of benzene (absolute) was added after 10 min of incubation at room temperature. After separating the benzene layer, equal volume of 0.1N NaOH was added. The alkaline layer was decanted and heated at 80°C for 10 min. The orange–yellow color was developed, and it was measured at 450 nm in a spectrophotometer. The percentage MAO activity was subsequently calculated using Eq. (1).

7.2.9 QUANTIFICATION OF ALKALOID COMPOUNDS BY GAS CHROMATOGRAPHY COUPLED WITH FLAME IONIZATION DETECTOR (GC-FID)

Alkaloid compound quantification was carried out using the modified method by Ngounou et al. [29]. For the GC-FID detection, a

Hewlett-Packard 6890 gas chromatograph (Hewlett-Packard, Palo Alto, CA, USA) was used, along with HP Chem station Rev. A09.01[1206] software equipped with a derivatized, nonpacked injection liner and an Rtx-5MS (5% diphenyl-95% dimethyl polysiloxane) capillary column (30 m length, 0.25 μm film thickness), and detected with an FID. The following conditions were employed: PA separation; injector temperature of 23°C; ramp temperature of 80°C for 5 min and then ramped to 250°C at 30°C/min; detector temperature of 320°C. Standard alkaloids were used.

7.2.9.1 Data Analysis

The results of three replicate experiments were pooled and expressed as mean ± standard deviation (SD). One-way analysis of variance (ANOVA) was used to analyze the mean, and the post-hoc treatment was performed using Duncan multiple test [44]. Significance was accepted at $P < 0.05$.

7.3 RESULTS AND DISCUSSION

The authors conducted this study to evaluate the beneficial effects of Utazi by investigating the inhibitory effect of aqueous polyphenols extracts obtained from Utazi (*G. latifolium*) on some enzymes linked to neurodegeneration in vitro [31]. In continuation of the previous research, an in vivo assay was conducted using an animal model, and alkaloid characterization was carried on the leaf samples to identify the specific bioactive metabolites [13], which were responsible for the observed results.

Eleyinmi [12] has reported the presence of alkaloid in Utazi plant. Figure 7.2 shows the general structure of alkaloid, which had been reported to possess neuroprotective properties by virtue of its interaction with the receptors at the nerve endings, thus making them therapeutic molecules for neurological disorders including AD [27]. Although AD has been considered as a multifactorial disease, oxidative stress has been considered as one of the important factors in the development of this neurodegenerative disease [4]. The animal model was pretreated with the supplemented diet of Utazi before induction of cyclophosphamide-oxidative stress as indicated by results. There was a significant increase in MDA produced and decrease

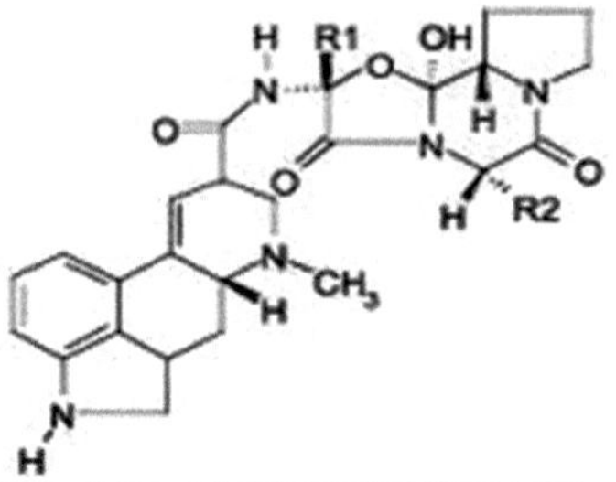

FIGURE 7.2 General structure of alkaloid.

in endogenous antioxidants such as vitamin C, SOD, and *GPx* when compared to the untreated positive control group I. However, the groups supplemented with Utazi had an increased level of antioxidants and reduced level of MDA as a result of presence of different alkaloids (Figures 7.3–7.5).

However, there was a significant difference ($p<0.05$) among all other groups and the negative control group without treatment. The groups that received treatment with Utazi and the positive control group (treatment with donepezil, a control drug for the treatment of mild to moderate AD) reduced the MDA formation, though there was no change among groups with only diet supplemented without induction when compared to the positive control group (Figure 7.6).

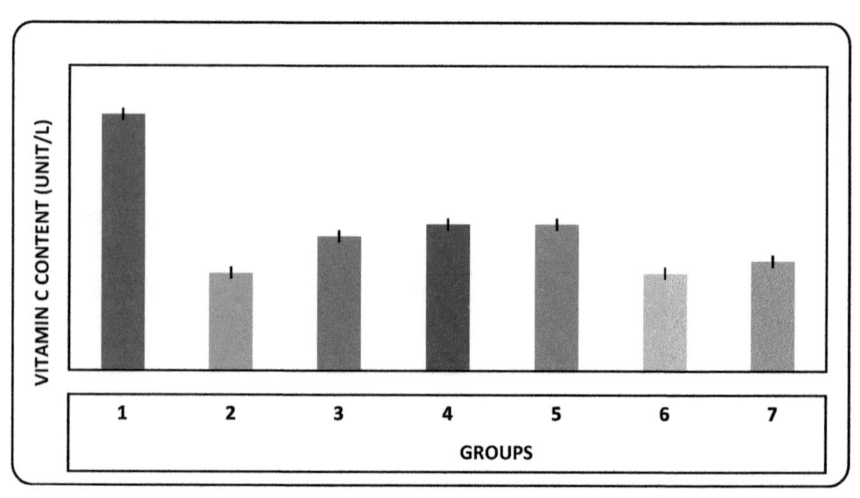

FIGURE 7.3 Effect of Utazi supplemented diet on vitamin C content in cyclophosphamide-induced oxidative stress in rats; values represent mean ± standard deviation ($n = 6$).

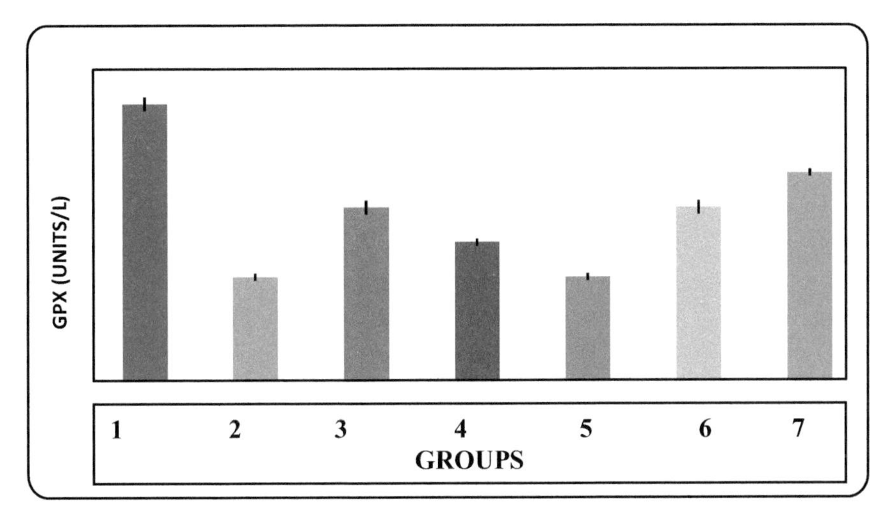

FIGURE 7.4 Effect of Utazi supplemented diet on glutathione peroxidase (GPx) content in cyclophosphamide-induced oxidative stress in rats; values represent mean ± standard deviation ($n = 6$).

This study clearly revealed that there is generation of free radical species, which can be controlled by phenolics and flavonoids in diet during the metabolic process. The restoration after oxidative stress by Utazi might be attributed to the presence of individual activity of

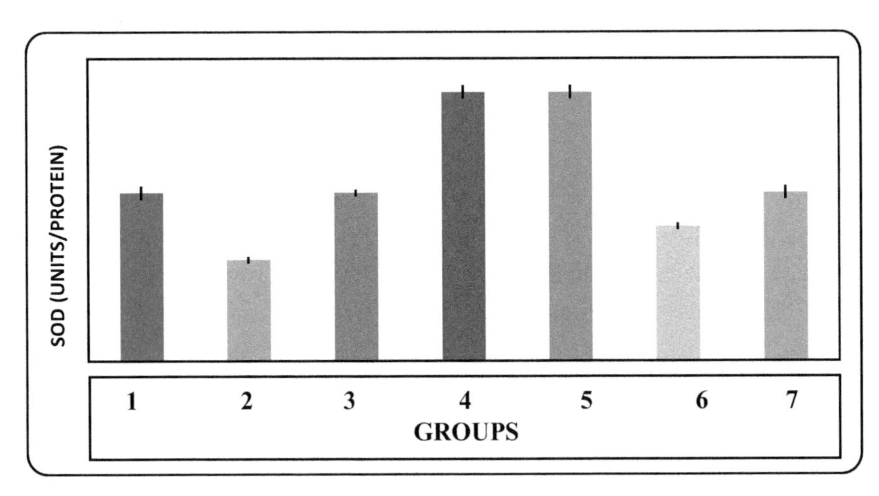

FIGURE 7.5 Effect of Utazi supplemented diet on superoxidase dismutase content in cyclophosphamide-induced oxidative stress in rats; values represent mean ± standard deviation ($n = 6$).

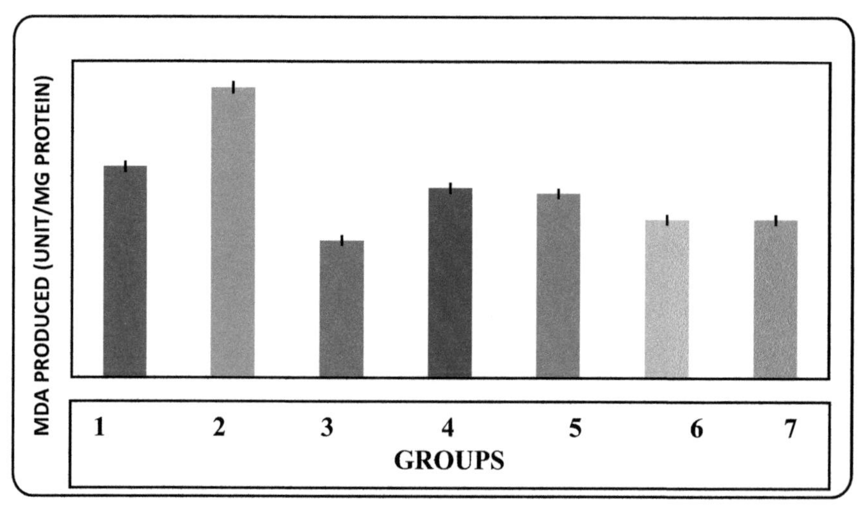

FIGURE 7.6 Effect of Utazi supplemented diet on MDA produced in cyclophosphamide-induced oxidative stress in rat brain; values represent mean ± standard deviation (n = 6).

alkaloid as revealed in the GC-MS analysis of the alkaloid. Table 7.1 shows its chromatogram peaks.

However, it was noted that excess intake of the supplement could deplete the plasma level of endogenous vitamin C (Figure 7.3). Increased activity of brain cholinesterases such as acetylcholinesterase (Ache) and butyryl cholinesterase (BChe) facilitates rapid hydrolysis of acetylcholine neurotransmitter, thus impairing the cholinergic neuronal function [16]. This has been shown to be major risk factor in the development and progression of dementia characteristic of AD [16]. As the loss of cognitive function in AD patients is strongly correlated with the reduction of cholinergic neurotransmission in the brain, rebalancing cholinergic input should theoretically increase memory and cognition in AD patients [8].

Cholinergic transmission is mediated by the neurotransmitter acetylcholine (ACh) through the activation of ionotropic nicotinic and metabotropic muscarinic acetylcholine receptors. AChE inhibitors can enhance cholinergic transmission by limiting degradation of ACh. Dietary sources of cholinesterase inhibitors such as alkaloids from fruits and vegetables could be a complementary approach at prevention/management of these diseases. The study in this chapter showed that there was increase in the activities of AChe and BChe of induced groups with cyclophosphamide-induced

TABLE 7.1 GC-MS Analysis of the Alkaloid Extract profiles of *G. latifolium*

Compounds	RT*	Utazil (mg/100 g)
1,3-Alphadrorhombifoline	11.269	$3.52109x10^{-5}$
6-Hydroxybuphanidrine	20.77	$5.09996x10^{-2}$
9-octadecenamide	12.921	$7.43151x10^{-5}$
Acronycine	20.924	$3.1490x10^{-3}$
Angustifoline	8.769	$7.91583x10^{-3}$
Augustamine	14.7	$7.16001x10^{-5}$
Buphanidrine	17.108	$5.74468x10^{-6}$
Caffeine	7.553	$1.96585x10^{-4}$
Choline	4.569	6.1964
Cinchonidine	16.885	$1.80206x10^{-5}$
Colchicine	28.831	$2.08469x10^{-2}$
Crinamidine	24.222	$5.31160x10^{-3}$
Crinane-3-alpha-ol	16.352	$1.73331x10^{-5}$
Cubebine	8.33	$3.44494x10^{-5}$
Dihydro-oxo-dimethoxyhaemanthamine	14.227	$9.00494x10^{-5}$
Dillapiole	8.445	$2.15890x10^{-5}$
Ellipcine	9.683	$2.17688x10^{-5}$
Emetine	29.618	$1.25372x10^{-3}$
Gingerdione	15.551	$7.96858x10^{-7}$
Lupanine	10.716	$8.88005x10^{-8}$
Monocrotaline	21.413	$4.67849x10^{-3}$
Myristicin	7.932	$2.06088x10^{-6}$
Nitidine	22.17	$4.91347x10^{-3}$
Paclitaxel	32.261	$3.24226x10^{-6}$
Shoqaol	15.132	$1.85755x10^{-6}$
Sparteine	9.298	$1.94664x10^{-5}$
Tetradrine	29.897	$5.24188x10^{-4}$
Thalicarpin	30.021	$1.5086x10^{-3}$
Theobromine	6.833	$9.24477x10^{-5}$
Theophylline	7.062	$1.41809x10^{-3}$
Trigonelline	5.616	$4.71773x10^{-4}$
Total		6.3001

RT* = Retention time (min).

oxidative stress in rats brain tissue when compared with positive control group. However, activities were reduced in the treated groups (Figures 7.7 and 7.8).

The diet was able to reduce the activity better in BChe than in Ache, which is in agreement with previous research [36], where plant extract

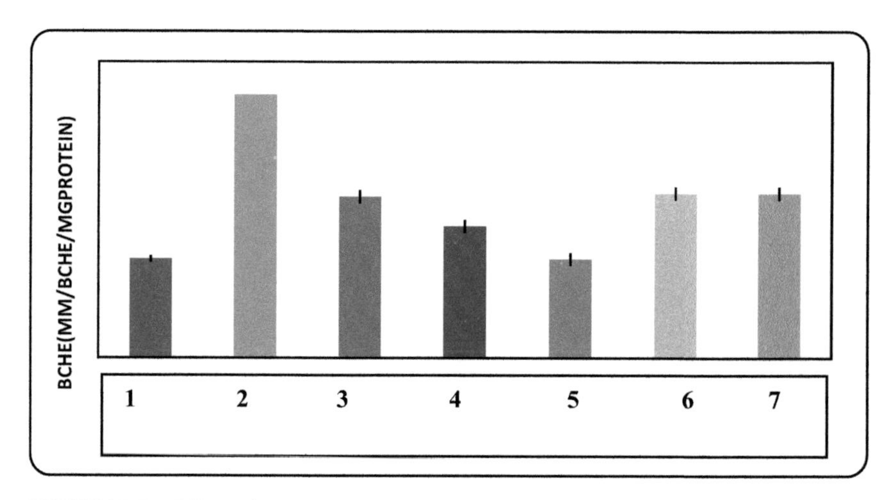

FIGURE 7.7 Effect of Utazi supplemented diet on BChe activity in cyclophosphamide-induced oxidative stress in rat brain; values represent mean ± standard deviation ($n = 6$).

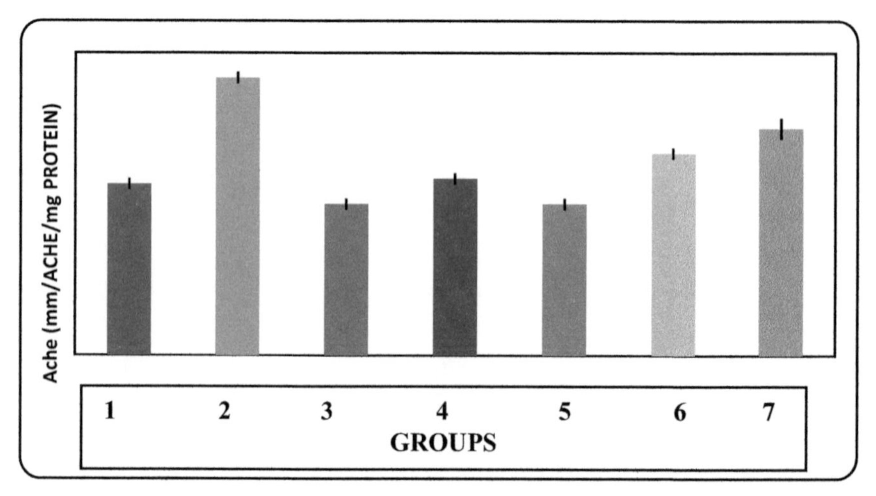

FIGURE 7.8 Effect of Utazi supplemented diet on AChe activity in cyclophosphamide-induced oxidative stress in rat brain; values represent mean ± standard deviation ($n = 6$).

inhibited BChe activity better than AChe activity in vitro. These effects have been largely linked to their constituent phytochemicals present such as choline, which is abundant in the Utazi plant. The amounts of acetylcholine released by physiologically active cholinergic neurons depend on the concentrations of choline available. In the absence of supplemented free choline, the neurons will continue to release constant quantities of the transmitter, especially when stimulated [20]. However, when choline is available (in concentrations bracketing the physiological range of 10–50 μM), a clear dose relationship has been observed between its concentration and acetylcholine release [6, 20]. When no free choline is available, then the source of the choline is used for acetylcholine synthesis in the cells' own membrane [6]. Membranes are very rich in endogenous phosphatidylcholine, and this phospholipid serves as a reservoir of free choline. It has been suggested that a prolonged imbalance between the amounts of free choline available to a cholinergic neuron and the amounts needed for acetylcholine synthesis might alter the dynamics of membrane phospholipids to the point of interfering with normal neuronal functioning ("autocannibalism") [5, 30], e.g., in patients with Alzheimer's disease. In that event, providing the brain with supplemented choline would serve two purposes: (1) It would enhance acetylcholine release from physiologically active neurons; and (2) it would replenish the choline-containing phospholipids in their membranes (Figure 7.9) [43].

MAO activity has shown to be upregulated in the basal ganglia of the brain of patients with neurodegenerative disease. The increase in MAO activity was correlated with the severity of the pathology [38]. Impairment of the monoaminergic neurotransmission through rapid oxidation of monoamine neurotransmitters by MAO has been implicated in the pathogenesis and progression of neurodegenerative diseases such as PD and AD [14]. Therefore, inhibition of MAO activity in brain tissue from this study with supplemented diet of alkaloid-rich vegetable of Utazi offers useful therapeutic strategy to manage these neurodegenerative diseases [19, 33].

Past findings have reported that alkaloids are potent inhibitors of MAO activity. The addictive or synergistic effect of the individual alkaloids could be responsible for the findings reported in this study (Figure 7.10).

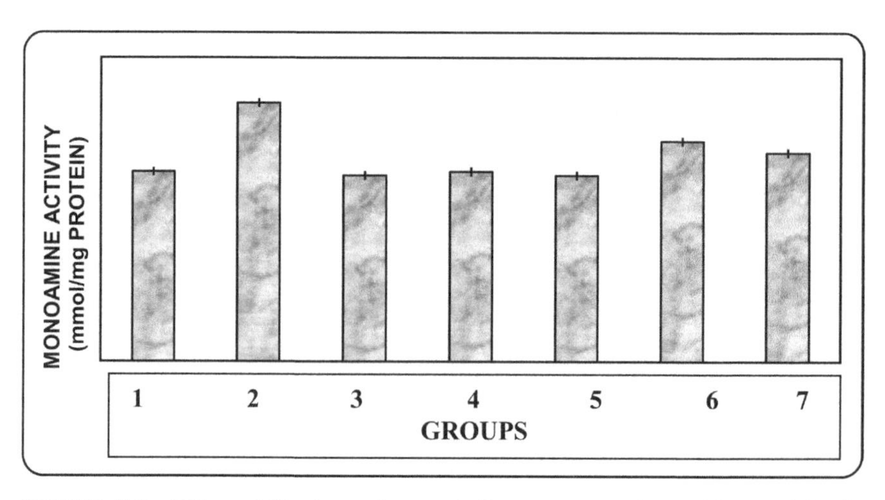

FIGURE 7.9 Effect of Utazi supplemented diet on monoamine oxidase activity in cyclophosphamide--induced oxidative stress in rat brain; values represent mean ± standard deviation (n = 6).

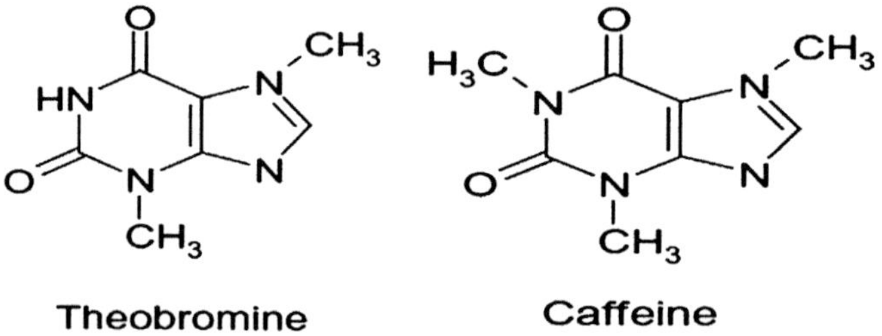

Theobromine

Caffeine

FIGURE 7. 10 Caffeine and structurally related methyl xanthines found in the alkaloid characterization of Utazi.

7.4 CONCLUSIONS

This study was able to reveal the in vivo ability of the vegetable Utazi (*Gongrenema latifolium*), which could modulate the activities of key enzymes such as AChe, BChe, and MAO and to increase the level of endogenous antioxidant relevant to neurodegeneration. This vegetable could offer possible dietary intervention to manage oxidative stress and neuronal pathology.

7.5 SUMMARY

The study in this chapter seeks to determine the neuroprotective ability of the vegetable *Gongronema latifolium* commonly called Utazi by Igbo *Arokeke* by Yoruba in Nigeria. It is a tropical rain forest plant from Ascelepindaceae family. Utazi formulated diet 2% and 4% was used to feed the Wistar male rat for 21 days in different groups and a known control drug Donepezil was used as a positive control. Cyclophosphamide (75 mg/kg wt) was used to induce damage in the brain of the rats for 24 h before termination of the experiment, leading to a neurodegenerative state. Cholinesterases (acetylcholinesterase (AChE) and butyryl cholinesterase (BChE)) and monoamine oxidase (MAO) activities were determined, together with endogenous antioxidant such as SOD, GPX, and Vitamin C. The ability of the diet supplement to scavenge thiobarbituric acid reactive species (TBARS) in brain tissue was assessed. Alkaloid profile was also determined using gas chromatography coupled with flame ionization detector (GC-FID).

The results showed that the diet was able to modulate the effect of cyclophosphamide-induced damage on the enzymes assayed linked to neurodegeneration in the brain when compared to the induced group. However, treatment with 2% inclusion showed the best results, and it was observed that 4% inclusion of the Utazi supplemented diet to normal rat brought about reduction in SOD and Vitamin C activity. Thirty-one different alkaloids were found from the GC-FID with a total of 6.30 mg/100 g, while Choline had the highest concentration of 6.19 mg/100 g. Thus, Utazi apart from serving as a type of vegetable goes beyond nutritional benefits to express useful pharmacological actions.

There are some molecular mechanisms underlying the possible neuroprotective properties of *Gongronema latifolium*. It was observed that choline was the highest alkaloid present, which can be isolated and used for the molecular studies of neuronal pathology. This is a future prospect for this vegetable.

KEYWORDS

- **acetylcholinesterase**
- **alkaloid profile**
- **antioxidants**
- **ascelepindaceae**
- **brain**
- **butyrylcholinesterase**
- **cholinesterases**
- **cyclophosphamide**
- **Donepezil**
- **gas chromatography coupled with flame ionization detector**
- *Gongronema latifolium*
- **monoamine oxidase**
- **Nigeria**
- **SOD**
- **thiobarbituric acid reactive species**
- **tropical rain forest**
- **Utazi**
- **Wistar male rat**

REFERENCES

1. Amos, S., Kolawole, E., Akah, P., Wambebe, C., & Gamaniel, K., (2001). Effects of aqueous extract of *Guiera Senegalenesis* in mice and rats. *Phyto Medicine*, *8*(5), 356–361.
2. Belle, N., Dalmolin, G., & Fonini, G., (2004). Polyamines reduce lipid peroxidation induced by different pro-oxidant agents. *Brain Research*, *1008*, 245–251.
3. Benderitter, M., Maupoil, V., Vergely, C., Dalloz, F., Briot, F., & Rochette, L., (1998). Studies by electron paramagnetic resonance of the importance of iron in hydroxyl scavenging properties of ascorbic acid in plasma: effects of iron chelators. *Fundamental Clinical Pharmacology*, *12*, 510–516.
4. Benzi, G., & Moretti, A., (1995). Are reactive oxygen species involved in Alzheimer's disease? *Neurobiology Aging*, *16*, 661–674.

5. Blusztajn, J. K., & Wurtman, R. J., (1983). Choline and cholinergic neurons. *Science, 221*, 614–620.
6. Blusztajn, J. K., Liscovitch, M., Richardson, U. I., & Wurtman, R. J., (1987). Phosphatidylcholine as a precursor of choline for acetylcholine synthesis. In: *Cellular and Molecular Basis of Cholinergic Function*, Dowdall, M. J., & Hawthorne, J. N., (eds.), Chichester (Sussex), UK: Ellis Harwood, pp. 341–346.
7. Conforti, F., Statti, G. A., Tundis, R., Loizzo, M. R., & Menichini, F., (2007). *In-vitro* activities of *Citrus medica* L., cv. Diamante (diamante citron) relevant to treatment of diabetes and Alzheimer's disease. *Phytotherapy Research, 21*, 427–433.
8. Craig, L. A., Hong, N. S., & Mcdonald, R. J., (2011). Revisiting the cholinergic hypothesis in the development of Alzheimer's disease. *Neuroscience and Biobehavioral Reviews, 35*(6), 1397–1409.
9. Crom, W. R., Glynn-Bamhart, A. M., & Rodman, J. H., (1987). Pharmacokinetics of Anticancer drugs in children. *Clinical Pharmacokinetics, 12*, 179–182.
10. Cyclophosphamide USP Drug information for the health care professional, (2002). 20th Ed. Englewood, Colorado: Micromedex, Inc., vol. *1*.
11. Danjuma, N. M., Zezi, A. U., Yaro, A. H., Musa, A. M., Ahmed, A., Sanni, H. A., & Maje, I. M., (2009). Residual aqueous fraction of stem bark extract of *Xeromphis nilotica* and behavioral effects in mice. *International Journal of Applied Research in Natural Products, 2*(3), 5–12.
12. Eleyinmi, A. F., (2007). Chemical composition and antibacterial activity of *Gongronema latifolium. J. Zhejiang Univ. Sci., 8*(5), 352–358.
13. Halliwell, B., (1992). Reactive oxygen species and central nervous systems. *J. Neurochem., 59*, 1609–1623.
14. Kaluderci, N., Carpi, A., Menabò, R., Di Lisa, F., & Paolocci, N., (2011). Monoamine oxidases (MAO) in the pathogenesis of heart failure and ischemia/ reperfusion injury. *Biochimica et Biophysica Acta (BBA) – Molecular Cell Research, 1813*(7), 1323–1332.
15. Kannappan, R., Gupta, S. C., Kim, J. H., Reuter, S., & Aggarwal, B. B., (2011). Neuroprotection by spice – derived nutraceuticals: You Are What You Eat! *Molecular Neurobiology, 44*(2), 142–159.
16. Klafki, H. W., Staufenbiel, M., Kornhuber, J., & Wiltfang, J., (2006). Therapeutic approaches to Alzheimer's disease. *Brain, 129*(11), 2840–2855.
17. Lawerence, R. A., & Burk, R. F., (1961). Glutathione peroxidase activity in selenium-deficient rat's liver. *Biochemical Biophysic. Research Communication, 71*, 952–958.
18. Lowry, O., H., Rosebrough, N. J., Farr, A. L., & Randall, R. J., (1951). Protein measurement with the Folin phenol reagent. *Journal of Biological Chemistry, 193*, 265–275.
19. Lühr, S., Vilches-Herrera, M., Fierro, A., Ramsay, R. R., Edmondson, D. E., & Reyes-Parada, M., (2010). The 2-Arylthiomorpholine derivatives as potent and selective monoamine oxidase B inhibitors. *Bioorganic Medical Chemistry, 18*(4), 1388–1395.
20. Maire, J. C., & Wurtman, R. J., (1985). Effects of electrical stimulation and choline availability on release and contents of acetylcholine and choline in super-fused slices from rat striatum. *Journal of Physiology (Paris), 80*, 189–195.
21. Marchetti, B., & Abbracchio, M. P., (2005). To be or not to be (inflamed): Is that the question in anti-inflammatory drug therapy of neurodegenerative disorders? *Trends in Pharmacology Sciences, 26*, 517–525.

22. Marksberry, W. R., & Lovell, M. A., (2007). Damage to lipids, proteins, DNA and RNA in mild cognitive impairment. *Archives of Neuroscience, 64*, 954–956.
23. Menken, M., Munsat, T. L., & Toole, J. F., (2000). The global burden of disease study: Implications for neurology. *Archives of Neuroscience, 57*, 418–420.
24. Meorebise, O., Fafunso, M. A., Makinde, J. M., Olajide, O. A., & Awe, E. O., (2002). Anti-inflammatory property of the leaves of *Gongronema latifolium. Phytotherapy Research*, 1675–1677.
25. Misra, H. P., & Fridovich, I., (1972). The generation of superoxide radical during the antioxidant of haemoglobin. *Journal of Biological Chemistry, 247*, 6960–6962.
26. Mukhtar, M. D., & Tukur, A., (2000). Biology of Pistiastratiotes and its toxicity effects in rat. *Journal of Applied Zoology in Environmental Biology, 49*(2), 39–49.
27. Mukherjee, P. K., Kumar, V., Mal, M., & Houghton, P. J. (2007). Acetylcholinesterase inhibitors from plants. *Phytomedicine, 14*, 289–300.
28. Murray, C. J., L., & Lopez, A. D., (1996). *The Global Burden of Disease*. In: Comprehensive assessment of mortality and disability from diseases, injuries, & risk factors in 1990 and projected to 2020, Harvard University Press: Cambridge, USA.
29. Ngounou, F. N., Manfouo, R. N., Tapondjou, L. A., Lontsi, D., Kuete, V., & Penlap, V., (2005). Antimicrobial diterpenoid alkaloids from *Erythrophleum suaveolens* (Guill. & Perr.). *Brenan Bulletin of the Chemical Society of Ethiopia, 19*, 221–226.
30. Nitsch, R. J., K., Blusztajn, A., Pittas, B. E., Slack, J. H., Growdon, D., & Wurtman, R. J., (1992). Evidence for a membrane defect in Alzheimer's disease brain. *Proceedings of the National. Academy of Sciences* (USA), *89*, 1671–1675.
31. Nwanna, E. E., Oyeleye, S. I., Ogunsuyi, O. B., Oboh, G., Boligon, A. A., & Athayde, M. L., (2016). In vitro neuroprotective properties of some commonly consumed green leafy vegetables in Southern Nigeria. *Nutrition and Food Science Journal, 2*, 19–24.
32. Oboh, G., & Ogunruku, O. O., (2010). Cyclophosphamide – induced oxidative stress in brain: Protective effect of hot short pepper (Capsicum frutescens L., var. abbreviatum). *Experimental Toxicology Pathology, 62*, 227–233.
33. Oboh, G. E., Nwanna, S., Oyeleye, I., & Tosin, A., (2016). In vitro neuroprotective potentials of aqueous and methanol extracts from *Heinsia crinita* leaves. *Food Science and Human Wellness, 5*, 95–102.
34. Ogundipe, O. O., Moody, J. O., Akinyemi, T. O., & Raman, A., (2003). Hypoglycemic potentials of methanolic extracts of selected plant foods in alloxanized mice. *Plant Foods Human Nutrition, 58*(3), 1–7.
35. Ohkawa, H., Ohishi, N., & Yagi, K., (1979). Assay for lipid peroxides in animal tissues by thiobarbituric acid reaction. *Analysis Biochemical, 95*, 351–358.
36. Orhan, I., Sener, B., Choudhary, M. I., & Khalid, A., (2004). Acetylcholinesterase and butyrylcholinesterase inhibitory activity of some Turkish medicinal plants. *Journal of Ethnopharmacology, 91*, 57–60.
37. Perry, N. S., Houghton, P. J., Theobald, A., Jenner, P., & Perry, E. K., (2000). In vitro activity of *Slavandulaefolia* Spanish sagedz relevant to treatment of Alzheimer's disease. *Journal of Pharmacology, 52*, 895–902.
38. Richards, G., Messer, J., Waldvogel, H. J., Gibbons, H. M., Dragunow, M., Faull, R. L., & Saura, J., (2011). Up-regulation of the isoenzymes MAO-A and MAO-B in

the human basal ganglia and pons in Huntington's disease revealed by quantitative enzyme radioautography. *Brain Res.*, *1370*, 204–214.

39. Schneider, L. J., (1998). Systematic review of the efficacy of rivastigmines for the patients with Alzheimer's disease. *Int. J. Geriatr. Psychopharmacol.*, *1*, S26.
40. Srinivasan, K., Viswanad, B., Asrat, L., Kaul, C. L., & Ramarao, P., (2005). Combination of high-fat diet-fed and low-dose streptozotocin – treated rat: A model for type-2 diabetes and pharmacological screening. *Pharmacology Research*, *52*, 313–320.
41. Turski, W., Turska, E., & Grossbel, M., (1973). Modification of spectrophotometric method of determination of monoamine-oxidase. *Enzyme*, *14*(4), 211–220.
42. Vladimir-Kneevic, S., Blaekovic, B., Kindl, M., & Vladic, J., (2004). Acetylcholinesterase inhibitory, antioxidant and phytochemical properties of selected medicinal plants of the *Lamiaceae* family. *Molecules*, *19*, 767–782.
43. Wurtman, J. J., Wurtman, R. J., Mark, S., Tsay, R., Gilbert, W., & Growdon, J., (1985). D-fenfluramine selectively suppresses carbohydrate snacking by obese subjects. *International Journal Eating Disorders*, *4*, 89–99.
44. Zar, J. H., (1984). *Biostatistical Analysis*. New Jersey: Prentice-Hall, pp. 620, ISBN-0-13-081542-X.
45. Zhang, J., Tian, Q., & Zhou, S., (2006). Clinical pharmacology of Cyclophophamide and ifosfamide. *Current Drug Therapy*, *1*, 55–84.

PART III

MEDICINAL PLANTS AND MANAGEMENT OF DIABETES MELLITUS

CHAPTER 8

THERAPEUTIC POTENTIALS OF SELECTED MEDICINAL PLANTS IN THE MANAGEMENT OF DIABETES MELLITUS: A REVIEW

FOLORUNSO ADEWALE OLABIYI, YAPO GUILLAUME ABOUA, and OLUWAFEMI OMONIYI OGUNTIBEJU

CONTENTS

8.1 INTRODUCTION

Since human existence on this planet, man has been surrounded by a wide array of plants that serve numerous purposes such as food and source of oxygen that sustain life. The importance of plants in the maintenance of health is increasingly recognized to the extent that the line between food and medicine could not, sometimes, be demarcated. For instance, *Allium sativum* (garlic) is a world-renowned medicinal plant species, but at the same time, it has been used for flavor in the food industry and is also one of the best natural antibiotics known in traditional medicine [121]. Plants play an important and unique role in disease prevention and management.

It has been scientifically demonstrated that eating certain vegetables with antioxidant properties, especially those that belong to the *Brassicaceae* group like cabbage (*Brassica oleracea L*) or certain Liliaceae such as garlic or onion has the capacity to resist the development of some degenerative diseases such as cancer or other diseases of the circulatory system [115]. Lycopene, a flavonoid that is present in tomatoes, has a great capacity to prevent or delay the advancement of cancers [64]. Similarly, apples (*Malus domestica* Borkh) and water melon (*Cucumis melo*), due to their antioxidant composition, guard the digestive system from colorectal cancer [43]. In addition, oranges (*Citrus sinensis*), lemon (*Citrus limonum*), and other citrus fruits are very rich in vitamin C and pectin, and have shown protective ability on the digestive system against diseases such as cancer [79].

Medicinal plants have been used to treat various diseases. In developed countries like United States, a significant number of Americans use medicinal plants to fight against certain diseases. Some years back, an estimated 25% of prescriptions contained plant-derived active ingredients, and the number of visits to providers of traditional medicine exceeded by far the number of visits to all primary care physicians [68, 134]. In Pakistan, it is estimated that about 52% of the population chose complementary and alternative medicine to take care of themselves of illnesses [118]. In Japan, 60–70% of allopathic doctors prescribe traditional medicines for their patients. In China, traditional medicine accounts for about 40% of all healthcare treatments [133, 135]. During 2014–2016, there has been a resurgence of interest in the field of herbal medicine known as phytotherapy. This is presumed to be due to better knowledge of plants, advancement in science and technology, whereby modern processes like high performance liquid chromatography (HPLC), gas chromatography (GC), thin layer chromatography (TLC), atomic absorption spectrophotometry (AAS), X-ray crystallography, and nuclear magnetic resonance spectroscopy (NMR) are now readily available to identify, quantify, and even determine the structure of the active principles in crude plant extracts. It could also be as a result of the increasing side effects of conventional drugs, the treatment failure rates, and high cost of these drugs that make people to seek alternative and/or complementary medicine.

This review focuses on the research on medicinal plants, with a view to elucidate the need for further empirical research, specifically on *Phyllanthus amarus* as the focus plant and provision of in-depth scientific knowledge, especially on its antidiabetic and antioxidant potentials. It will also, in no doubt, stimulate research initiatives into the production of affordable and effective remedy from this plant, for the management of diabetes mellitus (DM) among the low income and middle economic settings and countries in Africa [137].

8.2 DIABETES MELLITUS

DM has been described as a group of metabolic disorders of carbohydrate metabolism in which glucose produced is being underutilized, thereby resulting in excess glucose in the blood stream, a condition called hyperglycemia [22]. It has also been noted that if DM is left untreated, some individuals may experience acute life-threatening hyperglycemic episodes such as ketoacidosis or hyperosmolar coma. As the disease progresses, individuals are prone to come up with specific complications including retinopathy (potentially leading to blindness), renal failure, neuropathy (nerve damage), and atherosclerosis (which may result in stroke, gangrene, or coronary artery disease) [22]. Manifestations of DM include loss of body weight, excessive hunger (polyphagia), excessive thirst (polydipsia), and high frequency of urination (polyuria).

According to Karuna et al. [61], diabetes mellitus is characterized by chronic hyperglycemia, resulting from defects in insulin secretion and/or action, leading to disruptions in carbohydrate, lipid, and protein metabolism. Cardiovascular disease (CVD), a complication of diabetes mellitus, is the leading cause of morbidity and mortality in individuals with diabetes and 65% of deaths are traceable to heart disease or stroke [73]. The risk of myocardial infarction (MI) is high in patients with DM. Suggested possible mechanisms include insulin resistance, changes in endothelial functions, dyslipidemia, chronic inflammation and release of mediators of inflammation, procoagulability, and impaired fibrinolysis.

Growing evidence suggests that complications related to diabetes are associated with oxidative stress, induced by generation of free radicals [78]. The complications arising from chronic, prolonged hyperglycemia, as seen in uncontrolled diabetes mellitus, could cause damage to blood vessels and peripheral nerves, and could greatly increase the risk of heart attack. High blood cholesterol, triglycerides with increased atherogenic index has been implicated as an important risk factor for cardiovascular diseases particularly in diabetic patients [76]. Studies have also shown that patients with DM have increased risk of atherosclerotic vascular disease, and major advances in understanding its pathogenesis have been made [108]. There are suggestions that endothelial injury may be the initial event in the etiology of atherosclerosis, followed by platelet adhesion and aggregation at the site of injury.

It has been reported that in diabetes, evidence of endothelial dysfunction is present. Smooth muscle cell proliferation is an important pathological finding in atherosclerosis. Lipid accumulation in the area of the atherosclerotic lesion is mainly in the form of intracellular and extracellular esterified cholesterol. In uncontrolled diabetes, elevated plasma low-density lipoprotein cholesterol (LDL) levels and decreased plasma high density lipoprotein cholesterol (HDL) levels favor lipid deposition in large vessels. Evidence of a thrombotic state is seen in some patients. Together, these abnormalities of endothelial, platelet, smooth muscle, lipoprotein, and coagulation characteristics may be viewed as contributors to the problem of increased risk of atherosclerosis in diabetes. Furthermore, it is well documented that diabetic nephropathy (DN) is the chief cause of morbidity and premature mortality in patients with insulin–dependent DM [60].

Rao et al. [112] also identified DN to be the leading cause of chronic kidney disease and end-stage renal failure worldwide. They observed that DN patients experienced more oxidative stress than type 2 DM without DN. The authors concluded that elevated malondiadehyde and decreased glutathione peroxidase levels observed in DN are due to oxidative stress, while elevated levels of glycated hemoglobin (HbA1c) and micro-albuminuria seen in DN are due to prolonged hyperglycemia. The elevated total protein levels observed in DN may be due to increased protein catabolism, while the increased urea and creatinine levels seen in diabetic patients are due to renal dysfunctions.

8.2.1 PREVALENCE OF DIABETES MELLITUS

The report released by the World Health Organization (WHO) indicated that the worldwide prevalence of diabetes was estimated to be 2.8% in 2000; this was forecasted to increase to 4.4% by 2030. As such, the total number of people with diabetes was projected to rise from 171 million in 2000 to 366 million in 2030 [113]. Besides, Shaw et al. [119] projected that the number of adults affected from 285 (6.4%) million adults in 2010 would rise to about 439 million (7.7%) adults by 2030. In contrast, the estimated prevalence of diabetes in Africa is 1% in rural areas and ranges from 5% to 7% in urban sub-Saharan Africa [119].

Nigeria, with a population of about 158 million people, is the most populous country in Africa [136], with more than 1.56 million cases of diabetes in Nigeria in 2015 [51]. South Africa with a population of 53 million persons reported a total number of cases of adults (20–79 years) with diabetes at 2,286,000 in 2015. The highest prevalence of DM in South Africa is among the Indian population (11–13%) as this group is assumed to have strong genetic tendencies for diabetes. This is followed by 8–10% among the colored community, 5–8% among blacks, and 4% among the whites [25].

8.2.2 PATHOGENESIS OF DM

Most type 1 DM results from a cellular–mediated autoimmune destruction of the insulin-secreting cells of pancreatic β-cell [63], and this in turn, results in absolute insulin deficiency [36]. In most patients, the destruction is mediated by T-cells. Type 1 DM accounts for about 5–10% of all newly diagnosed DM cases [14]. On the other hand, insulin resistance and β-cell dysfunction are pathological defects in patients with type 2 diabetes. Insulin resistance, being a reduced ability of the insulin to act on the peripheral tissue, prevents glucose uptake. This is thought to be the main underlying pathological process. β-cell dysfunction is an inability of the pancreas to produce sufficient insulin to make up for the insulin resistance. It remains the most common form of DM and constitutes about 90–95% of all diabetic cases [14].

It is not clear whether type 2 diabetes is mainly due to a defect in β-cell secretion, peripheral resistance to insulin, or both. However, there are data to support the concept that insulin resistance is the main defect that comes before the derangement in insulin secretion and clinical diabetes. Although there is lack of agreement, it is clear that type 2 diabetes is an extremely heterogeneous disease and that no single cause is enough to explain the progression from normal glucose tolerance to diabetes [21].

8.2.3 DM VERSUS OXIDATIVE STRESS

Oxidative stress occurs when the production of reactive oxygen species (ROS) overpowers the antioxidant defense mechanisms, thereby leading to cellular damage [126]. With a view to enhance the understanding of the pathologic process, the role of oxidative stress in diabetic complication has been reported [91]. Furthermore, hyperglycemia has been reported to induce oxidative stress among diabetic subjects. During this process, ROS are produced by oxidative phosphorylation. More so, DM is known to increase the risk of developing cardiovascular disease nearly five-folds [22]. Consequently, hyperglycemia and insulin resistance are therefore noted to trigger oxidative stress in the diabetic myocardium that cannot adapt to ischemia reperfusion [15].

Table 8.1 summarizes the common African medicinal plants scientifically investigated for their antidiabetic activities in animal model and humans according to geographical regions.

8.3 MATERIALS AND METHODS

Google Scholar and PubMed databases were searched in order to obtain materials for this review on the therapeutic potential of selected medicinal plants for the management of diabetes mellitus between January 2000 and

TABLE 8.1 Common African Medicinal Plants Scientifically Investigated for Their Antidiabetic Activities in Animal Model and Human [Adapted from ref. 84.]

Scientific name	Common name	Parts of plants studied	References
West Africa			
Allium cepa	Onion	Bulb	[38, 106]
Allium sativum	Galic	Bulb	[38, 99]
Anacardium occidental	Cashew	Leaf	[39, 129]
Azadirachta indica	Neem	Leaf	[47, 67]
Carum carvi	Caraway	Fruit	[34, 35]
Gongronema latifolium	Amaranth globe, Utazi	Root/stem	[2, 92, 104]
Hibiscus sabdariffa L.	Red sorrel	Calyces	[3, 122]
Indigofera pulchra L.	Indigofera	Leaf	[125]
Moringa oleifera	Horseradish	Leaf	[27]
Nauclea latifolia	Bishop's head	Root/stem/ Leaf	[41, 138]
Occimum gratissimum	African/clove basil	Leaf	[28, 85, 102]
Parkia biglobosa jacq	African locust bean	Seed	[94]
Phyllanthus amarus	Stone breaker/gulf leaf flower	Root/stem/Leaf	[4, 48, 105,109]
Picralima nitida	Picralima	Pulp/seed	[9, 53]
Telfairia occidentalis Hook f.	Fluted pumpkin	Seed	[101]
Vernonia amygdalina	Bitter leaf	Leaf	[12, 18, 82]
Zingiber officinale	Ginger	Rhizome	[17, 85, 107]
North Africa			
Ajuga iva L.	Herb ivy	Whole plant	[46, 50]
Allium cepa	Onion	Bulb	[33, 124]
Balanites aegyptiaca	Desert date /Hegleg	Fruits	[33, 116]
Carum carvi	Caraway	Fruit/oil	[24, 26, 72]
Charmaemelum nobile	Charmomile	Aerial parts	[71]
Morus alba	White mulberry	Leaf/Root/bark	[31, 32]
Nigella sativa	Black seed	Seed	[20, 83]

TABLE 8.1 (Continued)

Scientific name	Common name	Parts of plants studied	References
Ziziphus spira Christi	Christ's Thorn Jujube	Leaf	[45, 81]
Southern Africa			
Artemisia afra jacq	African wormwood	Leaf	[7, 8]
Bryophyllum pinnatum Lam.	Good luck/ life plant	Leaf	[96]
Catharanthus roseus L.G Don	Madagascar periwinkle	Leaf	[98]
Hypoxis hemerocallidea	African potato	Corm	[75, 97]
Raphia gentiliana De wild	-	Fruit	[86]
Sclerocarya birrea A. Rich Hochst	Jelly plum	Stem/bark	[40, 88]
Sutherlandia frutescens R.	Cancer brush	Leaf	[23, 74]
Central Africa			
Bersama agleriana	Winged bersama	Leaf	[93, 132]
Dichrostechys glomerata chiov	Chinese lantern	Seed	[69]
Dracena arborea Wild	Dragon tree	Root	[131]
Kalanchoe crenata Andr Haw	Never ride	Whole plant	[58]
East Africa			
Aspilla pluriseta schweinf	Dwarf aspilia	Root	[90]
Catha edulis vahl	Bushman's tea	Root	[90]
Moringa stenopetala Baker F.	Cabbage tree	Leaf	[89]

March 2016. The following key words were used in the search: therapeutic potential, medicinal plants, diabetes mellitus, *Phyllanthus amarus*, antioxidant, and inflammation.

8.4 SELECTED AFRICAN MEDICINAL PLANTS WITH ANTIDIABETIC POTENTIALS

Several investigators have reported the antidiabetic activities of medicinal plants of African origin. Some of these have been highlighted and are presented in this section [62, 95].

8.4.1 AJUGA IVA L.SCHREIBER (MEDIT): LAMIACEAE

It has been shown that repeated oral administration of the water extract of *Ajuga iva* L. at a dose of 10 mg/kg resulted in a significant decrease in plasma glucose levels in normal rats 6 hours after administration and after 3 weeks of treatment. This reduction in glucose level continued to normality, which makes the investigators to conclude that *A. iva* (Figure 8.1) possesses a strong hypoglycemic effect in diabetic rats and supports its traditional use in DM control [49].

8.4.2 ALLIUM CEPA L. (ONION): LILIACEAE

Studies on different ether soluble portions as well as insoluble parts of *Allium cepa* powder show antihyperglycemic activity in diabetic rabbits. Onion (Figure 8.2) is also known to have antioxidant and hypolipidemic activity [62]. Administration of sulfur-containing amino acid, S-methyl cysteine sulfoxide (SMCS) @ 200 mg/kg for 45 days to alloxan-induced diabetic rats significantly reduced blood glucose as well as lipids in serum and tissues. The onion plant restores the activities of liver hexokinase, glucose 6-phosphatase, and HMG-coA reductase to their normal levels [70, 114]. Besides, when diabetic patients were administered with a single oral dose of 50 g of onion juice, it significantly controlled post-prandial glucose levels [77].

FIGURE 8.1 *Ajuga iva.*

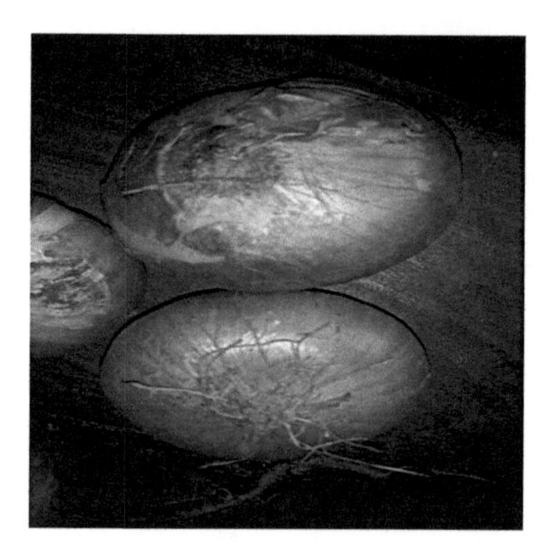

FIGURE 8.2 Onion bulb.

8.4.3 ALLIUM SATIVUM L. (GARLIC): LILACEAE

Garlic (Figure 8.3) is an all year-round herb that is widely grown in Africa and India [111]. A study showed that oral administration of garlic extract significantly lowered serum glucose, total cholesterol, triglycerides, urea, uric acid, creatinine, AST, and ALT levels, while it raised serum insulin levels in diabetic rats but not in normal rats. Further, the extract was found to be more potent than the standard, oral antidiabetic drug, glibenclamide. This made the plant a superior candidate of choice for future research on DM [29].

FIGURE 8.3 Garlic.

8.4.4 ALOE BARBADENSIS MILL.: LILIACEAE

Chronic treatment of exudates of *Aloe barbadensis* leaf showed the hypoglycemic effect in alloxanized diabetic rats. Single as well as chronic doses of the bitter active ingredient of the same plant also showed hypoglycemic effect in diabetic rats. This action is through stimulation of synthesis and/or release of insulin from pancreatic beta cells [10].

8.4.5 ALOE VERA (L) BURM.: ASPHODELACEAE

Aloe vera (Figure 8.4) gel at 200 mg/kg possesses significant anti-diabetic, cardioprotective activity, reduces the increased TBARS, maintains the superoxide dismutase and catalase activity up to normal levels, and increases reduced glutathione by four times in diabetic rats [55]. The leaf pulp extract showed hypoglycemic activity in insulin-dependent DM (IDDM) and noninsulin-dependent DM (NIDDM) rats, the effectiveness being enhanced for type 2 diabetes in comparison with glibenclamide [100].

FIGURE 8.4 Aloe vera plant.

8.4.6 *ARTEMISIA HERBA-ALBA ASSO (MED): ASTERACEAE*

Oral administration of an aqueous extract (0.39 g/kg) of the leaves or barks of *Artemisia herba-alba* (Figure 8.5) plant significantly reduced blood glucose level. However, the aqueous extract of roots and the methanolic extract of the aerial parts of the plant produced almost no significant reduction in blood glucose level. The extract of the aerial parts of the plant seems to have minimal adverse effect and high LD_{50} value [57]. In another study, it was reported that 15 patients with DM were treated with *Artemisia herba-alba* extract (AHE). Results showed that AHE caused considerable lowering of elevated blood sugar, and 14 out of 15 patients had good remission of diabetic symptoms with the use of AHE. It is concluded that AHE contains material capable of reducing raised blood sugar in DM [13].

8.4.7 *BRYONIA ALBA L.: CUCURBITACEAE*

Administration of trihydroxy actadecadienoic acids obtained from the roots of the native Armenian plant B. alba L. (@ 0.05 mg/kg/day for 15 days Lin.) restores the disordered lipid metabolism of alloxan-induced diabetic rats. Metabolic changes induced in diabetes significantly restores to their normal values with the exception of diminished triglyceride content of muscle, which was not restored. Thus, the extract can influence the profile of the formation of stable prostaglandin by actions downstream of prostaglandin endoperoxides [59].

FIGURE 8.5 *Artemisia herba alba.*

8.4.8 *CARUM CARVI L. (CC): APIACEAE AND CAPPARIS SPINOSA (CS): CAPPARIDACEAE*

After a single dose or 14 daily doses, oral administration of the aqueous CC and CS extract (20 mg/kg) produced a significant decrease in blood glucose levels in STZ-induced diabetic rats and reduced it to nearly normal after 2 weeks. However, there was no significant increase in blood glucose level or basal plasma insulin concentration in normal rats after both acute and chronic treatments with CS and CC [26].

8.4.9 *CORIANDRUM SATIVUM L.: APIACEAE*

Administration of coriander seed extract (200 mg/kg) significantly increased the activity of the beta cells in comparison with the diabetic control rats; decreased serum glucose in streptozotocin-induced diabetic rats; and released insulin from the beta cells of the pancreas [30]. The extract shows antihyperglycemic, insulin-releasing and insulin-like activity [42].

8.4.10 *CYAMOPSIS TETRAGONOLOBA (L) TAUBERT.: PAPILIONACEAE*

The aqueous extract of beans at 250 mg/kg of body weight significantly reduced blood glucose levels in alloxan-induced diabetic rats within 3 hours of administration. Continuation for 10 days produced significant reduction in the blood glucose level with marginal activity in normal and glucose-loaded rats [87].

8.4.11 *DIOSCOREA DUMETORUM (KUNTH) PAX.: DIOSCOREACEAE*

At a dose of 20 mg/kg, the fasting blood glucose in normoglycemic rabbits was reduced from 112 to 55 mg/dL after 4 hours. In alloxan-induced diabetic rabbits, the blood glucose was lowered from 52 to 286 mg/dL after the same time interval. The aqueous fraction of the methanol extract produced comparable effects at 100 mg/kg. The chloroform fraction elevated the fasting blood glucose of normal rabbits to 196 mg/100 mL after 6 hours. The hypoglycemic effects are comparable to those of tolbultamide [54].

8.4.12 *GARCINIA KOLA* HECKEL: CLUSIACEAE

Kolaviron (Figure 8.6), a mixture of c-3/c-8 linked bioflavonoids obtained from *Garcinia kola*, produces significant hypoglycemic effects. It has been reported that fractions obtained from kolaviron reduced blood glucose levels in streptozotocin-induced diabetic (STZ-diabetic) rats within 4 hours of oral administration and showed positive effects on the plasma lipid profile of diabetic animals [1]. At a dose of 100 mg/kg, the fasting blood glucose in normoglycemic rabbits reduced from 115 to 65 mg/dL after 4 hours. In alloxan-induced diabetic rabbits, the blood glucose was lowered from 506 to 285 mg/dL after 12 hours. It has also been documented that kolaviron (100 mg/kg) treatment significantly ameliorated hyperglycemia and liver dysfunction. In this study, it was noted that serum levels of hepatic marker enzymes were significantly reduced in kolaviron-treated diabetic rats. Besides, kolaviron prevented diabetes-induced increase in hepatic levels of pro-inflammatory cytokines, interleukin (IL)-1 beta, IL-6, tumor necrosis factor (TNF-α), and monocyte chemotactic protein (MCP-1) [19].

FIGURE 8.6 Kolaviron.

8.4.13 GONGRONEMA LATIFOLIUM END L.: ASCLEPIADACEAE

The aqueous extract of *G. latifolium* significantly increased the activities of hepatic hexokinase and decreased the activities of glucokinase but did not produce any change in the hepatic glycogen and glucose content of diabetic rats [127]. In this study, the effect of oral administration of aqueous and ethanolic extracts is shown to increase the activity of superoxide dismutase and the level of reduced glutathione. The aqueous extract further increases the activity of glutathione reductase, while the ethanolic extract caused a significant increase in the activity of glutathione peroxidase and glucose-6-phosphate dehydrogenase and a significant decrease in lipid peroxidation. These results suggest that the extract from *G. latifolium* leaves could exert their antidiabetic activities through their antioxidant properties [128].

8.4.14 HYPOXIS HEMEROCALLIDEA CONN. CORM (AFRICAN POTATO): HYPOXIDACEAE

At a dose of 800 mg/kg, the plant extract of *Hypoxis hemerocallidea* (Figure 8.7) has been reported to cause 30.20% and 48.54% reductions in the blood glucose concentration of fasted normal and STZ-induced diabetic rats, respectively, thus indicating hypoglycemic activity [75].

FIGURE 8.7 *Hypoxis hemerocallidea.*

8.4.15 *TELFARIA OCCIDENTALIS HOOK: CUCURBITACEAE*

The aqueous extract of this plant given orally in 1 g/kg to mice 60 mins before glucose administration reduced the blood glucose level from day two when compared with that of chlorpromide (200 mg/kg) under the same conditions. The results indicated that the aqueous extract of the leaves of *T. occidentalis* possess hypoglycemic activity [5].

8.4.16 *MORINGA OLEIFERA*

Moringa oleifera (Figure 8.8) is a popular food plant with much medicinal usefulness that includes the treatment of diabetes [80]. Various parts of the plant have shown to have antidiabetic potential [16]. In severely diabetic animals, 200 mg/kg aqueous leaf extract of *M. oleifera* reduced fasting blood glucose level by 69.2% after 3 weeks of treatment and also significantly reduced urine glucose [56]. The progression of diabetes was significantly reduced in STZ-induced diabetic rats treated with methanolic extract of *M. oleifera* pods [44].

8.4.17 *VERNONIA AMYGDALINA*

Alcoholic extract of *Vernonia amygdalina* (Figure 8.9) has been reported to significantly improve glucose tolerance in STZ-induced diabetic rats,

FIGURE 8.8 *Moringa oleifera* plant.

FIGURE 8.9 Bitter leaf.

to decrease fasting blood glucose, and to show a protective effect on pancreatic beta cells, thereby causing a slight increase in insulin level in STZ-induced diabetic rats [103]. The same authors also reported that *V. amygdalina* increased the expression of GLUT-4 in rat skeletal muscle and its translocation to the plasma membrane. Further, some investigators have reported that the extract of *Vernonia amygdalina* improved biochemical and hematological parameters in diabetic rats. Combination of extract with metformin at various ratios showed that the ratio of 1:2 (extract: metformin) caused the most significant ($P<0.05$) reduction in blood glucose (66.07%) compared to that in control [6, 11].

8.4.18 *PHYLLANTHUS AMARUS SCHUM AND THONN*

Phyllanthus amarus (PA) is a traditional Ayurveda herb used in Southern India but also commonly found in the Philippines, Cuba and southwestern Nigeria. This plant has been used in traditional medicine for more than 3,000 years [4, 123]. The herb belongs to the Euphorbiaceae family and has been used in folk medicine to treat quite a number of diseases. It is commonly called "carry me," "stone breaker," "wind breaker," "gulf leaf flower," or "gala of wind." This plant (Figure 8.10) possesses antidiabetic, antioxidant, and anti-inflammatory properties.

In recent time, there is a renewed interest in this medicinal plant, perhaps due to better knowledge of its potential benefits and advancement in science and technology. Increase in the side effects of conventional drugs and treatment failure rates that led to high cost of healthcare could also be another factor.

FIGURE 8.10 Freshly plucked leaves of *Phyllanthus amarus.*

8.4.18.1 Traditional Uses of *Phyllanthus amarus* schum and thonn

PA herb has been found to be traditionally useful in several health problems such as diarrhea, dysentery, dropsy, jaundice, intermittent fevers, urinogenital disorders (such as kidney problems, urinary bladder disturbances, gonorrhea), skin disorders (like wounds, scabies, sores, ulcers, itching, edema, ringworms, and tubercular ulcers, scabby), and crusty lesions [123]. Its effect in the excretory system is found to be due to its antiurolithic property and is used in the treatment of kidney or gallstone, other kidney-related problems, appendicitis, and prostate problems [65, 117, 130]. Because of its efficacy in the treatment of gastrointestinal disorders, it is used in the treatment of disorders such as dyspepsia, colic, diarrhea, constipation, and dysentery. The herb has been used in several gynecological conditions such as leucorrhoea, menorrhagia, and mammary abscess. In southwestern Nigeria, water decoction of the leaf and seed of PA is reputably used for the local management of DM, obesity, and hyperlipidemia [4]. It has also been reported that oral administration of ethanolic leaf extract (400 mg/kg) for 45 days resulted in a significant ($P<0.05$) decline in blood glucose and significant recovery in body weight of diabetic mice [20].

8.4.18.2 Antidiabetic and Antioxidant Activities of PA

A study shows that oral antihyperglycemic action of 150–600 mg/kg/day of ethanolic extract of PA were evaluated in normal and 10% sucrose-induced insulin resistance using indicators such as fasting blood glucose, insulin, and insulin resistance indices [4]. In this study, body weight, serum lipid profile, and atherogenic indices were also measured. The results showed that PA could effectively control glucose level in DM mediated via improvement in insulin resistance, thus supporting its ethno-medical use in the local management of DM.

Another study investigated PA in an experimental model, where fasted rats were induced diabetes by a single intraperitoneal injection of 120 mg/kg of alloxan monohydrate, and two doses of the aqueous and hydro-alcoholic extract of PA were then administered orally and were compared to the normal control group that received distilled water only. After 15 days treatment, the results demonstrated that the aqueous and hydro-alcoholic extract of PA decreased the blood glucose level significantly. Serum analysis of the PA-treated experimental animals showed an increase in insulin and a reduction in the malondialdehyde concentration, therefore demonstrating the potential anti-diabetic property of the aqueous and hydro-alcoholic extract of PA [37].

Kiran et al. [66] showed that the methanolic extract of PA inhibited lipid peroxidation and scavenged hydroxyl and superoxide radicals. Because free radicals are linked with diabetes, scavenging of free radicals could be one of the mechanisms of action [110]. Further experimental studies are, however, needed in order to isolate chemical constituents and their mechanisms of action. In a study that evaluated PA and phyllanthin, the DPPH free radical scavenging activity was seen to be concentration dependent and reaches maximum at a concentration of 20 mol/mL for phyllanthin and 300 g/mL for PA extract [123]. This finding suggests a strong antioxidant potential of PA.

8.5 FUTURE PROSPECT AND RESEARCH INITIATIVES

There is a general belief that because medicinal plants are "natural," they are safe and have no side effects and as such they are innocuous. In the opinion of authors of this chapter, the removal of extraneous substances (from the plants which might have been added, thereby causing their adulteration), right medicinal preparations and proper understanding of plants and drug interactions could reduce potential adverse reactions. Further, proper and extensive double-blind clinical trials are needed to determine safety and efficacy of each plant before they can be recommended for medical use. It has been observed that in sub-Sahara Africa, there has been very scanty empirical data on the effect of PA on male reproduction, inflammation, and apoptosis. Therefore, future studies will fill some gaps in the knowledge on male reproduction and on gene expression of diabetic rats that are administered with PA by studying the apoptotic pathway where expression of caspase-8, BAD, BAX, and Bcl-X_L as well as GLUTs would be assessed using western blot. Apart from the serum levels of inflammatory biomarkers such as MCP-1, vascular endothelial growth factor (VEGF) and interleukins (IL-1, IL-2, IL-6, and IL-8) could be evaluated. The following proteins could also be assessed using western blot: inducible nitrogen synthase (iNOS) and tumor necrotic factor-R1 (TNF-R1).

8.6 SUMMARY

Many synthetic, hypoglycemic agents or drugs have been developed for the treatment and management of various diseases, including diabetes mellitus. However, despite the advent of these various drugs, there have been reported cases of side effects, limited action, and increased treatment failure rates, which obviously could lead to high cost of treatment and loss of valuable manpower hours. In the light of these limitations, there is a need for cheaper but effective remedies that could either serve as an alternative and/or complimentary medication for the treatment and management of these diseases, especially in the upper middle, lower-middle, and low-income economies into which most sub-Sahara African countries come under.

Extracts from *Phyllanthus amarus* schum and thonn have been reported to possess antimicrobial, antidiabetic, anti-inflammatory, antioxidant, and other beneficial health properties. This review is therefore aimed at examining the research on this plant, with a view to elucidate the need for further empirical research and provision of in-depth scientific knowledge on its antidiabetic, anti-inflammatory, and antioxidant potentials. It is our opinion that it will stimulate research initiatives into the production of affordable and effective remedy for the management of diabetes mellitus.

KEYWORDS

- **alternative/complementary**
- **antidiabetic**
- **anti-inflammatory**
- **antioxidant**
- **apoptosis**
- **diabetes mellitus**
- **diseases**
- **empirical data**
- **euphorbiaceae**
- **free radicals**
- **Google scholar**
- **health**
- **herb**
- **hypoglycemic agent**
- **inflammation**
- **lower-middle economies**
- **low-middle economies**
- **male reproduction**
- **manpower hours**
- **medication**

- **medicinal plants**
- **oxidative stress**
- **pathogenesis**
- ***Phyllanthus amarus***
- **plant extract**
- **prevalence**
- **Pubmed**
- **remedies**
- **Sub-Saharan Africa**
- **therapeutic potential**
- **treatment**

REFERENCES

1. Adaramoye, O. A., & Adeyemi, E. O., (2006). Hypoglycemic and hypolipidemic effects of fractions from kolaviron, a biflavonoid complex from *Garcinia kola* in streptozotocin-induced diabetes mellitus rats. *Journal of Pharmacy and Pharmacology, 58*(1), 121–128.
2. Adebajo, A. C., Ayoola, M. D., Odediran, S. A., Aladesanmi, A. J., Schmidt, T. J., & Verspohl, E. J., (2013). Evaluation of ethnomedical claim III: anti-hyperglycemic activities of *Gongronema latifolium* root and stem. *Journal of Diabetes, 5*, 336–343.
3. Adedayo, O. A., & Ganiyu, O., (2013). Aqueous extracts of roselle (*Hibiscus sabdariffa* Linn.) varieties inhibit α-amylase and α-glucosidase activities in vitro. *Journal of Medicinal Food, 16*, 88–93.
4. Adeneye, A. A., (2012). The leaf and seed aqueous extract of *Phyllanthus amarus* improves insulin resistance diabetes in experimental animal studies. *Journal of Ethnopharmacology, 144*(3), 705–711.
5. Aderibigbe, A. O., Lawal, B. A., & Oluwagbemi, J. O., (1999). The Antihyperglycemic effect of *Telfaria occidentalis* in mice. *African Journal of Medicine and Medical Sciences, 28*, 171–175.
6. Adikwu, M. U., Uzuegbu, D. B., Okoye, T. C., Uzor, P. F., Adibe, M. O., & Amadi, B. V., (2010). Antidiabetic effect of combined aqueous leaf extract of *Vernonia amygdalina* and metformin in rats. *Journal of Basic Clinical Pharmacy, 1*(3), 197–202.
7. Afolayan, A. J., & Sunmonu, T. O., (2011). *Artemisia afra* Jacq. ameliorates oxidative stress in the pancreas of streptozotocin-induced diabetic Wistar rats. *Bioscience, Biotechnology & Biochemistry, 75*, 2083–2086.

8. Afolayan, A. J., & Sunmonu, T. O., (2013). Protective role of *Artemisia afra* aqueous extract on tissue antioxidant defense system in streptozotocin-induced diabetic rats. *African Journal of Traditional and Complementary Alternative Medicine, 10*, 15–20.
9. Aguwa, C. N., Ukwe, C. V., Inya-Agha, S. I., & Okonta, J. M., (2001). Antidiabetic effect of *Picralima nitida* aqueous seed extract in experimental rabbit model. *Journal of Natural Remedy, 1*, 135–139.
10. Ajabnoor, M. A., (1990). Effect of Aloes on blood glucose levels in normal and alloxan diabetic mice. *Journal of Ethnopharmacology, 28*, 215–220.
11. Akah, P. A., Alemji, J. A., Salawu, O. A., Okoye, T. C., & Offiah, N. V., (2009). Effects of *Vernonia amygdalina* on Biochemical and Haematological Parameters in Diabetic Rats. *Asian Journal of Medical Science, 1*(3), 108–113.
12. Akpaso, M. I., Atangwho, I. J., Akpantah, A., Fischer, V. A., Igiri, A. O., & Ebong, P. E., (2011). Effect of combined leaf extracts of *Vernonia amygdalina* (Bitter leaf) and *Gongronema latifolium* (Utazi) on the pancreatic *β*-Cells of streptozotocin induced diabetic rats. *British Journal of Medicine and Medical Research, 1*, 24–34.
13. Al-Waili, N. S., (1986). Treatment of diabetes mellitus by Artemisia herba-alba extract: preliminary study. *Clinical Experimental Pharmacology and Physiology, 13*(7), 569–573, PubMed PMID: 3791709.
14. American Diabetes Association (ADA), (2011). Diagnosis and Classification of diabetes mellitus. *Diabetes Care, 34*(1), 62–69.
15. Ansley, D. M., & Wang, B., (2013). Oxidative stress and myocardial injury in the diabetic heart. *Journal of Pathology*, *229*, 232–241.
16. Anwar, F., Latif, S., Ashraf, M., & Gilani, A. H., (2007). *Moringa oleifera*: a food plant with multiple medicinal uses. *Phytotherapy Research, 21*(1), 17–25.
17. Arikawe, A. P., Daramola, A. O., Olatunji-Bello, I. I., & Obika, L. F., (2013). Insulin, pioglitazone and *Zingiber officinale* administrations improve proliferating cell nuclear antigen immunostaining effects on diabetic and insulin resistant rat testis. *Journal of Experimental and Clinical Medicine*, *30*, 49–55.
18. Atangwho, I. J., Ebong, P. E., Eyong, E. U., Ashmawi, M. Z., & Ahmad, M., (2012). Synergistic antidiabetic activity of *Vernonia amygdalina* and *Azadirachta indica*: biochemical effects and possible mechanism. *Journal of Ethnopharmacology, 141*, 878–887.
19. Ayepola, O. R., Chegou, N. N., Brooks, N. L., & Oguntibeju, O. O., (2013). Kolaviron, a Garcinia biflavonoid complex ameliorates hyperglycemia-mediated hepatic injury in rats via suppression of inflammatory responses. *BMC Complementary and Alternative Medicine, 13*, 363.
20. Benhaddou-Andaloussi, A., Martineau, L., Vuong, T., Meddah, B., Madiraju, P., Settaf, A., & Haddad, P. S., (2011). The in vivo antidiabetic activity of *Nigella sativa* is mediated through activation of the AMPK pathway and increased muscle Glut4 content. *Evidence Based Complementary and Alternative Medicine*, DOI: 10.1155/2011/538671.
21. Brown, J. R., Brown, J. R., & Edwards, F. H., (2006). The diabetic disadvantage: historical outcomes measures in diabetic patients undergoing cardiac surgery – the pre-intravenous insulin era. *Semin Thoracic and Cardiovascular Surgery, 18*, 281–288.

22. Burtis, C. A., Ashwood, E. R., & Bruns, D. E., (2008). Carbohydrates. In: *Tietz Fundamentals of Clinical Chemistry*, Elsevier, New Delhi, vol. *22*, pp. 380.
23. Chadwick, W. A., Roux, S., Van de Venter, M., Louw, J., & Oelofsen, W., (2007). Anti-diabetic effects of *Sutherlandia frutescens* in Wistar rats fed a diabetogenic diet. *Journal of Ethnopharmacology, 109*, 121–127.
24. Dawidar, A. M., Abdel-Mogib, M., Abou-Elzahab, M. M., Berghot, M. A., Mahfouz, M., El-Ghorab, A. H., & Hussien, K. H., (2010). Effect of photo-oxygenation on biological activities of some commercial Egyptian essential oils. *Rev. Latinoam. Quim.*, *38*, 168–179.
25. Distiller, L., (2014). Prevalence of diabetes in South Africa. *Health24 Diabetes*. www.health24.com/Medical/Diabetes/about-diabetes/Diabetes-tsunami-hits-South-Africa-20130210. Accessed on October 08, 2016.
26. Eddouks, M., Lemhadri, A., & Michel, J. B., (2004). Caraway and caper: potential anti-hyperglycaemic plants in diabetic rats. *Journal of Ethnopharmacology*, *94*, 143–148.
27. Edoga, C. O., Njoku, O. O., Amadi, E. N., & Okeke, J. J., (2013). Blood sugar lowering effect of *Moringa oleifera* Lam in albino rats. *International Journal of Science and Technology, 3*, 88–90.
28. Egesie, U. G., Adelaiye, A. B., Ibu, J. O., & Egesie, O. J., (2006). Safety and hypoglycaemic properties of aqueous leaf extract of *Ocimum gratissimum* in streptozotocin induced diabetic rats. *Nigerian Journal Physiological Science, 21*, 31–35.
29. Eidi, A., Eidi, M., & Esmaeili, E., (2006). Antidiabetic effect of garlic *(Allium sativum L.)* in normal and streptozotocin – induced diabetic rats. *Phytomedicine, 13*(9–10), 624–629.
30. Eidi, M., Eidi, A., Saeidi, A., Molanaei, S., Sadeghipour, A., Bahar, M., & Bahar, K., (2009). Effect of coriander seed *(Coriander sativum L.)* ethanol extract on insulin-release from pancreatic beta cells in streptozotocin diabetic rats. *Phytotherapy*, *23*, 404–406.
31. El-Beshbishy, H. A., Singab, A. N., Sinkkonen, J., & Pihlaja, K., (2006). Hypolipidemic and antioxidant effects of *Morus alba* L. (Egyptian mulberry) root bark fractions supplementation in cholesterol-fed rats. *Life Sciences*, *78*, 2724–2733.
32. El-Sayyad, H. I. H., El-Sherbiny, M. A., Sobh, M. A., Abou-El-Naga, A. M., Ibrahim, M. A. N., & Mousa, S. A., (2011). Protective effects of *Morus alba* leaves extract on ocular functions of pups from diabetic and hypercholesterolemic mother rats. *International Journal of Biological Sciences*, *7*, 715–728.
33. El-Soud, N. A., & Khalil, M., (2010). Antioxidative effects of *Allium cepa* essential oil in streptozotocin induced diabetic rats. *Macedonia Journal of Medical Science, 3*, 344–351.
34. Ene, A. C., Bukbuk, D. N., & Ogunmola, O. O., (2006). The effect of different doses of black caraway (*Carum carvi* L.) oil on the serum creatinine of alloxan-induced diabetic rats. *Journal of Medical Science, 6*, 701–703.
35. Ene, A. C., Nwankwo, E. A., & Samdi, L. M., (2007). Alloxan-induced diabetic rats and the effects of black caraway (*Carum carvi* L.) oil on their body weight. *Research Journal of Medicine and Medical Science, 2*, 48–52.

36. Erejuwa, O. O., (2012). Oxidative stress in Diabetes mellitus: Is There a Role for Hypoglycemic Drugs and/or Antioxidants? In: Volodymyr Lushchak (Ed.). *Oxidative Stress and Diseases*. IntechOpen.com., pp. 217–246, ISBN: 978-953-51-0552-7.
37. Evi, P. L., & Degbeku, K., (2011). Antidiabetic activity of *Phyllanthus amarus* Schum and Thonn on Alloxan induced diabetes in male Wistar rats. *Journal of Applied Science, 11*(16), 2968–2973.
38. Eyo, J. E., Ozougwo, J. C., & Echi, P. C., (2011). Hypoglycaemic effects of *Allium cepa, Allium sativum* and *Zingiber officinalle* aqueous extracts on alloxan-induced diabetic *Rattus novergicus*. *Medical Journal of Islamic World Academic Science, 19*, 121–126.
39. Fagbohun, T. R., & Odufunwa, K. T., (2010). Hypoglycemic effect of methanolic extract of *Anacardium occidentale* leaves on alloxan-induced diabetic rats. *Nigerian Journal of Physiological Science, 25*, 87–90.
40. Gondwe, M., Kamadyaapa, D. R., Tufts, M., Chuturgoon, A. A., & Musabayane, C. T., (2008). *Sclerocarya birrea* [(A., Rich.) Hochst.] [Anacardiaceae] stem-bark ethanolic extract (SBE) modulates blood glucose, glomerular filtration rate (GFR) and mean arterial blood pressure (MAP) of STZ-induced diabetic rats. *Phytomedicine, 15*, 699–709.
41. Grace, S. E., Bob, I. A. M., Godwin, O. I., Justin, I. A., Eyong, U. E., & Patrick, E. E., (2013). Antioxidant enzymes activity and hormonal changes following administration of ethanolic leaves extracts of *Nauclea latifolia* and *Gongronema latifolium* in streptozotocin induced-diabetic rats. *European Journal of Medicinal Plants, 3*, 297–309.
42. Gray, A. M., & Flatt, P. R., (1999). Insulin–releasing and insulin-like activity of the traditional anti-diabetic plant *Coriandrum sativum* (coriander). *British Journal of Nutrition. 81*, 203–209.
43. Guaâdaoui, A., Bouhtit, F., Cherfi, M., & Hamal, A., (2015). The preventive approach of biocompounactives (2): A review in recent advances in common fruits. Recent advances in common fruits*: International Journal of Nutrition and Food Sciences, 4*(2), 189–207.
44. Gupta, R., Rajaj, V., & Marthur, M., (2012). Evaluation of antidiabetic and antioxidant activity of *Moringa oleifera* in experimental diabetes. *Journal of Diabetes*, *4*(2), 64–71.
45. Hala, M. H., Eman, M. E., & Aataa, A. S., (2006). Antihyperglycemic, antihyperlipidemic and antioxidant effects of *Ziziphus spina-christi* and *Ziziphus jujube* on alloxan induced diabetic rats. *International Journal of Pharmacology, 2*, 563–570.
46. Hamden, K., Ayadi, F., Jamoussi, K., Masmuodi, H., & Elfeki, A., (2008). Therapeutic effect of phytoecdysteroids rich extract from *Ajuga iva* on alloxan induced diabetic rats liver, kidney and pancreas. *Biofactors, 33*, 165–175.
47. Henry, D. A., Itemobong, S. E., Itoro, F. U., Patrick, E. E., & Isong, N. B., (2012). Effect of aqueous extract of *Azadirachta indica* (Neem) leaves on some indices of pancreatic function on alloxan-induced diabetic Wister rats. *Pharmacologia, 3*, 420–425.
48. Herbert, O. C., M., Clement, J., Idongesit, J., Godwin, E., Udeme, E., & Grace, E., (2011). Evaluation of the hypoglycemic effect of aqueous extract of *Phyllanthus amarus* in alloxan-induced diabetic albino rats. *International Journal of Pharmacy and Biomedical Research, 2*, 158–160.

49. Hilaly, J. E., & Lyoussi, B., (2002). Hypoglycaemic effect of the lyophilized aqueous extract of *Ajuga iva* in normal and streptozotocin diabetic rats. *Journal of Ethnopharmacology, 80*, 109–113.
50. Hilaly, J. E., Tahraoui, A., Israili, Z. H., & Lyoussi, B., (2007). Acute hypoglycemic, hypocholesterolemic and hypotriglyceridemic effects of continuous intravenous infusion of a lyophilised aqueous extract of *Ajuga iva* L., Schreber whole plant in streptozotocin-induced diabetic rats. *Pakistan Journal of Pharmaceutical Science, 20*, 261–268.
51. IDF, (2015). Diabetes in Nigeria. *www.idf.org/membership/afr/nigeria* Accessed on October 08, 2016.
52. IDF, (2015). Diabetes in South Africa. *www.idf.org/membership/afr/southafrica* Accessed on Sept. 01, 2016.
53. Igboasoiyi, A. C., Essien, E. E., & Eseyin, O. A., (2007). Ubam, G., Screening of the seed of *Picralima nitida* for hypoglycaemic activity. *Pakistan Journal of Biological Sciences, 10*, 828–830.
54. Iwu, M. M., Okunji, C. O., Ohaeri, G. O., Akah, P., Corley, D., & Tempesta, M. S., (1990). Hypoglycaemic activity of dioscoretine from tubers of *Dioscorea dumetorum* in normal and alloxan – induced diabetic rabbits. *Planta Medica, 56*, 264–267.
55. Jain, N., Vijayaraghavan, Pant, S. C., Lomash, V., & Ali, M., (2010). *Aloe vera* gel alleviates cardiotoxicity in streptozotocin–induced diabetic rats. *Journal of Pharmacy and Pharmacology, 62*, 115–123.
56. Jaiswal, D., Kumar, R. P., Kumar, A., Mehta, S., & Watal, G., (2009). Effects of *M., oleifera Lam.* leaves aqueous extract therapy on hyperglycaemic rats. *Journal of Ethnopharmacology, 123*(3), 392–396.
57. Kahzraji, S. M., Shamaony, L. A., & Twaij, H. A., (1993). Hypoglycaemic effect of *Artemisia herba alba*: Effect of different parts and influence of the solvent on hypoglycaemic activity. *Journal of Ethnopharmacology, 40*, 163–166.
58. Kamgang, R., Mboumi, Y. R., Fondjo, A. F., Tagne, M. A. F., Mengue N'dille, G. P., R., & Yonkeu, J. N., (2008). Antihyperglycaemic potential of the water-extract of *Kalanchoe crenata* (Crassulaceae). *Journal of Natural Medicine, 62*, 34–40.
59. Karageuzyan, K. G., Vartinanya, G. S., Agaadjanov, M. L., Panossian, A. G., & Hoult, J. R., (1998). Restoration of the disordered glucose-fatty acid cycle in alloxan induced diabetic rat by trihydroxyoctadecadienoic acids from *Bryonia alba*, a native Armenian medicinal plant. *Planta Medica, 64*, 417–422.
60. Karnib, H. H., & Ziyadeh, F. N., (2010). The cardiorenal syndrome in diabetes mellitus. *Diabetes research and clinical practice, 89*(3), 201–208.
61. Karuna, R., Bharathi, V. G., Reddy, S. S., Ramesh, B., & Saralakumari, D., (2011). Protective effects of *Phyllanthus amarus* aqueous extract against renal oxidative stress in steptozotocin – induced diabetic rats. *Indian Journal of Pharmacology, 43*(4), 414–418.
62. Kavishankar, G. B., Lakshmidevi, N., Mahadeva Murthy, S., Prakash, H. S., & Niranjana, S. R., (2011). Diabetes and medicinal plants- A review. *International journal of Pharmacy and Biomedical Science, 2*(3), 65–80.

63. Kawasaki, E., Abiru, N., & Eguchi, K., (2004). Prevention of type 1 diabetes: from the view point of β cell damage. *Diabetes Research and Clinical Practice, 66*(31), S27–S32.
64. Khan, N., Afaq, F., & Mukhtar, H., (2008). Cancer chemoprevention through dietary antioxidants: progress and promise. *Antioxidants & Redox Signaling.*, *10*(3), 475–510.
65. Khatoon, S., Rai, V., & Rawat, A., (2004). Comparative pharmacognostic studies of three *phyllanthus* species. *Journal of Ethnopharmacology*, *104*, 79–86.
66. Kiran, D., Rohilla, A., & Rohilla, S., (2011). *Phyllanthus amarus:* an ample therapeutic potential herb. *International Journal of Research in Ayurveda and Pharmacy*, *2*(4), 1099.
67. Koffour, G. A., Amoatengor, P., Okai, C. A., & Fiagbe, N. I. Y., (2011). Hypoglycemic effects of whole and fractionated *Azadirachta indica* (Neem) seed oil on alloxan-induced diabetes in New Zealand white rabbits. *Journal of Ghana Science Association, 13*, 34–39.
68. Kong, J. M., Goh, N. K., Chia, L. S., & Chia, T. F., (2003). Recent advances in traditional plant drugs and orchids. *Acta Pharmacologica Sinica*, *24*(1), 7–21.
69. Kuate, D., Etoundi, B. C., Ngondi, J. L., & Oben, E., (2011). Effects of *Dichrostachys glomerata* spice on cardiovascular diseases risk factors in normoglycemic and type 2 diabetic obese volunteers. *Food Research International, 44*, 1197–1202.
70. Kumari, K., Mathew, B. C., & Augusti, K. T., (1995). Anti-diabetic and hypolipidaemic effects of S-methyl cysteinesulfoxide isolated from *Allium cepa Linn. Indian Journal of Biochemistry and Biophysics*, *32*, 49–54.
71. Lemhadri, A., Burcelin, L., Sulpice, T., & Eddouks, M., (2007). *Chamaemelum nobile* L., aqueous extract represses endogenous glucose production and improves insulin sensitivity in streptozotocin-induced diabetic mice. *American Journal of Pharmacology and Toxicology*, *2*, 116–122.
72. Lemhadri, A., Eddouks, M., & Michel, J. B., (2006). Cholesterol and triglycerides lowering activities of caraway fruits in normal and streptozotocin diabetic rats. *Journal of Ethnopharmacology*, *106*, 321–326.
73. Lopez-Hernández, M. A., (2013). Hyperglycemia and diabetes in myocardial infarction. In: Oluwafemi O., Oguntibeju (ed.), *Diabetes Mellitus – Insights and perspectives*, Rijeka, *In. Tech. com.*, pp. 169–192.
74. MacKenzie, J., Koekemoer, T., Van de Venter, M., Dealtry, G., & Roux, S., (2009). *Sutherlandia frutescens* limits the development of insulin resistance by decreasing plasma free fatty acid levels. *Phytotherapy Research*, *23*, 1609–1614.
75. Mahomed, I. M., & Ojewole, J. A., (2003). Hypoglycaemic effect of *Hypoxis hemerocallidea conn.* (African potato) aqueous extract in rats. *Methods and Findings in Experimental Clinical. Pharmacology*, *25*, 617–623.
76. Martín-Timón, I., Sevillano-Collantes, C., Segura-Galindo, A., & Del Cañizo-Gómez, F. J., (2014). Type 2 diabetes and cardiovascular disease: Have all risk factors the same strength. *World Journal of Diabetes*, *5*(4), 444–70.
77. Mathew, P. T., & Augusti, K. T., (1975). Hypoglycaemic effect of onion, *Allium cepa Linn* on diabetes mellitus – A preliminary report. *Indian Journal of Physiology and Pharmacology, 19*, 213–217.

78. Matough, F. A., Budin, S. B., Hamid, Z. A., Alwahaibi, N., & Mohamed, J., (2012). The role of oxidative stress and antioxidants in diabetic complications. *Sultan Qaboos University Medical Journal, 12*(1), 5–18.
79. Mazed, M. A., Mazed, S., & Mazed Mohammad, A., (2009). Nutritional supplement for the prevention of cardiovascular disease, alzheimer's disease, diabetes, & regulation and reduction of blood sugar and insulin resistance. U.S. *Patent Application,* 12/390, 302.
80. Mbikay, M., (2012). Therapeutic potential of Moringa oleifera leaves in chronic hyperglycemia and dyslipidemia: a review. *Frontiers in Pharmacology, 3,* 24.
81. Michel, C. G., Nesseem, D. I., & Ismail, M. F., (2011). Anti-diabetic activity and stability study of the formulated leaf extract of *Ziziphus spina-christi* (L.) wild with the influence of seasonal variation. *Journal of Ethnopharmacology, 133,* 53–62.
82. Modu, S. A., Adeboye, E., Maisaratu, A., & Mubi, B. M., (2013). Studies on the administration of *Vernonia amygdalina* Del. (Bitter leaf) and glucophage on blood glucose level of alloxan – induced diabetic rats. *International Journal of Medicinal Plant and Alternative Medicine, 1,* 13–19.
83. Mohamed, A. M., EL-Sharkawy, F. Z., Ahmed, S. A. A., Aziz, W. M., & Badary, O. A., (2009). Glycemic control and therapeutic effect of *Nigella sativa* and *Curcuma longa* on rats with streptozotocin-induced diabetic hepatopathy. *Journal of Pharmacology and Toxicology, 4,* 45–57.
84. Mohammed, A., Ibrahim, M. A., & Islam, M. S., (2014). African medicinal plants with antidiabetic potentials: a review. *Planta Medica, 80*(5), 354–377.
85. Morakinyo, A. O., Akindele, A. J., & Ahmed, Z., (2011). Modulation of antioxidant enzymes and inflammatory cytokines: possible mechanism of anti-diabetic effect of ginger extracts. *African Journal of Biomedical Research, 14,* 195–202.
86. Mpiana, P. T., Masunda, T. A., Longoma, B. F., Tshibangu, D. S. T., & Ngbolua, K. N., (2013). Anti-hyperglycemic activity of *Raphia gentiliana* de wild (Arecaceae). *European Journal of Medicinal Plants, 3,* 233–240.
87. Mukhtar, H. M., Ansari, S. H., Ali, M., Bhat, Z. A., & Naved, T., (2004). Effect of aqueous extract of *Cyamopsistetragonoloba Linn.* beans on blood glucose level in normal and alloxan -induced diabetic rats. *Indian Journal of Experimental Biology, 42,* 1212–1215.
88. Musabayane, C. T., Gondwe, M., Kamadyaapa, D. R., Moodley, K., & Ojewole, J. A. O., (2006). The effects of *Sclerocarya birrea* [(a. rich.) hochst.] (Anacardiaceae) stem-bark aqueous extract on blood glucose, kidney and cardiovascular function in rats. *Endocrinology Abstracts, 12,* 36.
89. Nardos, A., Makonnen, E., & Debella, A., (2011). Effects of crude extracts and fractions of *Moringa stenopetala* (Baker f.) Cufodontis leaves in normoglycemic and alloxan-induced diabetic mice. *African Journal of Pharmacy and Pharmacology, 5,* 2220–2225.
90. Ngugi, M. P., Murugi, N. J., Kibiti, M. C., Ngeranwa, J. J., Njue, M. W., Maina, D., Gathumbi, K. P., & Njagi, N. E., (2011). Hypoglycemic activity of some Kenyan plants traditionally used to manage diabetes mellitus in Eastern province. *Journal of Diabetes and Metabolism, 2,* 1–6.

91. Niedowics, D. M., & Daleke, D. L., (2005). The role of oxidative stress diabetic complications. *Cellular Biochemistry and Biophysics, 43*(2), 289–330.
92. Nwanjo, H. U., Okafor, M. C., & Oze, G. O., (2006). Anti-lipid peroxidative activity of *Gongronema latifolium* in streptozotocin-induced diabetic rats. *Nigerian Journal of Physiological Sciences, 21*, 61–65.
93. Nyah, N. G., Watcho, P., Nguelefack, T., & Kamanyi, A., (2007). Hypoglycaemic activity of the leaves extracts of *Bersama engleriana* in rats. *African Journal of Traditional, Complementary and Alternative Medicine, 2*, 215–221.
94. Odetola, A. A., Akinloye, O., Egunjobi, C., Adekunle, W. A., & Ayoola, A. O., (2006). Possible antidiabetic and antihyperlipidaemic effect of fermented *Parkia biglobosa* (Jacq) extract in alloxan-induced diabetic rats. *Clinical and Experimental Pharmacology and Physiology, 33*, 808–812.
95. Oguntibeju, O. O., (2015). Medicinal plants: Potentials for the management and treatment of diabetes mellitus. *Inaugural Professorial Address at the Cape Peninsula University of Technology,* Bellville, South Africa, pp. 31–32.
96. Ojewale, J. A. O., (2005). Antinociceptive, anti-inflammatory and antidiabetic properties of *Bryophyllum pinnatum* (Crassulaceae) leaf aqueous extract. *Journal of Ethnopharmacology, 99*, 13–19.
97. Ojewale, J. A. O., (2006). Antinociceptive, anti-inflammatory and antidiabetic properties of *Hypoxis hemerocallidea* Fisch. & C. A., Mey. (Hypoxidaceae) corm (African potato) aqueous extract in mice and rats. *Journal of Ethnopharmacology, 103*, 126–134.
98. Ojewole, J. A. O., & Adewunmi, C. O., (2000). Hypoglycaemic effects of methanolic leaf extract of *Catharanthus roseus* (Linn.) G., Don (Apocynaceae) in normal and diabetic mice. *Acta Medical Biology, 48*, 55–58.
99. Ojo, R. J., Memudu, A. E., Akintayo, C. O., & Akpan, I. S., (2012). Preventive effect of *Allium sativum* on alloxan induced diabetic rat. *ARPN Journal of Agriculture and Biological Sciences, 7*, 609–612.
100. Okyar A., Can, A., Akev, N., Baktir, G., & Sutlupinar, N., (2001). Effect of *Aloe vera* leaves on blood glucose level in type 1 and type 2 diabetic rat models. *Phytotherapy Research, 15*, 157–161.
101. Olorunfemi, A. E., Patrick, E., Arit, E., Arnold, I., & Emmanuel, O., (2007). Hypoglycemic effect of seed extract of *Telfairia occidentalis* in rats. *Pakistan Journal of Biological Science, 10*, 498–501.
102. Onaolapo, A. Y., Onaolapo, O. J., & Adewole, S. A., (2011). Ethanolic extract of *Ocimum grattissimum* leaves (Linn.) rapidly lowers blood glucose levels in diabetic Wistar rats. *Macedonia Journal of Medical Science, 4*, 351–357.
103. Ong, K. W., Hsu, A., Song, L., Huang, D., & Tan, B. K., (2011). Polyphenol – rich *Vernonia amygdalina* shows anti-diabetic effect on streptozotocin – induced diabetic rats. *Journal of Ethnopharmacology, 133*(2), 598–607.
104. Orok, U. E., Eneji, E. G., Luke, O. U., Eyo, R. A., Sampson, E. V., Iwara, I. A., & Oko, O. M., (2012). Effect of ethanolic root and twig extracts of *Gongronema latifolium* (utazi) on kidney function of streptozotocin induced hyperglycemic and normal Wistar rats. *Journal of Medicine and Medical Sciences, 3*, 291–296.
105. Owolabi, O. A., James, D. B., Anigo, K. M., Iormanger, G. W., & Olaiya, I. I., (2011). Combined effect of aqueous extracts of *Phyllanthus amarus* and *Vitex do-*

niana stem bark on blood glucose of streptozotocin (STZ) induced diabetes rats and some liver biochemical parameters. *British Journal of Pharmacolology and Toxicology, 2*, 143–147.

106. Ozougwu, J. C., (2011). Anti-diabetic effects of *Allium cepa* (onion) aqueous extracts on alloxan-induced diabetic *Rattus novergicus*. *Journal of Medicinal Plants Research, 5*, 1134–1139.
107. Ozougwu, J. C., & Eyo, J. E., (2011). Evaluation of the activity of *Zingiber officinale* (ginger) aqueous extracts on alloxan-induced diabetic rats. *Pharmacology, 1*, 258–269.
108. Paneni, F., Beckman, J. A., Creager, M. A., & Cosentino, F., (2013). Diabetes and vascular disease: pathophysiology, clinical consequences, & medical therapy, part I., *European Heart Journal, 2*, 149.
109. Povi, L., Kwashie, E., Amegnona, A., Kodjo, A., Edmond, C., & Messanvi, G., (2011). Antidiabetic activity of *Phyllanthus amarus* Schum and Thonn (Euphorbiaceae) on alloxan induced diabetes in male Wistar rats. *Journal of Applied Science, 11*, 2968–2973.
110. Rahimi, R., Nikfar, S., Larijani, B., & Abdollahi, M., (2005). A review on the role of antioxidants in the management of diabetes and its complications. *Biomedicine & Pharmacotherapy, 59*(7), 365–373.
111. Rani, A., & Kumar, A., (2014). *Allium sativum*: A global natural herb with medical properties. *Journal Academica, 4*(1), 33–37.
112. Rao, P. P., Sajutha, C., Latha, N. M., Kumar, J. P., & Deepa, M., (2015). Association between microalbuminuria and oxidative stress in diabetic nephropathy. *International Journal of Clinical and Biomedical Research, 1*(2), 96–98.
113. Roglic, G., Unwin, N., Benneth, P. H., Mathers, C., Tuomilehto, J., Nag, S., Cannolly, V., & King, H., (2005). The burden of mortality attributable diabetes: realistic estimates for the year 2000. *Diabetes Care, 28*(9), 2130–2135.
114. Roman-Ramos, Flores-Saenz, J. L., & Alarco-Aguilar, F. J., (1995). Antihypoglycaemic effect of some edible plants. *Journal of Ethnopharmacology, 48*, 25–32.
115. Rupp, R., (2011). How carrots won the trojan war. In: *Curious (but True) Stories of Common Vegetables*, Deborah B. (Ed.) Storey Publishing, ISBN: 978-1-60342-968-9.
116. Samir, A. M. Z., Somaia, Z. A. R., & Mattar, A. F., (2003). Anti-diabetic properties of water and ethanolic extracts of *Balanites aegyptiaca* fruits flesh in senile diabetic rats. *Egyptian Journal of Hospital and Medicine, 10*, 90–108.
117. Sen, A., & Batra, A., (2013). The study of *in-vitro* and *in-vivo* antioxidant activity and total phenolic content of *Phyllanthus amarus* Schum and Thonn: A medicinally important plant. *International Journal of Pharmacy and Pharmaceutical Science, 5*, 947.
118. Shaikh, S. H., Malik, F., James, H., & Abdul, H., (2009). Trends in the use of complementary and alternative medicine in Pakistan: a population-based survey. *The Journal of Alternative and Complementary Medicine, 15*(5), 545–550.
119. Shaw, J. E., Sicree, R. A., & Zimmet, P. Z., (2010). Global estimates of the prevalence of diabetes for 2010 and 2030. *Diabetes Research in Clinical Practice., 87*(1), 4–14.

120. Shetty, A. A., Sanakal, R. D., & Kaliwal, B. B., (2012). Antidiabetic effect of ethanolic leaf extract of *Phyllanthus amarus* in alloxan induced diabetic mice. *Asian Journal of Plant Science Research.*, *2*(1), 11–15.
121. Simonetti, G., (1990). *Simon & Schuster's Guide to Herbs & Spices*. New York: Fireside, pp. 9.
122. Sini, J. M., Umar, I. A., & Inuwa, H. M., (2011). The beneficial effect of extract of *Hibiscus sabdariffa* calyces in alloxan-diabetic rats: Reduction of free radical load and enhancement of antioxidant status. *Journal of Pharmacognosy and Phytotherapy, 3*, 141–149.
123. Sonia, V., Sharma, H., & Garg, M., (2014). *Phyllanthus amarus*: A review. *Journal of Pharmacognosy and Phytochemistry, 3*(2), 18–22.
124. Taj Eldin, I. M., Ahmed, E. M., & Abd Elwahab, H. M., (2010). Preliminary study of the clinical hypoglycemic effects of *Allium cepa*(Red Onion) in type 1 and type 2 diabetic patients. *Environmental Health Insights, 4*, 71–77.
125. Tanko, Y., Abdelaziz, M. M., Adelaiye, A. B., Fatihu, M. Y., & Musa, K. Y., (2009). Effects of ethyl acetate portion of *Indigofera pulchra* leaves extract on blood glucose levels of alloxan-induced diabetic and normoglycemic Wistar rats. *Asian Journal of Medical Science, 1*, 10–14.
126. Tremellen, K., (2008). Oxidative stress and male infertility: a clinical perspective. *Human Reproduction Update*, *14*(3), 243–258.
127. Ugochukwu, N. H., & Babady, N. E., (2003). Antihyperglycaemic effect of aqueous and ethanolic extracts of *Gongronema latifolium* leaves on glucose and glycogen metabolism on livers of normal and streptozotocin–induced diabetic rats. *Life Sciences, 29*(73), 1925–1938.
128. Ugochukwu, N. H., & Babady, N. E., (2002). Antioxidant effects of *Gongronema latifolium*in hepatocytes of rat models of non-insulin-dependent diabetic mellitus. *Fitoterapia*, *73*, 612–618.
129. Ukwenya, V. O., Ashaolu, J. O., Adeyemi, D. O., Akinola, O. B., & Caxton-Martins, E. A., (2012). Antihyperglycemic activities of methanolic leaf extract of *Anacardium occidentale* (Linn.) on the pancreas of streptozotocin-induced diabetic rats. *Journal of Cell and Animal Biology*, *6*, 207–212.
130. Ushie, O., Neji, P., & Etim, E., (2013). Phytochemical screening and antimicrobial activities of *Phyllanthus amarus* stem bark extracts. *International Journal of Modern Biology and Medicine*, *3*, 101–112.
131. Wankeu-Nya, M., Watcho, P., Florea, A., Balici, S., Matei, H., & Kamanyi, A., (2013). *Dracena arborea* alleviates ultra-structural spermatogenic alterations in streptozotocin-induced diabetic rats. *BMC Complementary and Alternative Medicine*, *13*, 71.
132. Watcho, P., Anchountsa, J. H. G., Mbiakop, C. U., Nguelefack, T. B., Kamanyi, A., & Wankeu-Nya, M., (2012). Hypoglycemic and hypolipidemic effects of *Bersama engleriana* leaves in nicotinamide/streptozotocin-induced type 2 diabetic rats. *BMC Complementary and Alternative Medicine*, *12*, 264.
133. WHO, (1999). Consultation meeting on traditional medicine and modern medicine: harmonizing the two approaches. Geneva, World Health Organization, (document reference (WP)TM/ICP/TM/001/RB/98- RS/99/GE/32(CHN)).

134. WHO, (1999). Traditional, complementary and alternative medicines and therapies. Washington DC, WHO Regional office for the americas/pan american health organization (Working group OPS/OMS).
135. WHO, (2002). WHO traditional medicine strategy 2002–2005. World Health Organization, Geneva.
136. World Population Prospects. Population division of the department of economic and social affairs of the united nations secretariat. Available from: http //esa.un.org/wpp/documentation/pdf/WPP2012_Volume-II-Demographic-Profiles.pdf.
137. World Bank, (2016). http://data.worldbank.org/about/countries-and-lending-groups#Low_income.
138. Yessoufou, A., Gbenou, J., Grissa, O., Hitchami, A., Simonin, A., Tabka, Z., Moudachirou, M., Mountairou, K., & Khan, N., (2013). Anti-hyperglycemic effects of three medicinal plants in diabetic pregnancy: modulation of T-cell proliferation. *BMC Complementary and Alternative Medicine, 13*, 77.

CHAPTER 9

MEDICINAL ACTIVITIES OF *ANCHOMANES DIFFORMIS* AND ITS POTENTIAL IN THE TREATMENT OF DIABETES MELLITUS AND OTHER DISEASE CONDITIONS: A REVIEW

TOYIN DORCAS UDJE, NICOLE BROOKS,
and OLUWAFEMI OMONIYI OGUNTIBEJU

CONTENTS

9.1 INTRODUCTION

Medicinal plants are plants with potency to prevent and ameliorate specific ailment and pathological conditions [64, 65]. Apart from the well-known fact that plants serve as a source of food, they are also used for medicinal purposes [47]. Many of these plants are used as spices and local remedies and sometimes added to foods for supplementation or as treatment regimens [73].

Medicinal plants are of great importance to health of individuals and communities [31]. The medicinal value of plants is attributed to certain bioactive chemical substances present in those plants; which produce definite physiological actions in the human body [31]. Some of the bioactive constituents in plants that are of high importance include alkaloids, tannins, flavonoids, and phenolic compounds [86], and they elicit antioxidant, anticancer, anti-inflammatory, and antimalarial properties amongst many others [25]. These bioactive constituents are often referred to as phytochemicals.

This review is aimed at providing an overview of the medicinal properties and potentials of *Anchomanes difformis.*

Anchomanes difformis is a plant with many reported therapeutic properties [1, 70, 72, 74], and it is commonly used in traditional medicine to treat diseases with pathogenesis linked to oxidative stress among other factors [75]. *A. difformis* (Blume) is a species of flowering plants in the family Araceae, a herbaceous plant with prickly stem (up to 2 cm high) having huge divided leaf and spathe that arise from a horizontal tuber (which could be up to 80 cm long and 20 cm wide) growing as wild yam in the moist and shady places of tropical African forest [5]. It is prevalent in West African countries such as Ghana, Ivory Coast, Nigeria, Togo, Sierra Leone, Senegal, and Guinea [1] and is also found in southern-tropical Africa: Zambia and Angola [24]; Tanzania and Uganda. *A. difformis* has a wide range of local names, which is based on the location (Figure 9.1) [32]:

- in Ghana, *atõe, nyame kyin* (*FANTE*), *lukpogu* (*DAGBANI*);
- in Ivory Coast, *niamé kwanba* (*BAULE*), *eupé, niamatimi* (*AKAN-ASANTE*), *kohodié* (*ABURE*), *alomé* (*AKYE*), *tupain* (*ANYI*);
- in Nigeria ìgo *lángbòdó,* ògìrìòsákó (*YORUBA*), *oje, olumahi* (*IGBO*), *chakara, hántsàr gàdaá, hántsàr giawaá* (*HAUSA*), *eba ẹnàŋ* (*EFIK*);
- in Senegal, *éken* (*DIOLA*);
- in Sierra Leone, *a-thoŋbothigba* (*TEMNE*), *kalilugbo* (*MENDE-KPA*), *alatala-kunde-na (SUSU-DYALONKE);*

FIGURE 9.1 *Anchomanes difformis*: Left – leaves; Right – Rhizome.

- in Togo, *nau* (*TEM*);
- in Zambia, *Kabaka-kachulu* (*LUNDA*).

9.1.1 FOLKLORIC USES OF A. DIFFORMIS

Medicinal plants have been used for various traditional purposes varying from one community to another. Majority of people especially from the underdeveloped and developing countries rely on traditional plant usage for their day-to-day healthcare needs [32, 41], mostly due to the high cost of conventional health care or side effects of drugs. A. *difformis* has been reported for its wide range of traditional uses, some of which have been scientifically proven and confirmed. Different parts of the plants (leaves, stem, and tuber) are used to treat diverse ailments [37, 71].

The decoction from the leaves of *A. difformis* is traditionally used as an antibacterial agent particularly against *Staphylococcus aureus* in northern Nigeria [10]; to ameliorate pain and inflammation [1]; and in the treatment of cough, ulcer, and asthma in southern Nigeria [45]. The rhizome has been used for the treatment of many disease conditions in various parts of the world, and commonly in Africa. In Ivory Coast, it is considered to be a powerful purgative, and it is used to treat edema, control complications during child delivery, and as an antidote to poison as well as a strong diuretic for treating urethral discharge, jaundice, and kidney pains [8]. In Nigeria, the peeled tuber soaked in water is used for treating cases of dysentery and diarrhea [32]. In some parts of Africa such as Tanzania, the juice from the root tuber is used as eye drops in the treatment of river blindness [39]. The decoction from the tuber is used to treat cough, diabetes, dysentery, and throat-related conditions [75]. Most of the ethno-medicinal uses of *A. difformis* have been scientifically proven and established (Table 9.1).

9.1.2 NUTRITIVE VALUE OF A. DIFFORMIS

The leaves of *A. difformis* are consumed as a vegetable in some states in Nigeria [66, 71] and some other West African countries. Proximate

TABLE 9.1 Scientific Confirmation of Some Folkloric Uses of *A. difformis*

Folklore uses (scientifically proven)	Parts of plants used	Reference
Anti-asthma	Rhizome (aqueous fraction)	[70]
Anti-diabetes	Leaf (ethanolic extract)	[3]
	Rhizome (ethanolic extract)	[2]
Anti-inflammation	Leaf (ethanolic extract)	[1]
Anti-malarial	Rhizome (methanolic, aqueous and dichloromethane fractions)	[19]
Anti-microbial	Leaf and rhizome (Ethanol: methanol: water)	[32]
	Leaf and rhizome extracts	[7]
Anti-onchocercal (river blindness)	Rhizome (methanolic extract)	[63]
Anti-trypanosomal	Rhizome(dichloromethane extract)	[20]
	Aqueous and methanolic fractions	[14]
Anti-ulcer and gastro-protective	Rhizome (ethyl acetate extract)	[72]
	Leaf and rhizome (Ethanol: methanol: water)	[32]
Dysentery and diarrhea	Leaf and rhizome (Ethanol: methanol: water)	[32]
Pain killer/ analgesic	Leaf (ethanolic fraction)	[1]
	Rhizome (ethanolic fraction)	[38]

analysis carried out on the leaves revealed that it contains carbohydrate 58.63%, crude protein 30.55%, crude fiber 14.77%, fat and oil 0.49%, ash 11.71%, and moisture 12.39% [76]. The crude protein content (30.55%) of *A. difformis* leaves is found to be relatively higher when compared with some other vegetables such as *Momordica balsamina* (11.29%), *Moringa oleifera* (20.72%), *Lesianthera africana* leaves (13.10–14.90%), *Ocimum gratissimum* (8.00%), and *Hibiscus esculentus* (8.00%) [9, 68]. The high protein value in the leaves of *A. difformis* suggests a potential source of plant protein and therefore can be used as protein supplement in diet.

A. difformis is also considered to be rich in essential minerals [11]. Essential mineral elements play significant roles in human nutrition and metabolism [87], and their deficiencies could lead to anemia (iron), low glucose tolerance and liver necrosis (selenium), hypersensitivity (chloride), and reduced fertility (lithium) [92]. Certain macro elements play physiological roles such as structural (calcium in bones, iron in heme, phosphorus in phospholipids and nucleic acids), catalytic (magnesium) and signal transduction; and calcium in nerve and muscular cells [67]. Potassium, sodium, and chlorine are essential for the maintenance of osmotic balance between cells and the interstitial fluid [87, 92].

Other micro elements such as manganese and copper serve as co-factors of certain enzymes and are indispensable in numerous biochemical pathways that are essential for the normal functioning of the cell [87]. These include protein synthesis, carbohydrate metabolism, cell growth, and cell division. The mineral analysis of *A. difformis* (leaves) showed the presence of essential minerals such as potassium 1.74%, calcium 0.26%, sodium 0.026%, phosphorus 157.11 parts per million (ppm), magnesium 24525 ppm, iron 81.5 ppm, manganese 95.5 ppm, and copper 8.0 ppm [76]. This was supported by the output of mineral analysis carried out on the same plant [11].

The appreciable concentrations of minerals such as sodium, potassium, calcium, and phosphorus obtained from the plant showed that the plant has the potential of providing various secondary metabolites and mineral supply to enhance the curative process of ill-health [11, 76].

The nutritive value of the leaves and tuber of *A. difformis* has been compared and also used as a feed supplement to basal diet of West African dwarf sheep [13]. The proximate analysis indicated that the leaves contain more crude protein than the tuber, and nonfiber carbohydrate is higher in the tuber than in the leaves. The results illustrated that 10% inclusion of *A. difformis* into ruminant animal concentrate diet, increased nutrient digestibility coefficient in animals [13].

9.2 PHYTOCHEMICAL CONSTITUENTS OF *A. DIFFORMIS*

Phytochemicals are bioactive, non-nutrients compounds found in plants [49]. Medicinal plants contain numerous phytochemicals that are responsible for the plants' therapeutic potentials. Phytochemicals have been associated with amelioration of diseases and pathological conditions given that they possess several properties such as antioxidant, antidiabetic, anticancer, antimicrobial, anti-inflammatory, and gastro-protective [6, 42, 49, 50, 51, 62, 90]. The knowledge of phytochemical constituents of plants helps in the discovery of therapeutic agents and exploring new resources of such active chemical agents [60].

Ethanolic and methanolic extracts of leaves of *A. difformis* contain phytochemicals such as flavonoids, tannins, phlebotannins, cardiac glycosides, saponins, and reducing sugars [2, 11]. Conversely, aqueous extract from leaves indicates the presence of flavonoids, alkaloids, phenolics, and terpenoids, while cardiac glycosides and anthraquinones are absent [70]. A comparative study on the total phenolic content of root extract using three different solvents has been conducted by Aliyu et al. [47]. The results showed that n-butanol extract has significantly higher phenolic content than the methanol and acetone extracts. Other studies compared the presence of phytochemicals in the tuber and leaves of *A. difformis* [7, 32].

9.3 MEDICINAL BENEFITS OF *A. DIFFORMIS*

9.3.1 ANTIDIABETIC ACTIVITIES

Diabetes mellitus (Type I and II) is an endocrinological and metabolic disorder arising from insulin deficiency or impaired response of the body cells to insulin production [53, 81]. Diabetes mellitus is characterized by persistent hyperglycemia. Antihyperglycemic and antidiabetic properties of numerous plants have been explored, among which is *A. difformis* [2, 85]. The ethanolic extract of *A. difformis* tuber significantly reduced blood glucose concentration in alloxan monohydrate-induced diabetes in Wistar rats [3]. Similar results were found with the ethanolic extract of the leaves [2]. Conversely, the study carried out [3, 69] on aqueous extract of leaves of *A. difformis* revealed that it does not have any hypoglycemic effects in normal (nondiabetic) rats. The mechanism underlying the antihyperglycemic and antidiabetic effect of *A. difformis* has not been reported.

9.3.2 ANTIOXIDANT ACTIVITIES

Antioxidants and other bioactive components in plants have received much attention, especially in the medicinal field and food industry [82]. Natural antioxidants may function as reducing agents, free radical scavengers, complexers of pro-oxidant metals, or as quenchers of the formation of singlet oxygen [43]. The most common natural antioxidants are flavonoids (flavanols, isoflavones, flavones, catechins, flavanones), cinnamic acid derivatives, coumarins, tocopherols, and polyfunctional organic acids [48, 54].

The free radical scavenging activity, total antioxidant capacity, and reducing power assay have been measured in the acetone, n-butanol, and methanolic extracts of *A. difformis* tuber by Aliyu et al. [12], who showed that all extracts displayed strong radical scavenging activity, but n-butanol extract had the strongest reducing ability that was comparable to that of gallic acid at all the concentrations tested [12]. Similar antioxidant studies were carried out on methanolic leaf and root extracts of *A. difformis* by Agyare et al. [7]. Their report revealed that the methanolic root extract had higher antioxidant capacity than the leaves, and this is directly proportional to the total phenolic content. A study on the in vivo antioxidant capacity of aqueous fraction of *A. difformis* by Oghale and Idu [70] confirmed the plant as a potent inhibitor of free radicals and lipid peroxidation in guinea pigs.

9.3.3 ANTI-INFLAMMATORY ACTIVITIES

Inflammation plays a key role in the pathogenesis of certain disease conditions such as asthma, diabetes, arthritis, and cardiovascular diseases [1]. Elevated markers of inflammation are indicative of oxidative stress [97]. During oxidative stress, resident T-lymphocytes and activated macrophages around the β-cells release cytokines, particularly interleukin-1 (IL-1), which have been tagged as immunological molecules that inhibit secretion of insulin by the pancreatic β-cells and cause destruction of the β-cells [44, 58].

Many inflammatory diseases are associated with the synthesis of prostaglandins, which are responsible for a sensation of pain [51]. The primary enzyme responsible for prostaglandins synthesis is the membrane-associated cyclooxygenase (COX), which occurs in two isoforms: COX-1 and COX-2 [94]. COX-1 is constitutively expressed, while COX-2 is induced in the inflamed tissue. Modulation of the activity of the enzyme implies that the inflammation process can be modified.

Adebayo et al. [1] conducted a study to demonstrate the anti-inflammatory activity of *A. difformis*. This was done by inducing edema by injecting raw egg albumin (0.1 mL) on the left paw of the rats after 30 minutes of administration of *A. difformis* leaves (ethanolic extract). Edema size was measured and evaluated at intervals using a digital letica plethysmometer. The results showed that prior administration of *A. difformis* showed significant inhibition of edema (paw volume).

The anti-inflammatory studies performed by Agyare et al. [7] on leaf and tuber extracts of *A. difformis* in chicks have revealed that it possess significant anti-inflammatory activity and inhibitory effects on preformed mediators such as histamine and serotonin that are involved in the initial phase of the acute inflammatory process. Both extracts, at all the doses tested, demonstrated higher anti-inflammatory activity than aspirin. The anti-inflammatory activity could be due to the presence of steroids in the extracts, which exert their effects through switching-off multiple activated inflammatory genes and repressing NF-κB-regulated inflammatory genes, and it may also activate several anti-inflammatory genes and increase the degradation of mRNA encoding certain inflammatory proteins [16].

9.3.4 ANTINOCICEPTIVE ACTIVITIES

Nociception, which causes pain and injury or react to pain stimuli [26], can be induced by various methods such as tail-flick, hot plate, writhing, and injection of formalin among others depending on the nociceptors to be investigated [21]. In this study, 0.25% injection of formalin on the hind paw was used to induce pain in Wistar rats following the administration of ethanolic extract of *A. difformis* leaves and a standard drug aspirin [1]. The results indicated that formalin-induced paw nociception was significantly reduced, and this was dose-dependent. Therefore, *A. difformis* possesses anti-nociceptive ability as displayed by its inhibition of formalin-induced pain.

9.3.5 ANTIMICROBIAL ACTIVITIES

Microorganisms especially pathogens are of great medical and economical value [57, 80]. Due to side effects of antibiotics and the resistance built up by pathogenic microbes [61, 79] against these drugs, much attention has been shifted to plant extracts and their biological activity [29, 34]. Antimicrobial activity of plant origin shows great therapeutic potentials [79]. Medicinal plants have been explored for their antimicrobial properties [15, 19, 32, 63].

Reports on investigations of antimicrobial activities of *A. difformis* vividly showed that it is potent against common and drug-resistant microorganisms [7, 14, 32]. A comparative study on antimicrobial activities of the leaves and rhizome extracts of *A. difformis* using a ratio of different solvent has been reported by Eneojo et al. [32]. The microorganisms tested against were of public health importance, e.g., *Salmonella paratyphi, Salmonella typhi, Candida albicans, Proteus vulgaris, Staphylococcus aureus, Shigella flexneri, Shigella dysenterae, Pseudomonas aeruginosa*, and four different strains of *Escherichia coli.* The results revealed that *A. difformis* leaf and rhizome extracts demonstrated inhibition against all microorganisms except for *S. flexneri* and *P. aeruginosa* that conferred resistance against leaf and rhizome extracts, respectively [7, 32]. Another assessment of aqueous and methanolic extracts of *A. difformis* against *Trypanosoma brucei* presents trypanocidal activities of the plant [14].

In vitro antiplasmodial activity of *A. difformis* rhizome was tested against chloroquine-sensitive strain of *Plasmodium falciparum* (3D7). Roots of *A. difformis* exhibited moderate antiplasmodial activities compared with other plants used in the study [19]. In vivo study of antiplasmodial activities of *A. difformis* has been recommended given that inhibition or other mechanisms of action may take place at other stages of the plasmodium life cycle, such as the pre-erythrocytic development in the liver [19].

9.3.6 GASTRO-PROTECTIVE ACTIVITIES

Experimental peptic ulcer models can be induced using the following: acetic acid, ethanol, indomethacin, histamine, reserprine, pylorus ligation, ischemia-reperfusion, stress (hypothermic restraint, water-immersion), and diethyldithiocarbamate among others [4, 84, 99].

The research on Sprague-Dawley rats to assess the gastro-protective effect of ethyl acetate extract of *A. difformis* rhizome using different gastric ulcer models showed a positive result [72]. Ulcer conditions were induced using ethanol, indomethacin, and pylorus ligation in the animals after pre-treatment with rantidine. Although the mechanism through which it exerts its effect is not clear, the ethyl acetate fraction of *A. difformis* extract was shown to possess an antiulcer activity in all three tested models of ulcer [72]. The gastro-protective ability of *A. difformis* can be associated with the presence of flavonoids, which have proven to be gastro-protective [99]. It was found that flavonoids obtained from different plants were gastro-protective against ethanol-induced damage by causing increase in gastric microcirculation.

9.3.7 ANTIASTHMATIC ACTIVITIES

Bronchial asthma is one of the common syndromes of several respiratory diseases [78]. Antiasthmatic evaluation was carried out on leaves (aqueous extract) of *A. difformis* using guinea pigs as experimental animals [70]. The experimental groups were first sensitized with intraperitoneal administration of ovalbumin; 100 mg/kg body weight, intramuscular administration; 50 mg/kg body weight, 24 hours later. *A. difformis* leaf extract was administered for 7 days after which the animals were exposed to 0.2% histamine aerosol (bronchoconstrictor). The effect of *A. difformis* was assessed on the tracheal fluid volume and fluid viscosity against salbutamol, a standard drug for treating asthma [70]. The results of the study showed that *A. difformis* is effective against asthma as revealed by its bronchodilator properties similar to the effect of standard drug. Similar results have been reported by Boskabady et al. [23], who assessed the effect of *Nigella sativa* on airways of asthmatic patients.

9.3.8 ANTIONCHOCERCAL ACTIVITIES OF *A. DIFFORMIS*

Onchocerciasis or subcutaneous filariasis, popularly known as river blindness, is caused by filarial worm (*Onchocerca volvulus*) and transmitted by the black fly (*Simulium damnosum*) [52]. Onchocerciasis is one of the leading causes of blindness. It is prevalent in 37 countries of the world out of which 30 of these countries are in Africa [22]; and these account for 99% of the global burden of onchocerciasis and its related maladies, while the other 1% is confined to Yemen and some countries of Central and South America [91].

Mass drug administration (MDA) and vector control programs have been major approaches to control this disease [95]. Although onchocerciasis's control in Africa has lasted more than 40 years, the disease is still a public health concern in many African countries. Ivermectin, which is the main chemotherapeutic agent currently used, has limited micro-filaricidal efficacy [83]. Further, the fear of side effects and the duration of treatment (15–18 years) in order to terminate parasite transmission have led to a reduction in the ivermectin intake with serious epidemiological consequences [91]. These factors and the potential development of resistance to ivermectin have necessitated the need for new drugs in the treatment of onchocerciasis to achieve elimination of transmission.

The search for an efficient microfilaricide has led to exploration of certain plants such as: *Cyperus articulates Craterispermum laurinum, Morinda lucida, and Anchomanes difformis* [28, 59]. A study on the antionchocercal ability of *A. difformis* was performed on the methanolic extracts and on different fractionated extracts obtained at varying polarity using column chromatography [63]. The extracts and fractions were tested against *O. ochengi*: a strain very similar to *O. volvulus*. The methanolic extract and the four fractions obtained exhibited 100% inhibition against the microfiliariae [63], and this lends credence to its ethnobotanical use in the treatment of river blindness.

9.4 *A. DIFFORMIS*: OXIDATIVE STRESS AND DIABETES

Reactive oxygen species (ROS) and reactive nitrogen species (RNS) are produced in our body during normal metabolism and energy production [56]. They are produced to help normal healthy tissues to perform physiological roles such as signaling, regulation of signal transduction and gene expression, and activation of receptor and nuclear transduction [93].

Oxidative stress occurs as a result of imbalance between the systemic production of these free radicals (ROS and RNS) and the antioxidant capacity of the system to readily detoxify and eliminate the reactive intermediates or to repair the resulting damage [12]. Constant hyperglycemia is one of the major factors leading to oxidative stress. Hyperglycemia enhances the production of ROS, which is highly implicated in the development of oxidative stress [97].

Some of the mechanisms, by which hyperglycemia lead to oxidative stress, include the increased production of superoxide anion [27] that activates nuclear factor kappa-light-chain enhancer of activated beta cells (NF-κB); a transcription factor that leads to the increased expression of inducible nitric oxide synthase (iNOS) [88]. This increased iNOS results in enhanced production of nitric oxide (NO) that rapidly reacts with superoxide anion when present in high concentrations to form a strong oxidant (peroxynitrite) [18], which exerts its toxic effects through oxidation of proteins, initiation of lipid peroxidation and nitration of amino acids [18].

Triggering of the inflammatory response is another mechanism whereby hyperglycemia causes oxidative stress. Hyperglycemia contributes to increased glycation of lipids and proteins to form advanced glycation end products (AGEs) [40]. The AGEs bind to their receptors

(RAGE) on different cells and macrophages leading to intracellular generation of ROS [96, 98], which in turn activates NF-κB causing increased expression of a variety of cytokines such as TNF-α and TNF-ß (tumor necrosis factor-alpha and -beta), interleukins (IL-1, IL-6, IL-8, IL-18), and interferon-γ [17, 33].

Furthermore, hyperglycemia is strongly associated with glycation and glycoxidation of lipoproteins. ROS resulting from hyperglycemia contribute to initiation of lipid peroxidation [30] and ultimately lipoxidation to yield advanced lipoxidation end products (ALEs) [35, 36, 89]. Severe lipid peroxidation, protein oxidation, and nitration of proteins leads to oxidative stress that is an important risk factor in the development of diabetic complications such as retinopathy [77], nephropathy, liver injury, and cardiovascular diseases [46].

A. difformis has been clearly shown to possess antioxidant and anti-inflammatory properties [1]; and essential factors that minimize oxidative stress and ultimately ameliorate diabetes [2, 3] and possibly diabetic complications.

9.5 FUTURE PROSPECTIVE AND RESEARCH OPPORTUNITIES

Further investigations should be performed extensively on the leaves and roots of *A. difformis* to be able to understand the molecular mechanisms involved in its therapeutic ability against diabetes, inflammation, ulcer, dysentery, diarrhea, and microbial infections such as onchocerciasis and trypanosomiasis.

9.6 CONCLUSIONS

Useful products obtained from plants directly or indirectly demonstrate their importance. Medicinal plants are of great importance in healthcare and have been a common source of therapeutic agents due to presence of bioactive chemical substances (phytochemicals) that produce a definite physiological action on the human body. Some of the most important phytochemicals are alkaloids, tannins, flavonoids, and phenolic compounds. *Anchormanes difformis*, a rhizome used for medicinal purposes in Africa, has shown to elicit antioxidant, antidiabetic, antinociceptive, anti-inflammatory, antimalarial, antimicrobial, antionchocercal, and gastro-protective properties. These medicinal potentials and their roles in ameliorating diabetes-induced oxidative stress and its usefulness in other diseases have been considered in this chapter.

9.7 SUMMARY

A. difformis contains numerous phytochemicals that play significant roles in ameliorating disease conditions through certain mechanisms such as reducing or preventing oxidative stress. These phytochemicals include alkaloids, phenolics (flavonoids, tannins, and phlebotannins), cardiac glycosides, terpenoids and saponins. All parts of the plant (leaf, stem and tuber) contain most of these phytochemicals with just one or two absent in each part.

Certain folkloric and ethnobotanical uses of *A. difformis* have been scientifically proven and established such as its use in treating ulcer, diarrhea, pain, inflammation and diabetes, whilst other folklores are yet to be investigated. These include treatments against jaundice and kidney pains, cardiovascular disease and as hormonal modulator during child birth.

This review on *A. difformis* provided the opportunity to explore its activities against pathological conditions given its antidiabetic, antioxidant, anti-inflammatory, antihyperglycemic, antinociceptive, antimicrobial, antionchocercal, and gastro-protective properties.

KEYWORDS

- **analgesic**
- ***Anchomanes difformis***
- **anti-cancer**
- **anti-microbial**
- **antioxidant**
- **diabetes mellitus**

- **diabetic complications**
- **essential minerals**
- **extracts**
- **folklore**
- **free radicals**
- **gastro-protective**
- **glycation**
- **hyperglycemia**
- **inflammation**
- **mechanism**
- **molecular**
- **nociceptive**
- **nutritive**
- **onchocerciasis**
- **oxidative stress**
- **phytochemicals**
- **potentials**
- **reactive nitrogen species**
- **reactive oxygen species**
- **therapeutic**

REFERENCES

1. Adebayo, A. H., John-Africa, L. B., Agbafor, A. G., Omotosho, O. E., & Mosaku, T. O., (2014). Anti-nociceptive and anti-inflammatory activities of extract of *Anchomanes difformis* in rats. *Pakistan Journal of Pharmaceutical Sciences*, *27*(2), 265–270.
2. Aderonke, S. O., & Ezinwanne, A. J., (2015). Evaluation of the anti diabetic activity of ethanol extract of *Anchomanes difformis* (Araceae) leaves in albino rats. *International Research Journal of Pharmacy*, *6*(2), 90–93.
3. Adeyemi, O., Makinwa, T. T., & Uadia, R. N., (2015). Ethanol extracts of roots of *Anchomanes difformis* ENGL roots as an antihyperglycemic agent in diabetic rats. *Chemistry*, *1*(3), 68–73.

4. Adinortey, M. B., Ansah, C., Galyuon, I., & Nyarko, A., (2013). In vivo models used for evaluation of potential antigastroduodenal ulcer agents. *Ulcers,* 1–12.
5. Afolayan, M. O., Omojola, M. O., Orijajogun, J. O., & Thomas, S. A., (2012). Further physicochemical characterization of *Anchomanes difformis* starch. *Agriculture and Biology Journal of North America, 3*(1), 31–38.
6. Aggarwal, B. B., Kunnumakkara, A. B., Harikumar, K. B., Tharakan, S. T., Sung, B., & Anand, P., (2008). Potential of spice-derived phytochemicals for cancer prevention. *Planta Medica, 74*(13), 1560–1569.
7. Agyare, C., Boakye, Y. D., Apenteng, J. A., Dapaah, S. O., Appiah, T., & Adow, A., (2016). Antimicrobial and Anti-Inflammatory properties of *Anchomanes difformis* (Bl.) Engl. and *Colocasia esculenta* (L.) Schott. *Biochemistry & Pharmacology: Open Access,* ISSN: 2167–0501.
8. Akah, P. A., & Njike, H. A., (1990). Some pharmacological effects of rhizome aqueous extract of *Anchomanes difformis. Journal of Epidemiology, 61*, 368–370.
9. Akindahunsi, A. A., & Salawu, S. O., (2005). Phytochemical screening and nutrient-antinutrient composition of selected tropical green leafy vegetables. *African Journal of Biotechnology*, *4*(6).
10. Aliyu, A. B., Musa, A. M., Abdullahi, M. S., Oyewale, A. O., & Gwarzo, U. S., (2008a). Activity of plant extracts used in northern Nigerian traditional medicine against methicillin-resistant *Staphylococcus aureus* (MRSA). *Nigerian Journal of Pharmaceutical Sciences*, *7*(1), 1–8.
11. Aliyu, A. B., Musa, A. M., Oshanimi, J. A., Ibrahim, H. A., & Oyewale, A. O., (2008b). Phytochemical analyses and mineral elements composition of some medicinal plants of Northern Nigeria. *Nigerian Journal of Pharmaceutical Sciences*, *7*(1), 119–125.
12. Aliyu, A. B., Ibrahim, M. A., Musa, A. M., Musa, A. O., Kiplimo, J. J., & Oyewale, A. O., (2013). Free radical scavenging and total antioxidant capacity of root extracts of *Anchomanes difformis* Engl. (*Araceae*). *Acta Poloniae Pharmaceutica, 70*(1), 115–21.
13. Arigbede, M., Anele, Y., Südekum, K. H., Bolaji, O., Oni, A., Dele, P., & Sanusi, I., (2010). Chemical composition of different parts of wild Yam (*Anchomanes difformis)* and the nutritive potentials of its tuber for inclusion in small ruminant diet. In: *World Food System – A Contribution from Europe*. Tropentag, Zurich.
14. Atawodi, S. E., (2005). Comparative in vitro trypanocidal activities of petroleum ether, chloroform, methanol and aqueous extracts of some Nigerian savannah plants. *African Journal of Biotechnology*, *4*(2), 177.
15. Atawodi, S. E., & Atawodi, J. C., (2009). *Azadirachta indica* (neem): a plant of multiple biological and pharmacological activities. *Phytochemistry Reviews*, *8*(3), 601–620.
16. Barnes, P. J., (2006). How corticosteroids control inflammation: quintiles prize lecture 2005. *British Journal of Pharmacology*, *148*(3), 245–254.
17. Basta, G., Lazzerini, G., Massaro, M., Simoncini, T., Tanganelli, P., Fu, C., Kislinger, T., Stern, D. M., Schmidt, A. M., & De Caterina, R., (2002). Advanced glycation end products activate endothelium through signal-transduction receptor RAGE a mechanism for amplification of inflammatory responses. *Circulation, 105*(7), 816–822.

18. Beckman, J. S., & Koppenol, W. H., (1996). Nitric oxide, superoxide, & peroxynitrite: the good, the bad, & ugly. *American Journal of Physiology-Cell Physiology, 271*(5), 1424–1437.
19. Bero, J., Ganfon, H., Jonville, M. C., Frédérich, M., Gbaguidi, F., DeMol, P., Moudachirou, M., & Quetin-Leclercq, J., (2009). In vitro antiplasmodial activity of plants used in Benin in traditional medicine to treat malaria. *Journal of Ethnopharmacology, 122*(3), 439–444.
20. Bero, J., Hannaert, V., Chataigné, G., Hérent, M. F., & Quetin-Leclercq, J., (2011). In vitro antitrypanosomal and antileishmanial activity of plants used in Benin in traditional medicine and bio-guided fractionation of the most active extract. *Journal of Ethnopharmacology, 137*(2), 998–1002.
21. Björkman, R., (1994). Central antinociceptive effects of non-steroidal anti-inflammatory drugs and paracetamol. Experimental studies in the rat. *Acta Anaesthesiologica Scandinavica. Supplementum, 103*, 1–44.
22. Boatin, B. A., & Richards, F. O., (2006). Control of onchocerciasis. *Advances in Parasitology, 61*, 349–394.
23. Boskabady, M. H., Mohsenpoor, N., & Takaloo, L., (2010). Antiasthmatic effect of Nigella sativa in airways of asthmatic patients. *Phytomedicine, 17*(10), 707–713.

CHAPTER 10

SCREENING OF DIFFERENT EXTRACTS OF *AGERATUM CONYZOIDES* FOR INHIBITION OF DIABETES-RELATED ENZYMES

MUTIU IDOWU KAZEEM, OLUWATOSIN OGUNKELU, ADEMOLA OLABODE AYELESO, and EMMANUEL MUKWEVHO

CONTENTS

10.1 INTRODUCTION

Diabetes mellitus (DM) is a metabolic disorder characterized by chronic hyperglycemia as well as disturbance in carbohydrate, fat, and protein metabolism, due to absolute or relative lack of insulin [10, 11]. The number of people suffering from the disease worldwide is increasing at an alarming rate with 552 million people projected to have DM by 2035 as against 382 million estimated cases in 2014 [12]. Type 1 DM is characterized by an absolute deficiency of insulin caused by the destruction of pancreatic β-cells, while type 2 DM is due to a combination of peripheral resistance to insulin action as well as inadequate secretory response by pancreatic β-cells [21].

Most of the conventional drugs used for the management of diabetes have limited efficacy with various side effects such as hypoglycemia, weight gain, liver failure, and gastro-intestinal disorders [7, 27]. Thus, the search for more effective and safer drugs is essential to overcome these issues [20]. Several studies have reported the antidiabetic potential of medicinal plants [3, 4, 16]. However, there is dearth of information on the mechanism of action by which these plants elicit their hypoglycemic potential. One of the possible mechanisms of antidiabetic effect is the inhibition of carbohydrate hydrolyzing enzymes, pancreatic α-amylase and intestinal α-glucosidase.

The enzyme α-amylase, which belongs to the class of α-glucan-4-glucanhydrolases, is one of the important target enzymes for the conventional treatment of diabetes [1]. It catalyzes the initial steps in the hydrolysis of starch to disaccharides, which are then acted upon by α-glucosidase, which breaks them down into glucose that gets absorbed by the brush border epithelium of the intestine and enters the blood stream [26]. The pathological condition, which arises due to the excessive degradation of starch by α-amylase and α-glucosidase, is referred to as postprandial hyperglycemia.

Ageratum conyzoides is an annual herb, which grows approximately 1 m in height. The leaves and stem are covered with white fine hairs, while the leaves are ovate and up to 7.5 cm long [23]. It is commonly called white weed, goat weed, and billy goat weed. It is locally called "Otogo" in the Middle-belt region [9] and "Imi-eshu" in Southwestern region of Nigeria [24]. *A. conyzoides* is used in the treatment of pneumonia, wounds, and burns in Africa. It is also used as bacteriocide and antidysenteric agent in several regions of the world [23]. This plant has been reportedly used in the treatment of diabetes mellitus, infertility, and HIV/AIDS in Benue State, Nigeria [9].

Several studies have reported the hypoglycemic and antihyperglycemic potentials of the extracts of this plant [6, 22], but none has reported its mode of hypoglycemic action. In view of this, this study is aimed at evaluating the inhibitory potential of *A. conyzoides* leaf extracts on α-amylase and α-glucosidase activities, as a possible mode of eliciting its antidiabetic potentials.

10.2 MATERIALS AND METHODS

10.2.1 PLANT MATERIAL

The leaf of *Ageratum conyzoides* was obtained from Egan village, Igando Area of Lagos, Nigeria, in February 2013. It was identified and authenticated at the Department of Botany, University of Lagos, Nigeria, and voucher specimen (LUH 4727) was deposited in the University herbarium.

10.2.2 CHEMICALS AND REAGENTS

Porcine pancreatic α-amylase, rat intestinal α-glucosidase and paranitrophenyl – glucopyranoside were obtained from Sigma-Adrich Co., St Louis, USA, while starch soluble (extra pure) was purchased from J. T. Baker Inc., Phillipsburg, USA. Other chemicals and reagents were of analytical grade, and water used was glass-distilled.

10.2.3 PREPARATION OF PLANT EXTRACTS

Fresh leaves of *Ageratum conyzoides* were cut and washed with water to remove all contaminants; they were dried at room temperature and ground to powder using Warring blender. Equal quantity (20 g) each of the dried powdered material was extracted in 200 mL acetone, ethanol or water, with constant shaking on Labcon platform shaker (Laboratory Consumables, PTY, Durban, South Africa) for 24 h. The extracts were centrifuged (Hermle Laboratory Centrifuge, Lasec, South Africa) and later filtered using a Whatman No. 1 filter paper. Acetone and ethanol extracts were concentrated using rotary evaporator (Cole Parmer SB 1100, Shangai, China), while the aqueous extract was freeze-dried using Virtis Bench Top (SP Scientific Series, USA) freeze dryer. The extracts were dissolved in dimethyl sulphoxide (DMSO) to give stock solutions of 5.0 mg/mL each, and different concentrations (0.32, 0.63, 1.25 and 2.50 mg/mL) of the extracts were prepared using the serial dilution method with distilled water. All extracts were stored at 4°C *prior* further analysis.

10.2.4 α-AMYLASE INHIBITORY ASSAY

This assay was done using a modified procedure by McCue and Shetty [19]. Each extract (250 μL) at different concentrations (0.32–2.50 mg/mL) was added to 250 μL of 0.02 M sodium phosphate buffer (pH 6.9) containing α-amylase. This solution was pre-incubated at 25°C for 10 min, after which 250 μL of 1% starch solution in 0.02 M sodium phosphate buffer (pH 6.9) was added at timed intervals and then further incubated at 25°C for 10 min. The reaction was terminated by adding 500 μL of dinitrosalicylic acid (DNSA) reagent. The tubes were then incubated in boiling water for 5 min and cooled to room temperature. The reaction mixture was diluted with 5 mL distilled water, and the absorbance was measured at 540 nm using spectrophotometer. The control was prepared using the same procedure replacing the extract with distilled water, while the activity of the standard was tested by replacing the extract with acarbose. The α-amylase inhibitory activity was calculated as percentage inhibition:

$$\% \text{ Inhibition activity} = [(Abs_{control} - Abs_{extracts})/Abs_{control}] \times 100 \quad (1)$$

Concentrations of extracts resulting in 50% inhibition of enzyme activity (IC_{50}) were determined graphically.

10.2.5 MODE OF α-AMYLASE INHIBITION

The mode of inhibition of the enzyme by the leaf extract was conducted using the extract with the lowest IC_{50} according to the modified method described by Ali et al. [2]. Briefly, 250 μL of the (2.50 mg/mL) extract was pre-incubated with 250 μL of α-amylase solution for 10 min at 25°C in one set of tubes. In another set of tubes, α-amylase was pre-incubated with 250 μL of phosphate buffer (pH 6.9). Then, 250 μL of starch solution at increasing concentrations (0.32–2.50 mg/mL) was added to both sets of reaction mixtures to start the reaction. The mixture was then incubated for 10 min at 25°C, and then boiled for 5 min after the addition of 500 μL of DNS to stop the reaction. The amount of reducing sugars released was determined spectrophotometrically using a maltose standard curve and converted to reaction velocities. A double reciprocal plot (1/v versus 1/[S]) was plotted, where v is reaction velocity and [S] is substrate concentration. The type (mode) of inhibition of the enzyme by the extract was determined by double reciprocal plot using Michaelis-Menten kinetics.

10.2.6 α-GLUCOSIDASE INHIBITORY ASSAY

The effect of plant extracts on α-glucosidase activity was determined according to the method described by Kim et al. [15] using α-glucosidase from rat intestinal powder. The substrate solution p-nitrophenyl glucopyranoside (pNPG) was prepared in 20 mM phosphate buffer (pH 6.9). A total of 100 μL of α-glucosidase (E.C. 3.2.1.20) was pre-incubated with 50 μL of the different concentrations of the extracts (0.32–2.50 mg/mL) for 10 min. Then, 50 μL of 3.0 mM (pNPG) as a substrate dissolved in 20 mM phosphate buffer (pH 6.9) was then added to start the reaction. The reaction mixture was incubated at 37°C for 20 min and stopped by adding 2 mL of 0.1 M Na_2CO_3. The α-glucosidase activity was determined by measuring the yellow colored para-nitrophenol released from pNPG at 405 nm. The control was prepared using the same procedure replacing the extract with distilled water while activity of the standard was tested by replacing the extract with acarbose. Percentage inhibition was then calculated using Eq. (1). Concentrations of extracts resulting in 50% inhibition of enzyme activity (IC_{50}) were determined graphically.

10.2.7 MODE OF α-GLUCOSIDASE INHIBITION

The mode of inhibition of the enzyme by the extract was determined using the extract with the lowest IC_{50} according to the modified method described by Ali et al. [2]. Briefly, 50 μL of the (2.50 mg/mL) extract was pre-incubated with 100 μL of α-glucosidase solution for 10 min at 25°C in one set of tubes. In another set of tubes, α-glucosidase was pre-incubated with 50 μL of phosphate buffer (pH 6.9). Further, 50 μL of pNPG at increasing concentrations (0.32–2.50 mg/mL) was added to both sets of reaction mixtures to start the reaction. The mixture was then incubated for 10 min at 25°C, and 500 μL of Na_2CO_3 was added to stop the reaction. The amount of reducing sugars released was determined spectrophotometrically using a paranitrophenol standard curve and converted to reaction velocities. A double reciprocal plot (1/v versus 1/[S]) was plotted, where v is reaction velocity and [S] is substrate concentration. The type (mode) of inhibition of the enzyme by the crude extract was determined by double reciprocal plot using Michaelis-Menten kinetics.

10.2.8 STATISTICAL ANALYSIS

All analyses were performed in triplicates unless otherwise stated. Data were expressed as mean ± SEM and statistical significance was considered at $p < 0.05$. The IC_{50} values were obtained from percentage inhibitions using Microsoft Excel software. One-way analysis of variance (ANOVA) was used to assess differences in the percentage inhibitions and IC_{50} values of the extracts as well as standard. Modes of inhibition of the enzymes were determined by linear regression using GraphPad Prism statistical package (GraphPad Software, USA).

10.3 RESULTS

Among all concentrations tested, there were no significant differences ($p > 0.05$) in percentage inhibition of α-amylase by various extracts of *Ageratum conyzoides*, except at 1.25 mg/mL, where there was a significant difference ($p < 0.05$) between the values obtained for ethanol compared to the other extracts (Figure 10.1). Percentage inhibition of the enzyme by acetone and aqueous extracts were also significantly different at 0.63 mg/mL. Consequently, the mode of inhibition of α-amylase by the aqueous

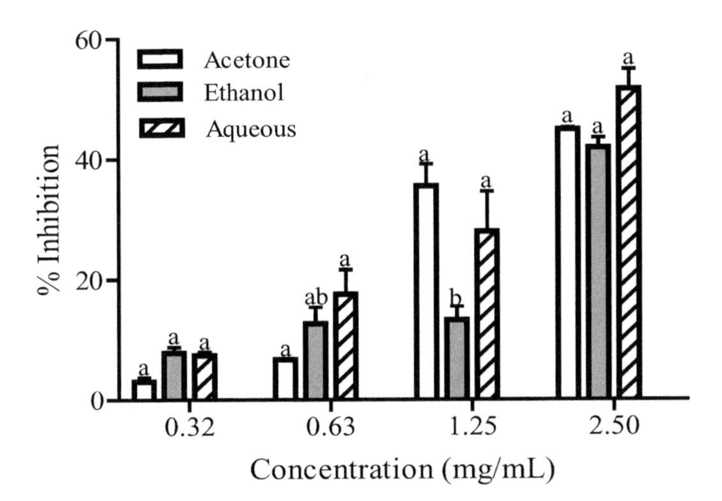

FIGURE 10.1 Percentage inhibition of α-amylase by different extracts of *Ageratum conyzoides*. Values are expressed as means ± SEM of triplicate determinations. Bars carrying different letters at the same concentration are significantly different ($p < 0.05$).

extract of *A. conyzoides* was determined, and it showed a characteristic mixed non-competitive inhibition of the enzyme (Figure 10.2).

The percentage inhibition of α-glucosidase by all the extracts of *A. conyzoides* is shown in Figure 10.3. At low concentrations (0.32 and 0.63 mg/mL), there were significant differences ($p < 0.05$) between aqueous extract and the other extracts. However, at high concentrations (1.25 and 2.5 mg/mL), the percentage inhibition of the enzyme by the acetone extract was significantly lower than that of aqueous and ethanolic extracts. The aqueous extract of *A. conyzoides,* which had the lowest IC_{50} value (1.05 mg/mL) compared to other extracts, was used to determine the mode of inhibition of this enzyme, and it also showed a mixed non-competitive inhibition (Figure 10.4).

Means down vertical column not sharing a common letter are significantly different ($p < 0.05$) from each other. Acarbose is the standard α-amylase and α-glucosidase inhibitor.

The IC_{50} value calculated showed that aqueous extract of *A. conyzoides* displayed the lowest IC_{50} (Table 10.1) for both α-amylase (2.13 mg/mL)

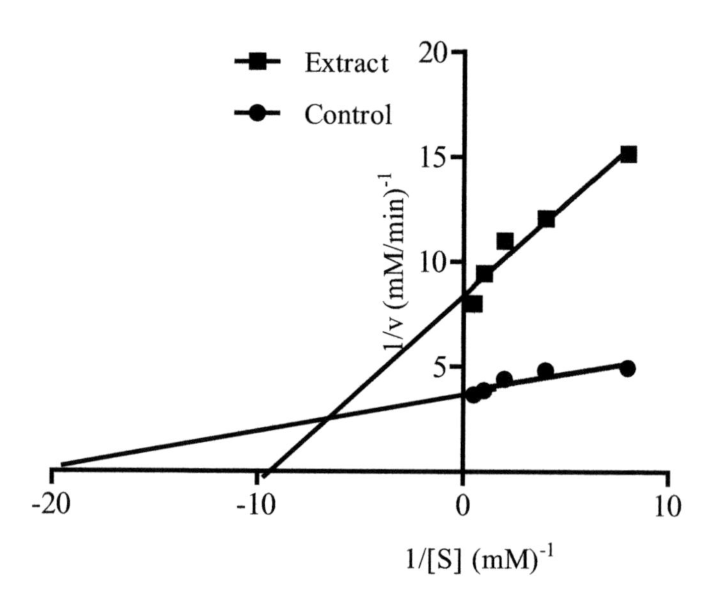

FIGURE 10.2 Mode of inhibition of α-amylase by the aqueous extract of *Ageratum conyzoides*.

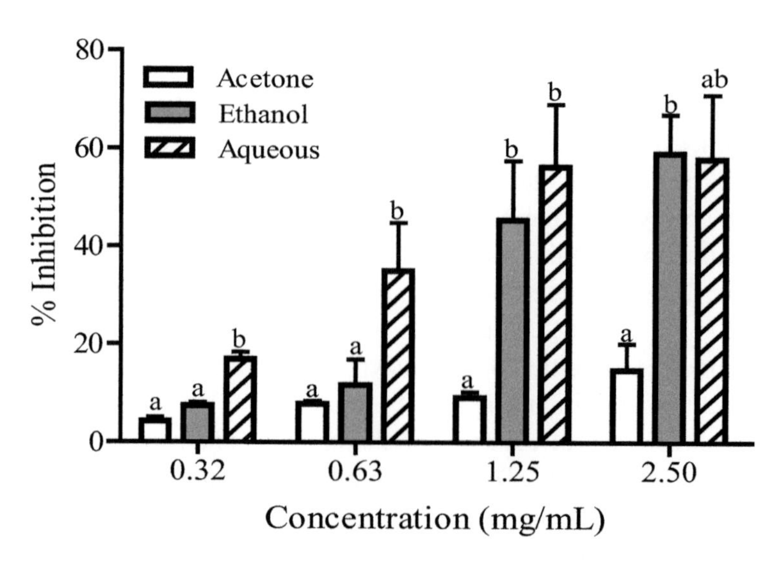

FIGURE 10.3 Percentage inhibition of α-glucosidase by different extracts of *Ageratum conyzoides*. Values are expressed as means ± SEM of triplicate determinations. Bars carrying different letters at the same concentration are significantly different ($p < 0.05$).

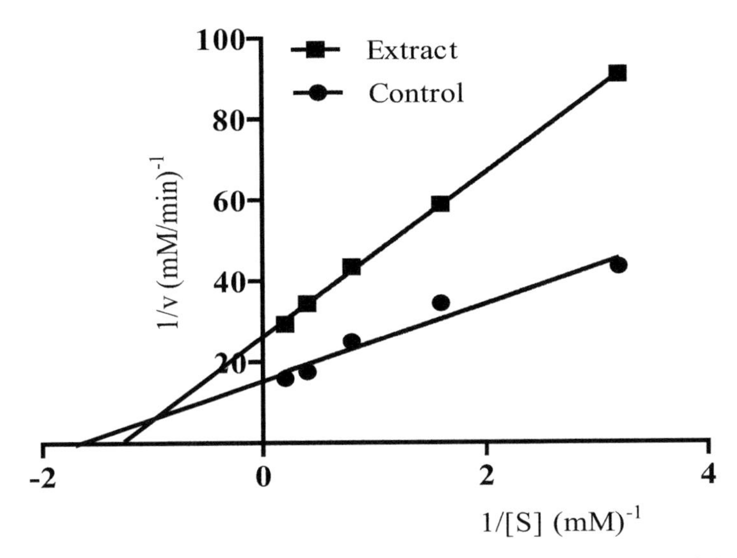

FIGURE 10.4 Mode of inhibition of α-glucosidase by the aqueous extract of *Ageratum conyzoides*.

TABLE 10.1 The IC_{50} values for the inhibition of α-amylase and α-glucosidase by extracts of *Ageratum conyzoides*

Extracts	IC_{50} (mg/mL)	
	α-amylase	α-glucosidase
Acetone	2.68 ± 0.12[a]	5.08 ± 0.80[a]
Aqueous	2.13 ± 0.10[b]	1.05 ± 0.05[c]
Ethanol	2.85 ± 0.29[a]	1.53 ± 0.07[b]
Acarbose	1.35 ± 0.03[c]	0.82 ± 0.01[c]

and α-glucosidase (1.05 mg/mL) inhibition compared to other extracts, though they were higher than acarbose. The values are expressed as mean ± SEM of triplicate determinations.

10.4 DISCUSSION

Hyperglycemia is a pathological state characterized by a rapid increase in blood glucose levels, and this is due to continuous hydrolysis of starch by pancreatic α-amylase and absorption of glucose by the intestinal α-glucosidases [5]. One of the therapeutic approaches for reducing postprandial hyperglycemia is to retard digestion of glucose by the inhibition of these carbohydrate hydrolyzing enzymes such as α-amylase and α-glucosidase, in the digestive tract [13]. Therefore, inhibition of these enzymes can significantly decrease postprandial hyperglycemia after a mixed carbohydrate diet and can be an important strategy in the management of diabetes mellitus [8].

In this chapter, the results of the α-amylase and α-glucosidase inhibitory assay revealed that the aqueous extract of *A. conyzoides* is a mild inhibitor of α-amylase and a strong inhibitor of α-glucosidase. This is attested to by the respective IC_{50} generated for the inhibition of the enzymes by the extract. This is in agreement with a previous report that any medicinal plant that may be used as an antidiabetic agent should be a mild inhibitor of α-amylase and strong inhibitor of α-glucosidase [17]. This has an edge over synthetic drugs such as acarbose, which strongly inhibits both α-amylase and α-glucosidase, leading to side effects, which include abdominal distention, flatulence, hypoglycemia, and liver failure [17, 25].

The mixed non-competitive inhibition displayed by the aqueous extract of *A. conyzoides* towards both α-amylase and α-glucosidase is in consonance with the report of Mayur et al. [18]. This suggests that the active components of the extract binds to a site other than the active site of the enzyme and combine with both the free enzyme and the enzyme-substrate complex, possibly interfering with the action of both, but has different affinity to both forms [18]. This affects the binding of the normal substrate to the enzyme due to conformational changes, thereby slowing down the breaking down of polysaccharides and disaccharides to glucose [14]. Therefore, the concentration of glucose in the blood is maintained, thereby controlling hyperglycemia and its associated effects.

10.5 SUMMARY

This study showed that out of the three extracts of *Ageratum conyzoides* tested, the aqueous extract displayed the most effective inhibition of α-amylase and α-glucosidase *in-vitro*; and the mode of inhibition of both enzymes is the mixed non-competitive one. Therefore, it can be concluded that one of the mechanisms by which *A. conyzoides* leaf exhibited its hypoglycemic properties may be due to inhibition of diabetes-related enzymes such as pancreatic α-amylase and intestinal α-glucosidase.

KEYWORDS

- ***Ageratum conyzoides***
- **diabetes mellitus**
- **enzymes**
- **hyperglycemia**
- **plant extracts**
- **α-amylase**
- **α-glucosidase**

REFERENCES

1. Akkarachiyasit, S., Charoenlertkul, P., Yibchok-Anun, S., & Adisakwattana, S., (2010). Inhibitory activities of cyanidin and its glycosides and synergistic effect with acarbose against intestinal α- glucosidase and pancreatic α -amylase. *International Journal of Molecular Sciences, 11*, 3387–3396.
2. Ali, H., Houghton P. J., & Soumyanath, A., (2006). Alpha-amylase inhibitory activity of some Malaysian plants used to treat diabetes, with particular reference to *Phyllanthus amarus. Journal of Ethnopharmacology, 107*, 449–455.
3. Bagri, P., Ali, M., Aeri, V., Bhowmik, M., & Sultana, S., (2009). Antidiabetic effect of *Punica granatum* flowers: Effect on hyperlipidemia, pancreatic cells lipid peroxidation and antioxidant enzymes in experimental diabetes. *Food Chemistry and Toxicology, 47*, 50–54.
4. Daisy, P., Eliza, J., Abdul Majeed, K., & Farook, M., (2009). A novel dihydroxy gymnemic triacetate isolated from *Gymnema sylvestre* possessing normoglycemic and hypolipidemic activity on STZ-induced diabetic rats. *Journal of Ethnopharmacology, 126*, 339–344.
5. Deshpande, M. C., Venkateswarlu, V., Babu, R. K., & Trivedi, R. K., (2009). Design and evaluation of oral bioadhesive controlled release formulations of miglitol, in-

tended for prolonged inhibition of intestinal alpha-glucosidases and enhancement of plasma glycogen like peptide-1 levels. *International Journal of Pharmacology, 380*, 16–24.

6. Egunyomi, A., Gbadamosi, I. T., & Animashahun, M. O., (2011). Hypoglycaemic activity of the ethanol extract of *Ageratum conyzoides* Linn. shoots on alloxan-induced diabetic rats. *Journal of Medicinal Plants Research, 5*, 5347–5350.
7. Fujisawa, T., Ikegami, H., Inoue, K., Kawabata, Y., & Ogihara T., (2005). Effect of two α-glucosidase inhibitors, voglibose and acarbose, on postprandial hyperglycemia correlates with subjective abdominal symptoms. *Metabolism, 54*, 387–390.
8. Hirsh, A. J., Yao, S. Y., Young, J. D., & Cheeseman, C. I., (1997). Inhibition of glucose absorption in the rat jejunum: A novel action of alpha-D-glucosidase inhibitors. *Gastroenterology, 113*, 205–211.
9. Igoli, J. O., Ogaji, O. G., Tor-Anyiin, T. A., & Igoli, N. P., (2005). Traditional medicine practice amongst the Igede people of Nigeria. *African Journal of Traditional, Complementary and Alternative Medicine, 2*, 134–152.
10. International Diabetes Federation, (2009). Diabetes Atlas, 4th Edition, Brussels, Belgium.
11. International Diabetes Federation, (2013). IDF Diabetes Atlas, 6th ed., Brussels, Belgium, pp. 12.
12. International Diabetes Federation, (2014). Global Diabetes Scorecard: Tracking progress for action. Brussels, Belgium, pp. 14.
13. Kazeem, M. I., Ayeleso, A. O., & Mukwevho, E., (2015). *Olax subscorpioidea* Oliv. leaf alleviates postprandial hyperglycaemia by inhibition of α-amylase and α-glucosidase. *International Journal of Pharmacology, 11*, 484–489.
14. Kazeem, M. I., Adamson, J. O., & Ogunwande, I. A., (2013). Modes of inhibition of α- amylase and α- glucosidase by aqueous extract of *Morinda lucida* Benth leaf. *Biomed Research International*, 1–6.
15. Kim, Y. M., Jeong, Y. K., Wang, M. H., Lee, W. Y., & Rhee, H. I., (2005). Inhibitory effects of pine bark extract on alpha-glucosidase activity and postprandial hyperglycemia. *Nutrition, 21*, 756–761.
16. Krisanapun, C., Peungvicha, P., Temsiririrkkul, R., & Wongkrajang, Y., (2009). Aqueous extract of *Abutilon indicum* Sweet inhibits glucose absorption and stimulates insulin secretion in rodents. *Nutrition Research, 29*, 579–587.
17. Kwon, Y. I., Apostolidis, E., & Shetty, K., (2008). Inhibitory potential of wine and tea against α-amylase and α-glucosidase for management of hyperglycemia linked to type 2 diabetes. *Journal of Food Biochemistry, 32*, 15–31.
18. Mayur, B., Sandesh, S., Shruti, S., & Sung-Yum, S., (2010). Antioxidant and α-glucosidase inhibitory properties of *Carpesium abrotanoides* L., *Journal of Medicinal Plant Research, 4*, 1547–1553.
19. McCue, P., & Shetty, K. (2004). Inhibitory effects of rosmarinic acid extracts on porcine pancreatic amylase in vitro. *Asian Pacific Journal of Clinical Nutrition, 13*, 101–106.
20. Moller, D. E., (2001). New drug targets for type 2 diabetes and the metabolic syndrome. *Nature, 414*, 821–827.
21. Motala, A. A., Omar, M. A., & Pirie, F. J., (2003). Diabetes in Africa: epidemiology of type 1 and type 2 diabetes in Africa. *Journal of Cardiovascular Risk, 10*, 77–83.

22. Nyunai, N., Manguelle-Dicoum, A., Njifutie, N. E., Abdennbi, E., & Gerard, C., (2010). Antihyperglycemic effect of *Ageratum conyzoides* L., fractions in normoglycemic and diabetic male wistar rats. *International Journal of Biomedical and Pharmaceutical Sciences*, *4*, 38–42.
23. Okunade, A. L., (2002). Review: *Ageratum conyzoides* L. (Asteraceae). *Fitoterapia*, *73*, 1–16.
24. Oladejo, O. W., Imosemi, I. O., Osuagwu, F. C., Oluwadara, O. O., Aiku, A., Adewoyin, O., Ekpo, O. E., Oyedele, O. O., & Akang, E. E., U., (2003). Enhancement of cutaneous wound healing by methanolic extracts of *Ageratum conyzoides* in the wistar rats. *African Journal of Biomedical Research*, *6*, 27–31.
25. Pinto, M. D. S., Kwon, Y. I., Apostolidis, E., Lajolo, F. M., Genovese, M. I., & Shetty, K., (2009). Potential of *Ginkgo biloba* L., leaves in the management of hyperglycemia and hypertension using in vitro models. *Bioresource Technolonogy*, 100, 6599–6609.
26. Smith, C., Marks, A., & Lieberman, M., (2005). *Mark's basic Medical Biochemistry: A Clinical Approach*, 2nd Ed., Lipincott Williams and Wilkins, Baltimore, Maryland, pp. 2120–2136.
27. Sriplang, K., Adisakwattana, S., Rungsipipat, A., & Yibchok-anun, S., (2007). Effects of *Orthosiphon stamineus* aqueous extract on plasma glucose concentration and lipid profile in normal and streptozotocin-induced diabetic rats. *Journal of Ethnopharmacology*, *109*, 510–514.

CHAPTER 11

POTENTIAL OF *CATHARANTHUS ROSEUS* AND *PUNICA GRANUTUM* IN THE MANAGEMENT AND TREATMENT OF DIABETES MELLITUS AND ITS COMPLICATIONS

MEDILINE GOBOZA, PRISCA KACHEPE,
YAPO GUILLAUME ABOUA,
and OLUWAFEMI OMONIYI OGUNTIBEJU

CONTENTS

11.1 INTRODUCTION

The adoption of westernized diet in Africa has been documented to have propelled the development of noncommunicable diseases such as type 2 diabetes mellitus (DM) and its complications. Type 2 DM remains incurable despite the availability of a number of orthodox drugs to manage it. The poor socio-economic factors in Africa make it difficult, if not impossible, for the poor to afford these expensive drugs. However, based on history, the intake of natural products as food or therapeutic agents has shown to enhance human health nutrition, leading to management of type 2 DM and its complications. The health promoting potentials of medicinal plants have been attributed to the presence of both nutritional and non-nutritional chemical compounds that can modify and modulate biological systems and thus eliciting therapeutic effects. A predilection for ethnopharmacological materials is on the increase owing to their safety, availability, and the cost of employing medicinal plants in research. In Africa, a number of medicinal plants have been traditionally used to treat diabetes with great success.

In this chapter, the authors have reviewed research studies on *Catharanthus roseus* and *Punica granutum* (Pomegranate) in the treatment of DM and its complications such as cardiovascular-related disorders, male infertility, and hepato-renal disorders. This chapter also presents results of comprehensive scientific investigations on medicinal plants for possible validation and development of cheap novel therapies against diabetes mellitus.

11.2 OVERVIEW OF DIABETES MELLITUS

DM is a potentially devastating, incurable but treatable lifelong metabolic disorder whose prevalence is rapidly increasing. It is a metabolic disorder that is marked by abnormal high glucose levels in the blood [2]. Diabetes was first described by a Greek physician named Areateus in the second century, for which he used the term *diabainein* referring to the excessive passing out of fluid that he constantly observed in diabetic patients. The Latin word "mellitus," which means "sweet-like honey," was later added to the word diabetes; both terms indicating the sweet taste of urine that was passed by diabetes patients (glucosuria) [5, 86]. Another rare form of diabetes identified as diabetes insipidus also exists. It is characterized by excessive urine output and fluid intake as a compensatory reflex action against fluid loss. Diabetes insipidus emerges due to the attenuated production of vasopressin, a hormone that controls water retention by reducing urine output. Although the two diabetic disorders exhibit the same symptoms, their diagnosis and treatment are not the same [2].

DM is classified into four main categories: type 1 DM (T1DM), type 2 DM (T2DM), gestational DM, and DM that occurs secondary to other conditions. The elevated levels of blood glucose in diabetes mellitus, also known as hyperglycemia, is implicated in the development of disorders that affect various organs such as the heart, kidneys, eyes, nerves, liver, and blood vessels [18]. This classical endocrine disease arises as a consequence of several pathogenic processes that include autoimmunity, genetic defects, and unhealthy lifestyle practices. These pathogenic processes predispose to defects in insulin secretion, insulin action, or both, resulting in the development of diabetic complications [1]. Insulin is a peptide exocrine hormone secreted by the pancreas and regulates glucose homeostasis. Deficiency or absolute lack of this hormone affects normal metabolism of macronutrients, leading to severe complications. DM is now seen as a growing global health crisis. It has become a common disease affecting people in both developing and developed countries. The rapid increase in the prevalence of DM is fueled by changes in the lifestyle patterns contributed by urbanization, obesity, physical inactivity, smoking, and sedentary practices [78, 70, 154]. Obesity has been labeled as a major risk factor determinant of DM development. Parallel escalations of overweight (body mass index between 25 kg/m$^{2)}$), obesity (BMI $\geq$ 30kg/m^2) and DM have been reported. According to World Health Organization (WHO) reports in 2014, 39% and 13% of the adult world population were overweight and obese, respectively. For these reasons, lifestyle changes are major therapeutic options for managing diabetes, involving regular exercises, medical nutrition, and weight control amongst others. Besides lifestyle management, pharmacological management of diabetes in the form of insulin therapy and the use of hypoglycemic drugs have improved the outlook of diabetic patients. However, it is expensive, harbors unfavorable side effects, and has achieved little to prevent the development of diabetes-related secondary complications.

11.3 THE ROLE OF INSULIN IN GLUCOSE METABOLISM

Insulin is the principal anabolic peptide hormone secreted by the β-cells of the pancreas, and plays a fundamental role in the regulation and maintenance of normal glucose and lipid levels [76]. The biosynthesis of insulin involves three main steps:

- The synthesis of preinsulin on the ribosomes of the rough endoplasmic reticulum.
- The conversion of preinsulin into proinsulin, which is then transported to the secretory granules of the Golgi apparatus.
- Cleavage of proinsulin into active insulin and C-peptide.

Insulin is released post-prandial by β-cells of the Islets of Langerhans in response to enhanced levels of glucose in blood circulation. Insulin performs its paracrine action on several tissues but mainly on the skeletal, adipose and the hepatic tissues [143]. These tissues possess a heterotetrameric transmembrane tyrosine kinase receptor composed of two extracellular α-chains and two intracellular β-chains. In response to increased glucose levels in circulation, insulin acts as a ligand, binding to the extracellular α-chains of its receptor. Binding of insulin to its receptor elicits auto-phosphorylation of the tyrosine site (Tyr^{960}) located on the intracellular domain of the receptor [34, 142]. The systemic phosphorylation of Tyr^{960} results in the formation of a recognition motif for the phosphorylation binding domain of the insulin receptor substrates (IRS). The IRS-1 protein is responsible to transmit signals that activate specific biological responses, e.g., glucose transport and synthesis of macromolecules [42]. IRS-1 binds to several proteins with SH-2 (Src homology 2) like p85, which is a regulatory subunit of phosphatidylinositol 3- kinase (PI3- kinase).

After IRS-1 binds to p85, a catalytic molecule p110 is recruited, causing a further activation of the 3-phosphoinositide-dependent kinases which as well phosphorylates Akt (a serine/threonine kinase) on Thr^{308}. The p13 kinase pathway coordinates the translocation process of GLUT-4, a glucose transporter molecule [33, 76, 143]. GLUT proteins migrate from the intracellular environment to the cell surface membrane, where their vesicles fuse with the cell membrane, allowing the channel for glucose entry to open, consequently lowering its levels in circulation [40, 155].

11.4 CLASSIFICATION OF DM

11.4.1 TYPE1 DM

Type-1 diabetes mellitus, previously referred to as insulin-dependent diabetes mellitus (IDDM), has a childhood onset. Type 1 DM emerges owing to the failure of insulin secreting β-cells of the islets of Langerhans to produce insulin as a consequence of auto-immune attack of these cells [120, 175]. The autoimmune destruction of the islets in diabetics is linked to the presence of islet cell-specific antibodies in the blood, excessive interleukin production by monokines (a deranged T-cell-mediated immunoregulatory system) and activation of antigens targeting the Islet cells [116]. Therefore, daily doses of insulin are given to these patients to maintain euglycemia and thus preventing life-threatening complications [117]. Type 1 DM accounts for 5–10% of all diabetic cases globally. Type 1 DM can also exist idiopathically with no known cause in the absence of autoimmune destruction of the β-cells [101].

11.4.2 TYPE 2 DM

Type 2 diabetes mellitus (T2DM) is a fast-growing global health threat promising to reach pandemic proportions if no appropriate measures are taken. T2DM is the most common type of DM and was previously known as non-insulin-dependent diabetes mellitus (NIDDM). It is responsible for almost 90% of the DM cases reported and its progression has been observed in the adult population with initial characteristics of high blood glucose levels, insulin resistance, and/or deficiency. A significant contribution of lifestyle and genetic factors have been documented in the predisposition to T2DM. In T2DM, the β-cells of the pancreas produce detectable amounts of insulin. However, the problem arises in the recognition of the insulin by its receptors in the target organs (insulin resistance), resulting in disturbed insulin intracellular signaling cascades [20, 118].

11.4.3 OTHER ETIOLOGICAL CLASSIFICATION

Other etiological classification of DM include gestational diabetes mellitus, genetic defects of β-cell function, genetic defects in insulin action, e.g., lipoatropic diabetes; diseases of the exocrine pancreas, e.g., pancreatitis, endocrinopathies, and drug and chemical-induced diabetes; infection-induced diabetes, e.g., by cytomegalovirus [50].

11.5 EPIDEMIOLOGY AND THE GLOBAL BURDEN OF T2DM

The World Health Organization (WHO) reported that approximately 422 million adults were suffering from DM in 2014, and it has been shown that the number of people living with DM in 2014 quadrupled when compared to 1980 global statistics [173]. In 2012, DM and its secondary disorders were root causes of 3.7 million deaths that occurred. In addition to that, 43% of these deaths occurred in individuals that were $\leq$ 70 years old. Based on the reported estimates, DM inflicts a substantial economic strain on the households way up to the nations at large. The burden of DM has an impact on healthcare systems owing to the direct medical costs and the attendant work absenteeism. DM, if not adequately controlled, can lead to severe visual impairment, lower limp amputations, kidney failure, male infertility, and cardiovascular associated disorders. In 2010, DM was responsible for the 4.5% of visual impairment and blindness cases. A constellation data from 54 countries revealed that 80% of the end-stage renal disease (ESRD) was as a result of DM. These complications directly contribute to the regression of economic growth.

The International Diabetes Federation [IDF, 80] recently published that in every 6 seconds, a person dies from DM and its complications, and

1 in 11 adults has DM. According to IDF reports [2016], more than two-thirds of people living in Africa have undiagnosed DM. Approximately 14.2 million people in sub-Saharan region have DM, with more than half of them residing in South Africa, Democratic Republic of Congo, Nigeria, and Ethiopia [80]. The management of DM and its complications in developing countries is difficult; therefore, adequate health resources are required to offer proper health care [164, 173].

11.6 OXIDATIVE STRESS IN DIABETES MELLITUS

Free radicals are produced during normal aerobic respiration in the mitochondria [125]. Free radicals including reactive oxygen species (ROS) and reactive nitrogen species (RNS) are described as reactive molecular species (RMS) that have unpaired electrons in their outer atomic orbitals. ROS is a broad term used to categorize molecules derived from incomplete oxygen reduction reactions whilst RNS comprises of products from nitrogen oxidation reactions [170]. Examples of ROS include superoxide ion ($O_2^{\cdot-}$), hydrogen peroxide (H_2O_2), hypochlorous acid (HOCl), and the hydroxyl radical ($HO^{\cdot}$), whereas nitric oxide (NO), nitroxyl (NO^{-}), S-nitrosothiol (RSNO), and peroxtnitrite ($OONO^{-}$) are types of RNS [21, 122]. Oxygen is the principal substrate that is utilized during cellular respiration. Not all of the oxygen inhaled is harnessed in the production of ATP; rather about 5% of the inhaled oxygen reacts with different molecules resulting in the formation of these ROS and RNS [32]. When produced during normal homeostasis, these free radicals participate in cellular signaling, cell growth, apoptosis, defense against pathogens and in regulatory pathways [104, 170].

If an imbalance arises between the production of ROS/RNS and the ability of antioxidants in biological systems to degrade or to remove these free radicals, where the imbalance favors free radical production, oxidative stress subsequently is resulted [21]. Oxidative stress is noted in various pathological states of different disorders displayed by the overwhelming production of free radicals and depleted production of antioxidants [146, 147]. In an event of overt oxidative stress, cellular components like proteins, lipids, and DNA are irreversibly modified, ultimately resulting in cellular and organ damage [122].

11.6.1 IMPLICATION OF OXIDATIVE STRESS IN THE PROGRESSION OF DIABETIC COMPLICATIONS

The chronic sustained hyperglycemic condition or state in diabetes is regarded as the major factor that triggers the development of both acute and long-term changes in the cellular metabolism of different molecules [65, 118]. Altered metabolism of macromolecules ultimately triggers excessive formation of free radicals via different pathways, resulting in the speeding up of development of diabetic complications [170]. ROS are incriminated in the disruption of glucose transport systems and in the defective receptor-ligand sensitivity between insulin and its receptor, giving rise to insulin resistance [123].

It has been strongly suggested that oxidative stress plays a key role in the development of diabetic vascular complications. Malignant transformations of tissues as a consequence of oxidative stress are mostly seen in the eyes, nervous system, and the kidneys [120]. Several pathways have been established that link the role of oxidative stress in promoting the pathogenesis of diabetic complications. The growing evidence has reported that hyperglycemia causes tissue injury via four major mechanisms:

- Increased glucose flux through the polyol pathway;
- Increased intracellular formation of AGEs (advanced glycation end products);
- Activation of the protein kinase (PK) C pathway; and
- Increased activity of the hexosamine pathway.

Scientific evidence has shown that all four mechanisms are triggered by a particular upstream event that points to the overproduction of reactive oxygen species by the mitochondria [65, 129].

11.6.2 FORMATION OF ADVANCED GLYCATION END PRODUCTS (AGES)

Progression of oxidative tissue damage occurs at a faster rate in patients with poor metabolic control of hyperglycemia, which in turn impels the formation of advanced glycation end-products (AGEs), which has been associated with the severity of diabetic complications [106].

AGEs is a heterogeneous group of molecules formed in the absence of enzymatic catalysis in a reaction termed the Maillard reaction [123]. The first step in the glycation pathway is the slow reversible nucleophilic reaction between the carbonyl groups (aldehydes/ ketones) of reducing sugars or monosaccharides with the amine groups of proteins, nucleic acid or lipids [94]. The outcome of the reaction is the formation of highly reversible products known as the Schiff bases [125, 151]. Spontaneous rearrangement reactions of the Schiff bases follows over a period of days to weeks forming more stable compounds termed Amadori products. Further rearrangement and oxidation of the latter compounds take place resulting in AGEs [6, 23, 94, 121, 125, 149].

The AGEs principally target long-lived connective tissue proteins like type-IV collagen, tubulin, plasminogen activator-1, and fibrinogen forming irreversible cross linkages among them [149]. The cross-links formed later develop into tough fibers, which stiffen the blood vessels, diminishing arterial and myocardial compliance as well as increasing diastolic dysfunction and systolic hypertension [23, 67]. Modification of the cellular matrix of the basement membranes by AGEs has also been implicated in the severity of tissue damage in diabetic nephropathy [121, 149].

Signaling of AGEs involves their interaction with their cell surface receptors called the receptor of advanced glycation end products (RAGE). Binding of AGEs to RAGE triggers the activation of NADPH oxidase system, cytokine release, and free radical activity further enhances tissue damage [157]. The production and translocation of NF-kβ via the Ras-MAPK pathway has been extensively studied as the product of the association of AGEs and RAGE. NF-kβ is a transcriptional factor that modulates the transcription of vascular endothelial growth factor (VEGF), which is a mitogen that stimulates the mitosis of endothelial cells therefore increasing their permeability and the progression of vascular diabetic complications [123]. Experimental evidence reported a direct proportional relationship between the levels of AGEs and the severity of diabetic complications. Furthermore, it has been indicated that AGEs, alter enzymatic activities, cause protein fragmentation, decrease ligand binding capacity and immunogenicity alterations [123]. Pentosidine, carboxymethyl-lysine (CML, a product of glycoxidation), imadazolones, and pyrraline are the predominant types of endogenous AGEs that have been studied extensively [6].

11.6.3 INCREASED POLYOL FLUX PATHWAY

The polyol pathway (also known as the aldose reductase (AR) pathway) is activated when there is an upsurge in the concentration of glucose in cells [98]. The pathway utilizes keto-reductase enzymes to catalyze the reduction of carbonyl compounds of sugars into their respective polyols [65]. In the course of normal metabolism, AR facilitates the reduction of reducing sugars into inactive alcohols, however if cells encounter excess levels of glucose, AR as the rate limiting enzyme converts glucose into sorbitol. The conversion depends on NADPH as the cofactor [37]. Oxidation of sorbitol into fructose by sorbitol dehydrogenase (SDH) follows with NAD^+ as the cofactor [65]. The most feasible explanation on how an increased polyol flux causes detrimental tissue damage is the depletion of NADPH in the aldo-keto reaction. NADPH is an important cofactor of glutathione reductase in reactions that trigger the generation of glutathione (GSH) from its oxidized form GSSH. Overconsumption of NADPH in the aldo-keto reaction deprives glutathione reductase (GR) of its cofactor resulting in attenuated GSH production, a critical antioxidant that scavenges free radicals and ROS.

Deficiency in GSH further depletes antioxidant defense systems leading to oxidative stress [37, 44, 65, 98]. In addition, phosphorylation of fructose to fructose-3-phosphate takes place, the phosphorylated fructose molecules undergo degradation resulting in the formation of 3-deoxyglucosone. These two products of fructose phosphorylation are strong glycosylating compounds that glycosylate molecules giving rise to AGEs precursors [44]. The oxidation reaction of sorbitol to fructose by SDH has been documented as the channel that favors ROS formation. This is because in the reaction, NAD^+ is reduced to NADH. NADH is a substrate of NADH oxidase in one of the pathways that forms ROS [44]. The increased polyol flux oxidative damage was greatly indicated in the retina of the diabetic mice that overexpressed AR. Overexpression of AR and sorbitol in diabetic mice also increased atherosclerosis and diabetic lesion, respectively [65].

11.6.4 MITOCHONDRIAL PRODUCTION OF THE SUPEROXIDE ANION

The mitochondria are cellular organelles housed in the cytoplasmic area that are important in the survival and normal functioning of cells [35]. The mitochondria are made up of two distinct areas that are surrounded by inner and outer membranes. These membranes are important in the maintenance of normal electrochemical gradients [53]. During normal respiration in the electron transport chain (ETC), electrons from reduced substrates pass through different redox centers grouped into four complexes:

- complex I (NADH dehydrogenase),
- complex II (succinate dehydrogenase),
- complex III (ubiquinone cytochrome c oxidoreductase), and
- complex IV (cytochrome oxidase).

The overall effect of the passage of electrons is the pumping of protons (H^+) into the inner membrane space. This movement of H^+ generates a proton motive force across the inner membrane that translocates the protons back into the matrix. As protons move back into the matrix, complex-V, also termed ATP synthase, is driven thus leading to the generation of ATP. Oxygen molecules are the final acceptors of these electrons resulting in the formation of H_2O and the superoxide anions [53, 120]. Although the superoxide anion is produced during normal respiration, yet it has been identified that evolution has mechanisms to counteract development of OS because of the availability of endogenous antioxidants such as manganese superoxide dismutase (MnSOD). MnSOD is a mitochondria-derived enzyme that dismutates superoxide anions into hydrogen peroxide (H_2O_2). Hydrogen peroxide is in turn converted to water and oxygen by the enzyme, glutathione peroxidase (GPx) [39, 53, 69, 84].

Prolonged exposure of cells to high glucose concentrations favors the oxidation of glucose in the form of pyruvate. Oxidation of pyruvate increases the influx of electron donors NADH and $FADH_2$ into the ETC increasing the voltage gradient across the mitochondrial membranes. The electrons being transferred bypass complex-III to coenzyme Q, which then donates them to oxygen molecules, generating excessive amounts of superoxide anions [65, 87]. Production of superoxide anions in this side reaction overpowers the ability of SOD to degrade them, thus empowering oxidizing ability of superoxide anions on proteins, DNA and lipids inflicting permanent oxidative changes. Moreover, hydrogen peroxide generated in the reaction has been associated with the production of hydroxyl radicals in the presence of ferrous ions. Hydroxyl radicals are powerful free radicals that potentially cause tissue damage [35, 63]. Despite the implication of mitochondrial superoxide production in diabetes, it has been implicated in the in the etiology of diseases such as cancer, aging, ischemia, obesity, and Alzheimer's and Parkinson's diseases [69, 140].

11.6.5 INCREASED ACTIVATION OF PROTEIN KINASE C

Protein kinase C (PKC) is an enzyme complex made up of 11 isoforms that belong to serine/threonine-related protein kinases family. The enzymes take part in the phosphorylation of the target proteins. Activation of PKC isoforms requires the presence of calcium ions, phosphatidylserine (PS) and diacylglycerol (DAG) [65]. Overactivity of the PKC pathway acts as an alternative route for the formation of hyperglycemia-induced ROS. Enhanced levels of glucose speeds up the synthesis of DAG via triose phosphate in which the glycolytic enzyme, glyceraldehyde-3-phosphate dehydrogenase (GAPDH), is inhibited by ROS. Inhibition of the GAPDH favors the formation of a DAG precursor [24, 65]. Hyperglycemia also increases the levels of dihydroxyacetone phosphate, which is reduced to glycerol-3-phosphate. Glycerol-3-phosphate conjugates with fatty acids thus increasing the *de novo* synthesis of DAG which further activates the PKCs again [138]. PKC isoforms control several cellular signals like the NADPH oxidase and NF-κβ, the two have major roles in the events that trigger ROS production. Activation of PKCs by glucose further contributes to the abnormal increase in vascular permeability as a consequence of the over-expression of vascular endothelial growth factor (VEGF), thickening of the basement membrane, apoptosis, and expansion of extracellular matrix [56, 59].

11.7 PHYSIOLOGICAL AND PATHOLOGICAL EFFECTS OF ROS IN THE DEVELOPMENT OF DIABETES COMPLICATIONS

11.7.1 MALE INFERTILITY

Uncontrolled production of ROS that exceeds the seminal plasma antioxidant capacity is a common phenomenon in diabetic patients that leads to male infertility [49, 92,146]. Antioxidants are crucial in the maintenance of free radicals within homeostatic levels that enable them to perform their homeostatic functions at the same time preventing pathological damage [52, 92]. Epidemiological and experimental studies have correlated diabetes-linked oxidative stress and male infertility [17, 92] as evidenced by testicular and epididymis dysfunction [75] in form of sperm cellular membrane lipid peroxidation [145], decreased sperm motility [38] and increased spermatic damage in diabetic patients [148].

At the cellular level, ROS attack can induce spermatozoal membrane lipid peroxidation, DNA fragmentation, and apoptosis. These effects have an overall ability to disrupt the motility of spermatozoa and negatively impact their ability to support normal embryonic development [3, 8–10, 12, 91]. At the testicular level, ROS-induced oxidative stress disrupts the steroidogenic capacity of Leydig cells as well as the capacity of the germinal epithelium to differentiate normal spermatozoa [111].

While seminal plasma antioxidants may help to prevent spermatozoal oxidative damage by ROS following ejaculation, they have no capacity to counter testicular and epididymal oxidative stress [75]. This is partly because during spermatogenesis and epididymal storage, spermatozoa are not in contact with seminal antioxidants and therefore must rely on epididymal/testicular antioxidants and their own intrinsic antioxidant capacity for protection against oxidative damage. Considering this, spermatozoa are vulnerable to oxidative damage during epididymal transit [38].

11.7.1.1 Mechanisms by which ROS Cause Male Infertility

While spermatozoa are vulnerable to oxidative stress, the presence of ROS appears to be crucial for the sperm to attain functional competence [4, 68, 134]. Under physiological conditions, normal metabolic processes are responsible for the generation of ROS, which are vital in the transduction of signals during complex spermatogenic biochemical cascades including processes of sperm maturation, capacitation, hyperactivation, acrosome reaction, and sperm-oocyte fusion [3, 7, 11, 49]. During the process of spermatozoal maturation, ROS contribute to the development of a keratinous-like protective coat called the membranous capsule around the mitochondria [132]. Hydrogen peroxide (H_2O_2) has been shown to contribute to the formation of this capsule by oxidizing hydroperoxide glutathione peroxidase (HPGxP), a phospholipid selenoenzyme, to form a chemical intermediate capable of forming a selenadisulfide bond with reduced protein thiol groups of the capsule. The inactivation of PHGPx and oxidation of thiol groups ultimately results in the production of a complex, yet stable protein network of the mitochondrial capsule. The importance of this mitochondrial capsule lies in the fact that deletion of spermatic mitochondrial capsule selenoprotein, seriously affects sperm motility despite having normal sperm morphology [29]. The superoxide anion has been shown to facilitate spermatozoal hyperactive motility providing spermatozoa with the necessary propulsion to penetrate the cumulus oophorous of the female gamete [47, 48]. Hyperactive spermatozoa exhibit an overall nonlinear motility characterized by significant side-to-side displacement of the sperm head, asymmetric flagellar movements, and high amplitude [90, 113]. During spermatogenesis, capacitation ensures that only fertile spermatozoal cells reach, bind, and fertilize the female gamete [29]. The superoxide anion is hypothesized to oxidize an essential thiol group thereby activating the acrosome reaction. On the other hand, H_2O_2 may promote tyrosine phosphorylation indirectly through the activation of tyrosine kinases (TKs) and inhibition of PTPases [90].

Physiological levels of ROS have been hypothesized to enhance spermatozoal capacitation by activating TKs, stimulating adenyl cyclase (AC) and inhibiting the activities of PTPases. By doing so, ROS increases the amount of tyrosine phosphorylation, one of the final and critical steps in capacitation [90]. During capacitation, by inhibiting PTPase activity, ROS prevents dephosphorylation and deactivation of phospholipase-A_2 [PLA_2], which cleaves the secondary fatty acids from the triglycerol backbone of the membrane phospholipid and increases membrane fluidity [8]. In addition, by activating AC and downstream molecules during the acrosome reaction, ROS may activate PLA_2. After successful propulsion of hyperactivated spermatozoa past the cumulous oophorous, the capacitated spermatozoa must bind to the glycoprotein layer that surrounds the oocyte called the Zona Pellucida (ZP), where they initiate the exocytotic release of proteolytic enzymes [47]. Low concentrations of the superoxide anion (O_2^-), hydrogen peroxide (H_2O_2) and nitric oxide (NO^{--}) have been demonstrated to be essential for the acrosome reaction by activating AC that triggers cAMP to initiate the exocytotic release of the enzymes [77, 114]. Activities of ROS have also been shown to increase the affinity of sperm for the ZP by the phosphorylation of membrane proteins at the sperm head, namely the sperm adhesin family, fertilin-β and p47 [36]. After penetration of the ZP, a high degree of spermatic membrane fluidity is essential for successful fertilization of the oocyte [9]. ROS are hypothesized to be crucial in terms of increasing membrane fluidity at this point where they increase the rate of sperm-oocyte fusion [172].

11.7.1.2 Lipid Peroxidation

The plasma membrane of spermatozoa is composed of polyunsaturated fatty acid (PUFA), making them extremely susceptible to lipid peroxidation (LPO) [41, 66, 156]. The makeup of PUFA includes more than two carbon-carbon double bonds; hence these fatty acids maintain the fluidity of spermatozoal membranes [134]. The membrane lipid components are involved in the regulation of spermatogenesis, spermatozoal maturation, capacitation, acrosome reaction, and membrane fusion. Under the influence of ROS, PUFA suffer lipid peroxidation, which is a self-propagating, autocatalytic reaction [134] that leads to a significant reduction in membrane fluidity [22, 96], loss of membrane enzyme system function and increased non-specific permeability to ions [99]. As a consequence of lipid peroxidation, spermatozoa rapidly lose ATP causing axonemal damage in form of defects in mid-piece morphology. The OS has also been shown to decrease the viability of sperm and induce deleterious effects on sperm capacitation and the acrosome reaction [30]. In this manner, lipid peroxidation disturbs spermatozoal functions and in severe instances, completely inhibits the process of spermatogenesis. Once ROS exerts its effects on membrane lipids, peroxyl and alkyl lipid radicals are formed which continually propagate peroxidative damage until their effects are countered or quenched by seminal antioxidants [51]. In this manner, LPO of spermatozoa is considered as the key mechanism of ROS-induced spermatozoal damage leading to male infertility.

11.7.1.3 DNA Damage

ROS has the potential to damage both nuclear DNA (nDNA) and mitochondrial DNA (mtDNA), but by contrast, mtDNA is more susceptible to oxidative stress because its DNA is naked [163] and not bound by histones, thus making them vulnerable to the effects of ROS, which arises naturally from processes that occur within the mitochondria [137]. In contrast, nDNA is less vulnerable to OS as it is tightly packed and stabilized by disulfide bonds to form crystalline toroids [31]. Under normal circumstances, spermatozoal DNA is protected from oxidative effects by its compact organization and antioxidants in the seminal plasma. As a consequence of OS, mtDNA can suffer base modifications, base oxidation, and denaturation into single-stranded DNA breaks, DNA crosslinks, base free deletions, chromosomal rearrangements, chromosomal microdeletions, point mutations, and polymorphism [81, 148]. The mtDNA mutations may cause a defect of mitochondrial energy metabolism [43]. In this manner, low levels of mutant mtDNA can adversely compromise sperm motility in vivo [152]. Due to all these oxidative effects, spermatozoa lack the capacity to repair the damaging attacks of ROS on its DNA [45]. They do not have the necessary cytoplasmic enzyme system to repair ROS-induced DNA damage making their DNA prone to the effects of oxidative stress. On that note, spermatozoa solely depend on the oocyte repair system for the correction of less severe DNA damage including those that occur as a consequence of OS as seen in diabetic patients [10]. However, in cases of extensive DNA damage, the oocyte repair mechanisms are inhibited resulting in infertility. In this manner, the oocyte repair mechanism helps prevent the propagation of abnormal DNA across generations [10].

11.7.1.4 Apoptosis

ROS may also initiate apoptosis that in turn promote male infertility. Apoptosis is a natural physiologic process by which the body gets rid of old, senescent cells including abnormal germ cells thereby preventing their overgrowth during spermatogenesis [3, 133]. Research has shown that high levels of seminal ROS correlates with decreased sperm variables as seen in infertile men. In most cases, increased sperm damage by ROS is associated with higher levels of cytochrome-c and caspases 3 and 9, indicating that apoptosis is a significant factor of male infertility. High levels of ROS are capable of disrupting the inner and outer mitochondrial membranes thereby stimulating the release of cytochrome-c by the mitochondria. Cytochrome c in turn signals the release of caspases particularly caspase 3 and 9, which induces apoptosis. Levels of these caspases have been revealed to be significantly higher in infertile men. In this manner, the mitochondria play a significant role in oxidative stress-induced apoptosis of spermatozoa [46, 171].

11.7.2 DIABETIC KIDNEY DISEASES

Diabetic kidney disease (DKD), which is a term used interchangeably as diabetic nephropathy is a major complication of diabetes that affects approximately 20-40% of diabetic patients globally. DKD is categorized as a microvascular complication that develops into end stage renal disease (ESRD) and cardiovascular disease, and as a result, increasing the mortality rate of diabetes. Diabetic kidney disease exists in several phases of development that are initiated by different mechanisms. The initial marker of diabetic kidney disease in the absence of other renal diseases, is the leakage of albumin into the urine also referred to as urine albumin excretion (UAE).

11.7.2.1 Implication of OS in the development of DKD

It has been clearly demonstrated that OS plays a crucial role in impairing normal kidney function in diabetes. Hyperglycemia disturbs normal cellular processes through induction of OS and inflammation. Generation of ROS in the renal tissue has been shown to be related to the modification of the extracellular matrix, vasoconstriction, and overgrowth of the vascular smooth muscles, endothelial dysfunction and sodium imbalance [55]. The relationship between OS and cytokine production has been established as ROS act as second messengers that activate various transcription factors that encode for cytokine genes. For example NF-κβ is a transcription factor produced via the Ras-MAPK pathway as a result of the association of AGEs and RAGE. NF-κβ plays a key role in the events that lead mesangial cell activation [123, 55]. The overproduction of the AGEs in renal tissues has been reported as the chief pathway that exerts tissue damage as a result of the formation of crosslinks within the renal matrix. The extensive damage was ascribed to the role that the renal system plays in the final clearance of these AGEs.

Other studies reported that long-lived AGEs in renal tissue cripple the antioxidant defense functions of CuSOD and MnSOD thus empowering ROS damage [60]. In addition, the effects of abnormal mitochondrial electron transport chain, auto-oxidation of glucose, activation of the polyol, aldo-keto reductase and PKC pathways have also been strongly emphasized in the pathogenesis of diabetic nephropathy. Induction of apoptosis in renal tissue as a consequence of ROS production has been shown by different studies [60, 87]. Other studies done in streptozotocin (STZ)-induced animal subjects indicated that the inhibition of OS was successful in response to certain interventions employed to reverse manifestations of diabetic nephropathy [55].

11.7.3 DIABETIC LIVER DISEASE

The prevalence of diabetic hepatic disease in diabetic patients has been estimated to lie between 17 and 100%, and association of the disease has been reported in obese type-2 diabetic subjects [14]. Hepatic insulin resistance is the major metabolic factor reported in diabetics that leads to the development of non-alcoholic fatty liver disease (NAFLD), which describes different liver pathologies including benign conditions, hepatic glycogenosis, and nonalcoholic steatohepatitis. Diabetic hepatologies are frequently marked by the following clinical and metabolic features: abnormal liver function tests, abdominal discomfort, nausea, vomiting, hepatomegaly and increased glycogen storage in hepatocytes [62].

In type-1 diabetes, insulin deficiency negatively influences glucokinase levels and hence lowering the glucose uptake in hepatic cells. In response to the disturbed glucose metabolism, hepatic glucagon levels increases thus initiating glycogenolysis and gluconeogenesis [14]. It has been shown that decreased levels of insulin will cause glucose output by the liver to increase. Overinsulization (as a result of insulin therapy) in type-1 diabetes in turn causes hepatic glycogenosis where there is a marked increase of glycogen molecules in hepatocytes. Hepatic glycogenosis is a reversible condition that is treated by glycemic control [83]. ROS have been implicated in the emergence of diabetic hepatopathies to a lesser extent. The involvement of ROS has been linked to lipid peroxidation of hepatocytes' membranes and apoptosis of hepatocytes [61].

11.8 MEDICINAL PLANTS AND THEIR HEALTH BENEFITS

Medicinal plants have been defined as naturally occurring substances that possess potent compounds that can be utilized for therapeutic reasons or synthetically modified to pharmacological drugs [167]. Plants are fundamental to human and animal lives as both food and medicine. For thousands of years, various cultures relied on plants as the primary source of health care. It is estimated that almost 80% of the developing countries' and 40–60% of the developed countries' populations still depend on medicinal plants for the management and treatment of ailments [57,136, 167]. Africa is a continent where the use of medicinal plants is greatly practiced for traditional and therapeutic purposes. The heavy dependence on medicinal plants is undeniably a consequence of their accessibility, safety and affordability [109]. It is documented that 122 pharmacological drugs have been developed from 94 plant species among the 250,000 plant species available globally, although more than 20,000 have been documented to have medicinal values [58, 153]. Unfortunately, the validation and inclusion of medicinal plants in the orthodox treatment regimens are still lagging especially in Africa.

Throughout history, medicinal plants have been used to alleviate symptoms of various diseases such as cancer, diabetes, hypertension, infertility, gastrointestinal problems, infections, headaches etc. [85]. Globally, information pertaining to certain plants' medicinal activities has been passed from generation to generation, although some of the claims have been scientifically proven to be erroneous [28]. Fortunately, scientific studies have been able to confirm some of these claims and established the importance of medicinal plants in health care. Plant-based diets composed of fruits, vegetables, spices, teas, and seeds have been linked to the decreased risks of developing noncommunicable diseases (NCDs) such as cancer and diabetes mellitus [126, 161].

In the past decades, an etiology of NCDs in Africa was generally low. However due to rapid globalization and behavioral changes, increased trends of NCDs are now common. Inclusion of genetically modified organisms was unusual in Africa, while consumption of natural products such as plant materials was common; thus explaining prior lower trends of NCDs [164]. The burden of diabetes mellitus in Africa is reported to be on the increase to such an extent that a 2.5 fold increase in its prevalence is expected by the years 2020–2025 [19]. Moreover, in sub-Saharan Africa, more than 10 million people are reported to have type-2 DM and if not urgently addressed, figures are expected to reach 18.7 million by 2025 [1, 97]. It has also been observed that hypertension imposes heavy constrains in sub-Saharan regions as highest prevalence of this condition was reported [15]. Without new approaches to redeem the escalating prevalence of NCDs and their secondary disorders in Africa, economic and health burdens will keep crippling livelihood.

No doubt scientific advances have played a major role in the management of human diseases, orthodox formulations however have been reported to have unfavorable effects, in addition to the high costs that make it impossible for the populace in developing countries to rely on [115]. This calls for comprehensive scientific investigations on medicinal plants for the development of cheap novel therapies. The nonnutritional chemical components of medicinal plants, also referred to as phytochemicals (including caretenoids, polyphenols, alkaloids, saponins, tannins, and vitamins), possess established antioxidant activities against free radicals that are chief factors implicated in the development of noncommunicable diseases [144]. Inclusion of exogenous antioxidants from dietary sources is important since there is consumption of endogenous antioxidants in biological systems in response to free radical attack [54].

In the 20th century, drugs such as ectoposide, metformin, vinblastine, vincristine, artemisinin etc. have been developed from medicinal plants. This indicates clearly how medicinal plants have made a significant contribution to modern therapeutics in the management and treatment of various diseases. It is important to note that medicinal plants undisputedly play a key role in the management of human health. The goal for optimizing the usage of medicinal plants to prevent/ cure NCDs is thus imperative.

11.8.1 ROLE OF ANTIOXIDANTS IN PREVENTING OXIDATIVE STRESS

An antioxidant is defined as "any substance, which when present in low concentration compared to that of the oxidizable substrate, significantly delays or attenuates the oxidation of that substrate and the final outcome is its oxidation" [174]. Antioxidant is a broad term to characterize enzymes that degrade free radicals (catalase, SOD, and glutathione peroxidase) and proteins that have the ability to bind to metals to initiate free radical generation (transferring that bind iron) and exogenous antioxidants (vitamin C, vitamin E, flavonoids, carotenoids, alpha-tocopherol) [124, 165]. The antioxidant properties of these abovementioned compounds are derived from their ability to:

- break the reaction chains of free radicals e.g. during lipid peroxidation.
- sequestrate transition metal ions.
- repair damaged molecules.
- scavenge radicals (uric acid, ascorbic acid and albumin) [72, 112, 165].

Constant replenishing of the endogenous antioxidant pool is essential because they are inevitably consumed during attack by free radical. The shortage is counterbalanced by antioxidant supplementation from plant materials. A growing body of evidence has shown that plants possess phytochemicals that exhibit strong antioxidant activities [53]. The antioxidant compounds extracted from plants that have been associated with high antioxidant powers include: flavanoids, isoflavones, flavones, anthocyanins, coumarins, lignans, catechins, alkaloids, terpernoids, polyphenols, and isocatechins [112]. For this reason, adequate intake of exogenous antioxidants plays a fundamental role in the prevention and treatment of various pathologies, which include diabetes, neurodegenerative diseases, pulmonary disorders, aging-related diseases and cancer [141, 168]. Studies conducted to compare the antioxidant levels of non-diabetics and type-1 diabetic subjects reported a 16% lower level in type-1 patients [124].

Accumulation of reactive oxygen species (ROS) and their ability to oxidize biological molecules is known to contribute to an increased rate of cell shrinkage, chromatin condensation, DNA fragmentation as well as compromised signaling mechanisms [73]. Further, the lack of appropriate defense strategies to revert the uncontrolled generation of ROS leads to the development of adverse diabetic complications. The availability of adequate supply of antioxidants greatly intercepts the oxidizing effects of free radicals. In this chapter, authors examined two medicinal plants (*Catharanthus roseus* and *Punica granatum*) and their therapeutic activities in respect to diabetes and associated complications.

11.9 *CATHARANTHUS ROSEUS* IN THE TREATMENT OF DM

Catharanthus roseus is a popular ornamental shrub that has its origin in Madagascar; hence, its alternative name is "Madagascar periwinkle." It has been used traditionally in several countries to treat diseases such as malaria, Hodgkins disease, skin diseases and diabetes [110, 127]. *C. roseus* grows vertically up to 100 cm and is easily identified with its white, dark pink, or purple petal like flowers and its oval leaves, which are glossy and hairless [105]. The roots, leaves, and flowers of this plant have been used throughout ancient times in the treatment of various ailments. The medicinal values of *C. roseus* were ascribed to the presence of more than 100 alkaloids. Among these alkaloids, vincristine and vinblastine have been successfully isolated and used as anticancer agents because of its anti-mitotic properties [159].

The antidiabetic activities of *C. roseus* have been extensively studied. It was reported that its leaves are consumed as tea to treat DM [127]. Various scientific evidences documented the increase in glucose utilization in both in vivo and in vitro setups following administration of *C. roseus* [158].

11.9.1 ROLE OF C.ROSEUS TO INCREASE THE ACTIVITIES OF ENZYMES OF CARBOHYDRATE METABOLISM

Singh et al. [150] evaluated the effects of dichloromethane-methanol (DCMM) leaf and twig extract of *C. roseus* on the activities of the enzymes involved in the metabolism of carbohydrates in STZ-induced Sprangue Dawley rats. From their findings; 500 mg/kg body weight (B.W.) of DCMM leaf and twig extract of *C. roseus* reduced blood glucose levels of diabetic rats by 57.6% and 48.6% in a treatment period of 15 and 7 days, respectively. They also reported increase in the activity of glucose-6-phosphate dehydrogenase enzyme in treated diabetic rats when compared to the untreated diabetic controls. Glucose-6-phosphatase dehydrogenase is a regulatory enzyme that plays an important role in the pentose phosphate pathway. The higher activity observed in the treated diabetic animals indicated improvement in glucose utilization. In the same study, it was reported that treatment of diabetic rats with the DCMM extract of *C. roseus* increased the activity of glucokinase. The group additionally measured levels of malate dehydrogenase, an enzyme that releases oxaloacetate required in the generation of citrate in the citric acid cycle. In this reaction, oxaloacetate has to react with acetyl-CoA to produce malate. Malate is a reactant needed in the cytosolic gluconeogenic pathway. The study showed significant increase in the levels of malate dehydrogenase in the liver and plasma of treated diabetic rats.

11.9.2 ROLE OF *C. ROSEUS* TO ENHANCE EXPRESSION OF GLUCOSE TRANSPORTER GENES

In a study that assessed the effects of *C. roseus* on the expression of glucose transporter genes (GLUT-2 and GLUT-4), the investigators reported that the mechanism through which *C. roseus* mitigates hypoglycemia is by increasing the expression of GLUT-4 and GLUT- 2 genes. GLU-4 and GLUT-2 molecules are insulin sensitive molecules that recruit glucose from the extracellular environment into the cytosol during insulin signaling. The liver and muscle cells are the main sites of action.

The effect of two different concentrations of ethanolic extract of *C. roseus* leaves in STZ-induced rats was evaluated on the expression of GLUT genes. Treatment of diabetic rats with ethanolic extracts of *C. roseus* leaves was shown to enhance the expression of GLUT genes. Based on the findings, it was concluded that in untreated diabetic rats, insulin deficiency may contribute to the down regulation of GLUT gene expression [16].

11.9.3 ROLE OF *C. ROSEUS* TO PREVENT THE DEVELOPMENT OF CVDs

Administration of the methanolic extract of *C. roseus* leaves for 14 days in STZ-induced diabetic rats was found to significantly reduce the levels of serum lipids. Improvement in the reduction of serum lipid parameters may contribute to the delay or prevention of macrovascular complications of DM. In the same study, a higher dose of 400 mg/kg B.W. significantly reduced the levels of other serum lipids while it elevated the levels of HDL. Investigations on the integrity of pancreatic β-cells in treated diabetic rats *versus* the untreated diabetic controls showed that *C. roseus* can rejuvenate the loss of β-cells. Loss of β-cells in the pancreas is a pathologic feature observed in diabetes patients. The findings of this study strongly suggest a need for further evaluation of *C. roseus* in higher animals or clinical trials [89]. Similar findings were also reported by Ghosh and Suryawansh [64], who measured the effects of the extracts of *C. roseus* leaf and flower on diabetic rats and normal controls for 7 days. The serum triglycerides, cholesterol, and free fatty acids were reported to be elevated in diabetic controls, whereas treatment with either the flower or leaf extract reduced the serum lipids to near normal.

11.9.4 ROLE OF C. ROSEUS TO PREVENT DIABETES-INDUCED OXIDATIVE STRESS

As previously highlighted, the development of DM complications is attributed to the uncontrolled production of ROS. Zhang et al. [176] investigated the effect of *C. roseus* alkaloid mixture against DM in high fat diet-fed diabetic rats. Induction of DM in the Wistar rats resulted in reduction in the activities of antioxidant endogenous enzymes in different tissues. Treatment of diabetic rats with the alkaloid mixture significantly increased the activity of SOD in the heart, liver and kidney tissues by 130%, 108% and 71%, respectively when compared to the untreated diabetic controls. The activities of catalase (CAT) in the same tissues were reported to have increased by 82.52%, 65.95%, and 25.79%, respectively. In addition, their findings showed the reversal of TBARS levels to near normal in treated diabetic group. The increase in the activities of the endogenous antioxidant enzymes in diabetic rats after treatment indicated that the alkaloids derived from *C. roseus* may potentially reduce glycation of enzymes and/or decrease the levels of ROS. The results observed in this study concluded that *C. roseus* might prevent diabetes-induced oxidative tissue damage and therefore the risk of diabetic complications.

Additional antidiabetic research on *C. roseus* is summarized in Table 11.1.

11.10 *PUNICA GRANATUM* (POMEGRANATE)

Pomegranate (*Punica granatum*) is a medicinal plant that is native to the Himalayas, northern India, and Iran, but has been cultivated and consumed as a fresh fruit or in beverage form since ancient times over the entire Mediterranean region [130, 131]. Based on scientific evidence, pomegranate extracts have powerful antioxidant capacity [108] that is superior to red wine and equal to or better than green tea due to their free

TABLE 11.1 Summary of antidiabetic research on *C. roseus*

Drug used to induce diabetes	Extract of *C. roseus* used for treatment	Treatment	% blood glucose reduction in treated diabetic groups	Other parameters improved in treated diabetic groups	Comment	Reference
Alloxan	Aqueous leaf	Treatment groups received 100mg/kg of the leaf extract for 4 weeks	51%	Increased total cholesterol, LDL and VLDL-cholesterol, and triglycerides were reduced 2.plasma HDL-cholesterol increased to near normal	*C. roseus* may reduce the risk of CVDs	[105]
Alloxan		0.5, 0.75 and 1.0 mL/kg body weight of C.roseus leaf juice	1.0 mL/kg caused 31.9% reduction (20 hr post treatment, $p < 0.01$)		*C. roseus* served as a good adjuvant in the presence f of anti-diabetic drugs	[107]
Strepto-zotocin	Fraction of ethanolic extracts	150mg/kg of ethanolic extract given to diabetic treated group for once in a 24hr test period	48.34% reduction in petroleum ether fraction while 40.68% was observed at 24th hour for ethyl acetate	significant reduction in the levels of total cholesterol and triglycerides	*C. roseus* may reduce the risk of CVDs	[82]

radical scavenging properties [131]. The antioxidant activities of pomegranate rely on the high polyphenol content, in particular ellagitinnins, anthocyanins, and condensed tannins in pomegranate juice and other components of this fruit. Scientific studies have revealed that pomegranate juice has the ability to ameliorate diabetes-related oxidative stress as evidenced by improved glycemic control, reduced lipid peroxidation, improved antioxidant enzymatic status as well as the ability to prevent various secondary complications of diabetes [130, 131]. Based on the current scientific knowledge of the beneficial effects of pomegranate, there is no doubt that this fruit is a valuable medicinal plant with great potential in the management of diabetes mellitus and its associated secondary complications. Table 11.2 summarizes the antidiabetic research on pomegranate.

11.10.1 POSSIBLE THERAPEUTIC EFFECTS OF POMEGRANATE

The wide range therapeutic benefits of pomegranate may be attributable to its several bioactive compounds and most research has focused on its antioxidant, anticarcinogenic, antibacterial, antimicrobial, and anti-inflammatory properties.

11.10.1.1 Pomegranate and Diabetes

Pomegranate juice (PJ) consumption significantly reduced oxidative stress in the diabetic patients as evidenced by a 56% reduction in lipid peroxides and a 28% reduction in TBARS compared to baseline serum levels. In addition, a 39% decrease in uptake of oxidized LDL by human monocyte-derived macrophages (an early development in foam cell formation and atherogenesis) was observed in diabetic patients after PJ consumption. Researchers concluded that despite the sugars naturally present in pomegranate juice, consumption did not adversely affect diabetic parameters but had a significant effect on atherogenesis *via* reduced oxidative stress. [131].

TABLE 11.2 Summary of Antidiabetic Research on Pomegranate

Drug used to induce diabetes	Extract of Pomegranate used for treatment	Treat ment	% blood glucose reduction in treated diabetic groups	Other parameters improved in treated diabetic groups	Comment	Ref.
Alloxan in rats	Aqueous peel extract	Treatment groups received 0.43 g/kg of the aqueous peel extract for 4 weeks	57%	Elevated levels of β pancreatic cells 2. significant increase in insulin release	Pomeg-ranatepeel extract may reduce blood glucose levels through generation of β pancreatic cells	[88]
Genetic model of T2DM rats	methanolic extract of pomegranate flower	500 mg/kg of the methanolic extract were given to the obese Zucker rats daily for 6 weeks		Enhanced expression of the cardiac PPAR-γ mRNA. 2 Increased GLUT-4 mRNA expression	Pomegranate flower extract may Improve the sensitivity of the insulin receptor	[79]
Non diabetic obese mice tested	Aqueous leaf extract	800 mg/kg of (aq) leaf extract was fed daily for 5 weeks in obese mice		Supple-mentation with Pomegranate (aq) leaf extract reduced serum levels of glucose, triglycerides and total cholesterol	Pomegranate leaf extract may reduce the development dyslipidaemia, a major risk factor of T2DM in obese patients	[95]

11.10.1.2 Pomegranate and Antioxidant Capacity

In vitro assays demonstrated that pomegranate juice and seed extracts have 2–3 times the antioxidant capacity of either red wine or green tea. Pomegranate extracts have shown to scavenge free radicals and decrease macrophage oxidative stress and lipid peroxidation in animals [131] and increase plasma antioxidant capacity in elderly humans [71].

11.10.1.3 Pomegranate and Anticarcinogenic Properties

Various pomegranate extracts (juice, seed oil, peel) have also been demonstrated to inhibit prostate cancer cell invasiveness and proliferation, cause cell cycle disruption, induce apoptosis, and inhibit tumor growth, in vitro. These studies also demonstrated that combinations of pomegranate extracts from different parts of the fruit were more effective than any single extract [13, 93]. These results indicate that pomegranate may reduce prostate cancer because of its anti-proliferative, apoptotic, antioxidant, and possibly anti-inflammatory effects [119]. Recent research also indicates that pomegranate constituents inhibit angiogenesis *via* down regulation of vascular endothelial growth factor in MCF-7 breast cancer and human umbilical vein endothelial cell lines [160].

11.10.1.4 Anti-inflammatory Properties of Pomegranate

Cold-pressed pomegranate seed oil (CPSO) has shown to inhibit cyclooxygenase and lipoxygenase enzymes, which play major roles in the formation of inflammation mediators, in vitro [13]. Flavanoids extracted from CPSO have been reported to inhibit cyclooxygenase by 31–44%, while 69–81% inhibition of lipoxygenase was observed. By comparison, the pomegranate fermented juice (PFJ) extract in the same study resulted in a 23.8% inhibition of lipoxygenase in vitro [139].

11.10.1.5 Antimicrobial Activity of Pomegranate

Numerous in vitro studies [108, 169] and two human trials [102, 166] demonstrated the antimicrobial activities of pomegranate extracts. The growth of *Staphylococcus aureus*, *Streptococcus pyogenes*, *Diplococcus pneumoniae*, *Escherichia coli* O157:H7, and *Candida albicans* was inhibited *via* direct bactericidal or fungicidal activities.

11.10.1.6 Antihypertensive Properties of Pomegranate

The PJ consumption by hypertensive patients inhibited serum angiotensin converting enzyme activity (ACE: a catalyst for the conversion of angiotensin I to angiotensin II, a potent vasoconstrictor), thereby reducing systolic blood pressure [25] and potentially protecting against cardiovascular disease. Animal studies have revealed three possible hypoglycemic mechanisms for *Punica granatum* extracts. Pomegranate flower extract (PFLE) improved insulin sensitivity and lowered glucose levels in rats as early as 30 minutes post-glucose loading. PFLE also inhibited alpha-glucosidase in vitro, thereby decreasing the conversion of sucrose to glucose [79]. PFLE demonstrated significant hypoglycemic activity in diabetic rats *via* enhanced insulin levels and regeneration of pancreatic β-cells [88].

11.10.1.7 Pomegranate and Male Infertility

Research in rats demonstrated that PJ consumption is capable of improving epididymal sperm concentration, spermatogenic cell density, diameter of seminiferous tubules, and sperm motility, and it decreased the number of abnormal sperms compared to control animals. An improvement in antioxidant enzyme activity in both rat plasma and sperm was also noted [162].

11.10.1.8 Erectile Dysfunction

A study using the rabbit model of arteriogenic erectile dysfunction (ED) measured the effect of PJ concentrate on intracavernous blood flow and penile erection. Azadzoi *et al.* [26] found that eight weeks daily administration of 3.87 mL PJ concentrate (112 μmol polyphenols) significantly increased intravenous blood flow and smooth muscle relaxation, an effect probably exerted *via* its antioxidant property to enhance NO preservation and bioavailability [26].

11.10.1.9 Pomegranate and Obesity

PFLE (400 or 800 mg/kg/day) given to obese hyper-lipidemic mice for 5 weeks caused significant decrease in body weight, percentage of adipose pad weights, energy intake, and serum cholesterol, triglyceride, glucose, and total cholesterol/HDL ratios. Decreased appetite and intestinal fat absorption were also observed, together with improvements mediated in part by inhibition of pancreatic lipase activity [95].

11.11 BENEFICIAL EFFECTS OF SOME SELECTED MEDICINAL PLANTS ON MALE REPRODUCTIVE SYSTEM IN DIABETES-INDUCED SUBJECTS

Table 11.3 summarizes effects of selected medicinal plants on male reproductive parameters.

11.12 CONCLUSIONS

The adoption of westernized diet in Africa has been documented to propel the development of non-communicable diseases such as type-2 diabetes mellitus and its complications. The health promoting potentials of medicinal plants have been attributed to the presence of both nutritional and non-nutritional chemical compounds that can modify and modulate biological systems and thus indorsing therapeutic effects. A recent upsurge of ethno-pharmacology predilection has been linked to safety, availability, and the cost of employing medicinal plants in research. In Africa, a number of medicinal plants have been traditionally used to treat diabetes. In this chapter, authors reviewed the studies were conducted on two plants *Pomegranate* and *Catharanthus roseus* in the treatment of diabetes mellitus and its complications such as cardiovascular-related disorders, male infertility and hepato-renal

TABLE 11.3 Effects of Selected Medicinal Plants on Male Reproductive Parameters Drug

Used to induce diabetes	Plant species used for treatment	Treatment	Testicular parameters improved	Reference
Alloxan	*Ajuga iva*	One diabetic experimental group received orally the aqueous extract of *Ajuga iva* at the dose of 50 mg/kg for three weeks. Another experimental group received the same extract in the same concentration for 3 weeks before diabetes induction	Sperm count and motility, 17-β estradiol levels, serum testosterone, lipid peroxidation, antioxidant enzymes activity, total protein	[75]
	Cnidoscolus aconitifolius	Three diabetics experimental groups received orally doses of ethanol extract of *Cnidoscolus aconitifolius* in different concentrations (100 mg/kg, 500 mg/kg and 1000 mg/kg) daily for 28 days	Motility, number, morphology and viability of sperm	[27]
	Phoenix dactylifera	Two diabetic experimental groups were orally treated with solutions of the component 1 (diosmetin 7-O-β-L-arabinofuranosyl (1→2) β-D-apiofuranoside) and component 2 (diosmetin 7-O-β-D-apiofuranoside) isolated from the acetone extract of *Phoenix dactylifera*	Serum testosterone	[103]
	Hyphaene thebaica	One diabetic experimental group received orally an aqueous solution of the soluble fraction of *Hyphaene thebaica* (20 mg/kg) daily for 30 days. A second diabetic group received orally a component 5 solution of soluble aqueous fraction of *Hyphaene thebaica* (20 mg/kg) daily for 30 days		[135]

Used to induce diabetes	Plant species used for treatment	Treatment	Testicular parameters improved	Reference
Streptozotocin	*Momordica charantia*	One healthy and one diabetic experimental group received orally the extract of seeds of *Momordica charantia* at the dose of 10 mL/kg daily for 14 days	Endogenous peroxidase activity, localization of GST alpha	[128]
	Musa paradisiaca, Tamarindus indica, Eugenia jambolana and *Coccinia indica*	One diabetic experimental group was orally treated with a compound of specific parts of the species *Musa paradisiaca, Tamarindus indica, Eugenia jambolana Coccinia indica* at the dose of 60 mg/0.5 mL olive oil/100 g body weight/twice daily for 14 days	Gonadosomatic index, serum testosterone, sperm count and viability, testicular glucose level, activities of testicular catalase and glutathione peroxidase, lipid peroxidation, quantification of germ cells at stage VII in seminiferous epithelial cycle, seminiferous tubules diameter, Leydig cell nuclear area	[99]
	Musa paradisiaca and *Coccinia indica*	One diabetic experimental group was orally treated with hexane fraction of hydro-methanolic extract of root of *Musa paradisiaca* and leaf of *Coccinia indica* at the dose of 2 mg/0.2 mL olive oil/100 g body weight for 45 days	Serum testosterone, sperm count and viability, antioxidant enzymes activity, lipid peroxidation, incidence of apoptosis in seminiferous tubules, quantification of germ cells at stage VII in seminiferous epithelial cycle and giant cell in testicular section	[100]

TABLE 11.3 (Continued)

Used to induce diabetes	Plant species used for treatment	Treatment	Testicular parameters improved	Reference
	Mucuna pruriens	One healthy and one diabetic experimental group received orally the ethanolic extract of seeds of *Mucuna pruriens* at the dose of 200 mg/kg once a day for 60 days	Sperm count, morphology, motility, DNA integrity, quantitative assessment of DNA damage, estimation of lipid peroxidation and antioxidant	[156]
	Musa paradisiaca, Tamarindus indica, Eugenia jambolana and Coccinia indica	One diabetic experimental group was orally treated with a compound of specific parts of solvent fractions of *Musa paradisiaca*, *Tamarindus indica*, *Eugenia jambolana* and *Coccinia indica* at the dose of 10 mg/0.5 mL of 2% Tween 80/100 g body weight twice daily for 28 days	Gonadosomatic index, sperm count and motility, seminal vesicular fructose and testicular cholesterol levels, serum testosterone, antioxidant enzymes activity, estimation of lipid peroxidation, Bax-α protein expression, histological evaluation of the testis	[41]
	Eugenia jambolana	One diabetic experimental group was orally treated with the ethyl acetate fraction of the hydromethanolic extract of *Eugenia jambolana* at a dose of 20 mg/0.5 mL of distilled water/100 g body weight for 60 days	Gonadosomatic index, seminiferous tubules diameter, sperm count, seminal vesicular fructose level, quantification of germ cells at stage VII in seminiferous epithelial cycle, antioxidant enzymes activity, serum testosterone, analysis of Bax and Bcl-2 gene expression	[64]

Used to induce diabetes	Plant species used for treatment	Treatment	Testicular parameters improved	Reference
	Chlorophytum borivilianum	Two experimental diabetic group orally treated with aqueous *Chlorophytum borivilianum* extract at a dose of 250 mg/kg and 500 mg/kg for 28 days	Sperm count, morphology, motility and viability, sperm's flagella membrane integrity, lipid peroxidation, hydrogen peroxide and nitric oxide levels, sperm total antioxidant capacity, activity levels of endogenous antioxidant enzymes, expression of sperm caspase-3	[66]
	Morus alba	One diabetic experimental group received orally *Morus alba* extract at a dose of 1 g/kg per day for 8 weeks	Lipid peroxidation, antioxidant enzymes activity, mRNA expression level of StAR and P450scc	[74]

disorders. This review calls for comprehensive scientific investigations on medicinal plants on the validation and development of cheap novel therapies.

11.13 SUMMARY

Medicinal plants have been in use since ancient times to treat various diseases and ailments. They have undoubtedly played very significant roles in the maintenance of human health and livelihood. Throughout history, the intake of plants as food has been shown to enhance human health nutrition.

Previous studies on these two medicinal plants reported on their effectiveness against diabetes-induced oxidative stress, inflammation, hyperlipidemia, hyperglycemia, and male infertility. The antioxidant activities of these plants could possibly be responsible for the delay and/ prevention of the development of DM complications. These plant species therefore appear to contain promising antidiabetic agents that need further and extensive study.

KEYWORDS

- **advanced glycation end products**
- **alkaloids**
- **antioxidants**
- **apoptosis**
- **catalase**
- **diabetes mellitus**
- **diabetic complications**
- **DNA damage**
- **free radicals**
- **hyperglycemia**
- **lipid peroxidation**

- **macrovascular complications**
- **male infertility**
- **medicinal plants**
- **non-communicable diseases**
- **orthodox drugs**
- **oxidative stress**
- **phytochemicals**
- **phytotherapy**
- **reactive oxygen species**
- **streptozotocin**
- **superoxide dismutase**
- **tissue damage**
- **traditional medicines**
- **type-1 diabetes mellitus**
- **type-2 diabetes mellitus**

REFERENCES

1. Abo, K. A., Fred-Jaiyesimi, A. A., & Jaiyesimi, A. E. A., (2008). Ethnobotanical studies of medicinal plants used in the management of diabetes mellitus in South Western Nigeria. *Journal of Ethnopharmacology, 115*(1), 67–67.
2. Afolayan, A. J., & Sunmonu, T. O., (2010). Studies of in vivo antidiabetic plants used in South African herbal medicine. *Journal of Clinical Biochemistry and Nutrition, 42*(2), 98–106.
3. Agarwal, A., Abdelrazik, H., & Sharma, R. K., (2008). Oxidative stress measurement in patients with male or female factor infertility. Chapter 10, In: *Handbook of Chemiluminescent Methods in Oxidative Stress Assessment,* Popov, I and Lewin, G. (Eds.), Transworld Research Network, India, pp. 195–218.
4. Agarwal, A., & Allamaneni, S., (2006). Oxidative stress and human reproduction. Chapter 3, In: *Oxidative Stress, Disease and Cancer*, Singh, K. K., Ed., Imperial College Press, New York, pp. 687–703.
5. Ahmed, A. M., (2002). History of diabetes mellitus. *Saudi Medical Journal, 23*(4), 373–378.
6. Ahmed, N., (2004). Advanced glycation end-products: Role in pathology of diabetic complications. *Diabetes Research and Clinical Practice, 67*(1), 3–21.

7. Aitken, R. J., (1997). The cell biology of fertilization. *Advances in Experimental Medicine and Biology*, *424*, 291–299.
8. Aitken, R. J., Paterson, M., Fisher, H., Buckingham, D. W., & Van Duin, M., (1995). Redox regulation of tyrosine phosphorylation in human spermatozoa and its role in the control of human sperm function. *Journal of Cell Sci.*, *108*(5), 2017–2025.
9. Aitken, R. J., & De Iuliis, G. N., (2007). Origins and consequences of DNA damage in male germ cells. *Reproduction Biomed Online*, *14*(6), 727–33.
10. Aitken, R. J., & Krausz, C., (2001). Oxidative stress, DNA damage and the Y chromosome. *Reproduction*, *122*(4), 497–506.
11. Aitken, R. J., & Sawyer, D., (2003). The human spermatozoon – not waving but drowning. *Advances in Experimental Medicine and Biology*, *518*, 85–98.
12. Aitken, R. J., Harkiss, D., & Buckingham, D., (1993). Relationship between Iron-Catalyzed Lipid-Peroxidation Potential and Human Sperm Function. *Journal of Reproduction and Fertility*, *98*(1), 257–265.
13. Albrecht, M., Jiang, W., Kumi-Diaka, J., et al., (2004). Pomegranate extracts potently suppress proliferation, xenograft growth, & invasion of human prostate cancer cells. *Journal of Medicinal Food*, *7*(3), 274–283.
14. Al-Hussaini, A. A., Sulaiman, N. M., Al-Zahrani, M. D., Alenizi, A. S., & Khan, M., (2012). Prevalence of hepatopathy in type-1 diabetic children. *Biomedical Central Pediatrics*, *12*(1), 160.
15. Alinde, O. B. L., Esterhuyse, A. J., & Oguntibeju, O. O., (2014). Potential role of parkia biglobosa in the management and treatment of cardiovascular diseases. Chapter 15, In: *Antioxidant-Antidiabetic Agents and Human Health*, Oguntibeju, O. O., Ed., *In. Tech.* Croatia, pages 350–369.
16. Al-Shaqha, W. M., Khan, M., Salam, N., Azzi, A., & Chaudhary, A. A., (2015). Antidiabetic potential of *Catharanthus roseus* Linn and its effect on the glucose transport gene (GLUT-2 and GLUT-4) in streptozotocin induced diabetic Wister rats. *BMC Complementary and Alternative Medicine*, *15*(379), 1–8.
17. Alves, M. G., & Oliveira, P. F., (2013). Diabetes Mellitus and male reproductive function: where we stand? *International Journal of Diabetology & Vascular Disease Research*, *1*(1), 101–103.
18. American Diabetes Association, (2014). Standards in Medical care in diabetes. *Diabetes Care*, *37*(1), S14-S80.
19. Amine, E. K., Baba, N. H., Belhadj, M., Derenber-Yap, M., Djazayery, A., Forrester, T., Galusk, A., et al., (2003). Diet, nutrition and the prevention of chronic diseases report of a Joint WHO/FAO Expert Consultation. *916*, pp. 1–13.
20. Amod, A., Ascott-Evans, B. H., & Berg, G. I., (2012). The 2012 SEMDSA guideline for the management of type-2-diabetes. *Journal of Endocrinology, Metabolism and Diabetes of South Africa (JEMDSA)*, *17*(1), S1–S95.
21. Araki, E., & Nishikawa, T., (2010). Oxidative stress: A cause and therapeutic target of diabetic complications. *Journal of Diabetes Investigation*, *1*(3), 90–96.
22. Armstrong, J. S., Rajasekaran, M., Chamulitrat, W., Gatti, P., Hellstrom, W. J., & Sikka, S. C., (1999). Characterization of reactive oxygen species induced effects on human spermatozoa movement and energy metabolism. *Free Radical Biology and Medicine*, *26*(7–8), 869–880.

23. Arsov, S., Graaff, R., Van Oeveren, W., Stegmayr, B., Sikole, A., Rakhorst, G., & Smit, A. J., (2014). Advanced glycation end-products and skin auto-fluorescence in end-stage renal disease: A review. *Clinical Chemistry and Laboratory Medicine, 52*(1), 11–20.
24. Avignon, A., & Sultan, A., (2006). PKC-ε inhibition: A new therapeutic approach for diabetic complications. *Diabetes & Metabolism, 32*(3), 205–213.
25. Aviram, M., & Dornfeld, L., (2001). Pomegranate juice consumption inhibits serum angiotensin converting enzyme activity and reduces systolic blood pressure. *Atherosclerosis, 158*(1), 195–198.
26. Azadzoi, K., M., Schulman, R. N., Aviram, M., & Siroky, M. B., (2005). Oxidative stress in arteriogenic erectile dysfunction: prophylactic role of antioxidants. *Journal of Urology, 174*(1), 386–393.
27. Azeez, O. I., Oyagbemi, A. A., Oyeyemi, M. O., & Odetola, A. A., (2010). Ameliorative effects of *Cnidoscolus aconitifolius* on alloxan toxicity in Wister rats. *African Health Sciences, 10*(3), 283–291.
28. Bailey, C. J., & Day, C., (1989). Traditional plant medicines as treatments for diabetes. *Diabetes Care, 12*(8), 553–564.
29. Baker, M. A., & Aitken, J., (2004). The importance of redox regulated pathways in sperm cell biology. *Molecular and Cellular Endocrinology, 216*(1–2), 47–54.
30. Bansal, A. K., & Bilaspuri, G. S., (2007). Effect of ferrous ascorbate on in vitro capacitation and acrosome reaction in cattle bull spermatozoa. *Animal Science Report, 1*(2), 69–77.
31. Bennetts, L. E., & Aitken, R. J., (2005). A comparative study of oxidative DNA damage in mammalian spermatozoa. *Molecular Reproductive Development, 71*(1), 77–87.
32. Berk, B. C., (2007). Novel approaches to treat oxidative stress and cardiovascular diseases. *Transactions of the American Clinical and Climatological Association, 118*, 209–214.
33. Bevan, P., (2001). Insulin signaling. *Journal of Cell Science, 114*(8), 1429–1430.
34. Bjornholm, M., & Zierath, J. R., (2005). Insulin signal transduction in human skeletal muscle: identifying the defects in type II diabetes. *Biochemical Society Transactions, 33*(2), 354–357.
35. Brand, M. D., Affourtit, C., Esteves, T. C., Green, K., Lambert, A. J., Miwa, S., & Parker, N., (2004). Mitochondrial superoxide: production, biological effects, & activation of uncoupling proteins. *Free Radical Biology and Medicine, 37*(6), 755–767.
36. Breitbart, H., & Naor, Z., (1999). Protein kinases in mammalian sperm capacitation and the acrosome reaction. *Reviews of Reproduction, 4*(3), 151–159.
37. Brownlee, M., (2005). The pathobiology of diabetic complications. *Diabetes, 54*(6), 1615–1625.
38. Bucak, M. N., Sarıozkan, S., Tuncer, P. B., Ulutaş, P. A., & Akçadağ, H. I., (2009). Effect of antioxidants on microscopic semen parameters, lipid peroxidation and antioxidant activities in Angora goat semen following cryopreservation. *Small Ruminant Research, 81*(2–3), 90–95.
39. Buettner, G. R., (2011). Superoxide dismutase in redox biology: the roles of superoxide and hydrogen peroxide. *Anti-cancer Agents in Medicinal Chemistry, 11*(4), 341–346.

40. Byon, J. C. H., Kusari, A. B., & Kusari, J., (1998). Protein-tyrosine phosphatase-1B acts as a negative regulator of insulin signal transduction. *Molecular and Cellular Biochemistry*, *182*(1–2), 101–108.
41. Chatterjee, K., Ali, K. M., De, D., Bera, T. K., Jana, K., Maiti, S., Ghosh, A., & Ghosh, D., (2013). Hyperglycemia induced alteration in reproductive profile and its amelioration by the polyherbal formulation MTEC (modified) in streptozotocin induced diabetic albino rats. *Biomarkers and Genomic Medicine*, *5*(1–2), 54–66.
42. Cheng, A., Dube, N., Gu, F., & Tremblay, M. L., (2002). Coordinated action of protein tyrosine phosphatases in insulin signal transduction. *European. Journal of Biochemistry, 269*(4), 1050–1059.
43. Chinney, P. F., & Turnbull, D. M., (2000). Mitochondrial DNA mutations in the pathogenesis of human disease. *Molecular Medicine Today, 6*(11), 425–432.
44. Chung, S. S. M., Ho, C. C., M., Lam, K. S. L., & Chung, S. K., (2003). Contribution of polyol pathway to diabetes-induced oxidative stress. *Journal of the American Society of Nephrology, 14*(3), S233–S236.
45. Cocuzza, M., Sikka, S. C., Athayde, K. S., & Agarwal, A., (2007). Clinical relevance of oxidative stress and sperm chromatin damage in male infertility: An evidence based analysis. *International Brazilian Journal of Urology*, *33*(5), 603–621.
46. De Iuliis, G. N., Newey, R. J., King, B. V., & Aitken, R. J., (2009), Mobile phone radiation induces reactive oxygen species production and DNA damage in human spermatozoa in vitro. *PLoS One*, *4*(7), 1–9.
47. De Lamirande, E., & Gagnon, C., (1993). A positive role for the superoxide anion in triggering hyperactivation and capacitation of human spermatozoa. *International Journal of Andrololog*y, *16*(1), 21–25.
48. De Lamirande, E., & O'Flaherty, C., (2008). Sperm activation: role of reactive oxygen species and kinases. *Biochim Biophys Acta.*, *1784*(1), 106–115.
49. Desai, N., Sharma, R., Makker, K., Sabanegh, E., & Agarwal, A., (2009). Physiologic and pathologic levels of reactive oxygen species in neat semen of infertile men. *Fertility and Sterility*, *92*(5), 1626–1631.
50. Diabetes American Association (DAA), (2008). Diagnosis and classification of diabetes mellitus. *Diabetes Care, 3*(1), S55–S60.
51. Dorota, S., & Kurpisz, M., (2004). Reactive oxygen species and sperm cells. *Reproductive Biology and Endocrinology*, *2*(12), 1–7.
52. Du Plessis, S. S., Kashou, A. H., Benjamin, D. J., Yadav, S. P., & Agarwal, A., (2011). Proteomics: a subcellular look at spermatozoa. *Reproductive Biology Endocrinology, 9*(36), 10.
53. Echtay, K. S., (2007). Mitochondrial uncoupling proteins: what is their physiological role? *Free Radical Biology and Medicine, 43*(10), 1351–1371.
54. Edeoga, H. O., Okwu, D. E., & Mbaebie, B. O., (2005). Phytochemical constituents of some Nigerian medicinal plants. *African Journal of Biotechnology*, *4*(7), 685–688.
55. Elmarakby, A. A., & Sullivan, J. C., (2012). Relationship between oxidative stress and inflammatory cytokines in diabetic nephropathy. *Cardiovascular Therapeutics*, *30*(1), 49–59.
56. Evcimen, N. D., & King, G. L., (2007). The role of protein kinase C activation and the vascular complications of diabetes. *Pharmacological Research*, *55*(6), 498–510.

57. Farnsworth, N. R., Akerele, O., Bingel, A. S., Soejarto, D. D., & Guo, Z., (1985). Medicinal plants in therapy. *Bulletin of the World Health Organization, 63*(6), 965–981.
58. Fennell, C. W., Lindsey, K. L., McGaw, L. J., Sparg, S. G., Stafford, G. I., Elgorashi, E. E., Grace, O. M., & Van Staden, J., (2004). Assessing African medicinal plants for efficacy and safety: pharmacological screening and toxicology. *Journal of Ethnopharmacology, 94*(2), 205–217.
59. Fernández-Mejía, C., Lazo-de-la-Vega, & Monroy, M. L., (2013). Oxidative stress in diabetes mellitus and the role of vitamins with antioxidant actions. Chapter 9, In: *Oxidative Stress and Chronic Degenerative Diseases: A Role For Antioxidants*, Morales-Gonzalez, J. A., Ed., *In. Tech.*, Mexico, vol. *9*, pp. 209–232.
60. Forbes, J. M., & Cooper, M. E., (2012). Glycation in diabetic nephropathy. *Amino Acids, 42*(4), 1185–1192.
61. Francés, D. E., Ronco, M. T., Monti, J. A., Ingaramo, P. I., Pisani, G. B., Parody, J. P., & Carnovale, C. E., (2010). Hyperglycemia induces apoptosis in rat liver through the increase of hydroxyl radical: new insights into the insulin effect. *Journal of Endocrinology, 205*(2), 187–200.
62. Fridell, J. A., Saxena, R., Chalasani, N. P., Goggins, W. C., Powelson, J. A., & Cummings, O. W., (2007). Complete reversal of glycogen hepatopathy with pancreas transplantation: two cases. *Transplantation, 83*(1), 84–86.
63. Genova, M. L., Ventura, B., Giuliano, G., Bovina, C., Formiggini, G., Parenti, G., & Lenaz, C. G., (2001). The site of production of superoxide radical in mitochondrial Complex I., is not a bound ubisemiquinone but presumably iron sulfur cluster N2. *FEBS Letters, 505*(3), 364–368.
64. Ghosh, S., & Roy, T., (2013). Evaluation of antidiabetic potential of methanolic extract of Coccinia indica leaves in streptozotocin induced diabetic rats. *International Journal of Pharmaceutical Sciences and Research, 4*(11), 4325–4328.
65. Giacco F., & Brownlee, M., (2010). Oxidative stress and diabetic complications. *Circulation Research, 107*(9), 1058–1070.
66. Giribabu, N., Kumar, K. E., Rekha, S. S., Muniandy, S., & Salleh, N., (2014). *Chlorophytum boviriliarum* (Safed Musli) root extract prevents impairment in characteristics and elevation of oxidative stress in sperm of streptozotocin induced adult male diabetic rats. *BMC Complementary and Alternative Medicine, 14*(291), 377.
67. Goh, S. Y., & Cooper, M. E., (2008). The Role of advanced glycation end products in progression and complications of diabetes. *Journal of Clinical Endocrinology and Metabolism, 93*(4), 1143–1152.
68. Gonc-Alves, F., Barretto, L. S., S., Arruda, R. P., Perri, S. H .V., & Mingoti, G. Z., (2010). Effect of antioxidants during bovine in vitro fertilization procedures on spermatozoa and embryo development. *Reproduction in Domestic Animals, 45*(1), 129–135.
69. Grivennikova, V. G., & Vinogradov, A. D., (2006). Generation of superoxide by the mitochondrial Complex I., *Biochimica. et. Biophysica. Acta., 175*(5–6), 553–561.
70. Guariguata, L., Whiting, D. R., Hambleton, I., Beagley, J., Linnenkamp, U., & Shaw, J. E., (2014). Global estimates of diabetes prevalence for 2013 and projections for 2035. *Diabetes Research and Clinical Practice, 103*(2), 137–49.

71. Guo, C., Wei, J., & Yang, J., (2008). Pomegranate juice is potentially better than apple juice in improving antioxidant function in elderly subjects. *Nutrition Research, 28*(2), 72–77.
72. Gupta, R. K., Patel, A. K., Shah, N., Kumar, A., Chaudhary, U. K. J., Yadav, U. C., & Pakuwal, U., (2014). Oxidative stress and antioxidants in disease and cancer: *Asian Pacific Journal of Cancer Prevention, 15*(11), 4405–4409.
73. Habib, S. L., (2013). Diabetes and renal tubular cell apoptosis. *World Journal of Diabetes, 4*(2), 27–30.
74. Hajizadeh, M., Eftekhar, E., Zal, F., Jafarian, A., & Mostafavi-Pour, Z., (2014). Mulberry leaf extract attenuates oxidative stress mediated testosterone depletion in streptozotocin induced diabetic rats. *Iranian Journal of Medical Sciences, 39*(2), 123–129.
75. Hamden, K., Carreau, S., Jamoussi, K., Ayadi, F., Garmazi, F., Mezgenni, N., & Elfeki, A., (2008). Inhibitory effects of 1alpha, 25dihydroxyvitamin D3 and Ajuga iva extract on oxidative stress, toxicity and hypofertility in diabetic rat testes. *Journal of Physiology and Biochemistry, 64*(3), 231–239.
76. Henriksen, E. J., Diamond-Stanic, M. K., & Marchionne, E. M., (2011). Oxidative stress and the etiology of insulin resistance and type-2 diabetes. *Free Radical Biology and Medicine, 51*(5), 993–999.
77. Herrero, M., B, De Lamirande, E., & Gagnon, C., (2003). Nitric oxide is a signaling molecule in spermatozoa. *Current Pharmacology Design, 9*(5), 419–425.
78. Hu, F. B., (2011). Globalization of diabetes: Role of diet, lifestyle and genes. *Diabetes Care, 34*(6), 1249–1257.
79. Huang, T. H., Peng, G., Kota, B. P., Li, G. Q., Yamahara, J., Roufogalis, B. D., & Li, Y., (2005). Antidiabetic action of *Punica granatum* flower extract: activation of PPAR-gamma and identification of an active component. *Toxicology Applied Pharmacology, 207*(2), 160–169.
80. International Diabetes Federation (IDF), (2016). Atlas, 7th edition.
81. Irvine, D. S., (1996). Glutathione as a treatment for male infertility. *Reviews of Reproduction, 1*(1), 6–12.
82. Islam, M. A., Akhtar, M. A., Islam, M. R., Hossain, M. S., Alam, M. K., Wahed, M. I., Rahman, B. M., Anisuzzaman, A. S., Shaheen, S. M., & Ahmed, M., (2001). Antidiabetic and hypolipidemic effects of different fractions of Catharanthus roseus (Linn.) on normal and streptozotocin-induced diabetic rats. *Journal of Scientific Research, 1*(2), 334–44.
83. Jardim, J., Trindade, E., Carneiro, F., & Dias, J. A., (2013). Hepatomegaly in Type-1 Diabetes Mellitus: When to suspect of glycogenic hepatopathy? *Journal of Medical Cases, 4*(11), 726–728.
84. Jezek, P., & Hlavata, L., (2005). Mitochondria in homeostasis of reactive oxygen species in cell, tissues, & organism. *The International Journal of Biochemistry & Cell Biology, 37*(12), 2478–2503.
85. Joshi, A. R., & Joshi, K., (2000). Indigenous knowledge and uses of medicinal plants by local communities of the Kali Gandaki Watershed Area, Nepal. *Journal of Ethnopharmacology, 73*(1), 175–183.

86. Karamanou, M., Protogerou, A., Tsoucalas, G., Androutsos, G., & Poulakou-Rebelakou, E., (2016). Milestones in the history of diabetes mellitus: The main contributors. *World Journal of Diabetes*, *7*(1), 1–7.
87. Kashihara, N., Haruna, Y., Kondeti, V. K., & Kanwar, Y. S., (2010). Oxidative stress in diabetic nephropathy. *Current Medicinal Chemistry*, *17*(34), 4256–4269.
88. Khalil, E. A., (2004). Antidiabetic effect of an aqueous extract of pomegranate (Punica granatum L) peels in normal and alloxan diabetic rats. *Egyptian Journal of Hosp. Med.*, *16*(1), 92–99.
89. Khan, A., (2015). A comparative study of antidiabetic activity of *catharanthus roseus* and *catharanthus alba* flower extracts on alloxan induced diabetic rats. *World Journal of Pharmacy and Pharmaceutical Science*, *5*(2), 527–543.
90. Khothari, S., Thompson, A., Agarwal, A., & du Plessis, S. S., (2010). Free radicals: Their beneficial and detrimental effects on sperm function. *Indian Journal of Experimental Biology*, *48*(5), 425–435.
91. Kumar, T. R., (1999). Muralidhara. Male-mediated dominant lethal mutations in mice following prooxidant treatment. *Mutation Research*, *444*(1), 144–149.
92. La Vignera, S., Calogero, A. E., Condorelli, R., Lanzafame, F., Giammusso, B., & Vicari, E., (2009). Andrological characterization of the patient with diabetes mellitus. *Minerva Endocrinology*, *34*(1), 1–9.
93. Lansky, E. P., Jiang, W., & Mo, H., (2005). Possible synergistic prostate cancer suppression by anatomically discrete pomegranate fractions. *Invest New Drugs*, *23*(1), 11–20.
94. Lehmann, R., & Schleicher, E. D., (2000). Molecular mechanism of diabetic nephropathy. *Clinica. Chimica. Acta.*, *297*(1), 135–144.
95. Lei, F., Zhang, X. N., Wang, W., et al., (2007). Evidence of anti-obesity effects of the pomegranate leaf extract in high-fat diet induced obese mice. *International Journal of Obesity (London)*, *31*(6), 1023–1029.
96. Lenzi, A., Culasso, F., Gandini, L., Lombardo F., & Dondero, F., (1993). Placebo-controlled, double-blind, cross-over trial of glutathione therapy in male infertility. *Human Reproduction*, *8*(10), 1657–1662.
97. Levitt, N. S., (2008). Diabetes in Africa: Epidemiology, management and healthcare challenges in global burden of cardiovascular diseases. *Heart*, *94*(11), 1376–1382.
98. Lorenzi, M., (2007). The polyol pathway as a mechanism for diabetic retinopathy: attractive, elusive, & resilient. *Journal of Diabetes Research*, *2007*, 1–10.
99. Mallick, C., Bera, T. K., Ali, K. M., Chatterjee, K., & Ghosh, D., (2010). Diabetes-induced testicular disorders vis-a-vis germ cell apoptosis in albino rat: Remedial effect of hexane fraction of root of *Musa paradisiaca* and leaf of *Coccinia indica*. *Journal of Health Science*, *56*(6), 641–654.
100. Mallick, C., Mandal, S., Barik, B., Bhattacharya, A., & Ghosh, D., (2007). Protection of testicular dysfunctions by MTEC, a formulated herbal drug, in streptozotocin induced diabetic rat. *Biological & Pharmaceutical Bulletin*, *30*(1), 84–90.
101. McLarty, D. G., Pollitt, C., & Swai, A. B., M., (1990). Diabetes in Africa. *Diabetic Medicine*, *7*(8), 670–684.
102. Menezes, S. M., Cordeiro, L. N., & Viana, G. S., (2006). *Punica granatum* (pomegranate) extract is active against dental plaque. *Journal of Herb Pharmacotherapy*, *6*(2), 79–92.

103. Michael, H. N., Salib, J. Y., & Eskander, E. F., (2013). Bioactivity of diosmetin glycosides isolated from the epicarp of date fruits, *Phoenix dactylifera,* on the biochemical profile of alloxan diabetic male rats. *Phytotherapy Research, 27*(5), 699–704.
104. Mittler, R., (2002). Oxidative stress, antioxidants and stress tolerance. *Trends in Plant Science, 7*(9), 405–410.
105. Muralidharan, L., (2014). *Catharanthus roseus* leaves as an anti-diabetic and hypolipidemic agents in alloxan-induced diabetic rats. *American Journal of Phytomedicine and Clinical Therapeutics, 2*(12), 1393–1396.
106. Musabayane, C. T., (2012). The effects of medicinal plants on renal function and blood pressure in diabetes mellitus: review article. *Cardiovascular Journal of Africa, 23*(8), 462–468.
107. Nammi, S., Boini, M. K., Lodagala, S. D., & Behara, R. B., (2003). The juice of fresh leaves of *Catharanthus roseus* Linn. reduces blood glucose in normal and alloxan diabetic rabbits. *BMC complementary and Alternative Medicine, 2*(3), 1–4.
108. Naqvi, S. A., Khan, M. S., & Vohora, S. B., (1991). Antibacterial, antifungal, & antihelminthic investigations on Indian medicinal plants. *Fitoterapia, 62*(3), 221–228.
109. Nasri, H., & Shirzad, H., (2013). Toxicity and safety of medicinal plants. *Journal of Herbal Medicine Plarmacology, 2*(2), 21–22.
110. Natarajan, A., Ahmed, K. S., Sundaresan, S., Sivaraj, A., Devi, K., & Kumar, B. S., (2012). Effect of aqueous flower extract of *Catharanthus roseus* on alloxan induced diabetes in male albino rats. *IJPSDR, 4*(2), 150–153.
111. Naughton, C. K., Nangia, A. K., & Agarwal, A., (2001). Pathophysiology of varicoceles in male infertility. *Human Reproduction Update, 7*(5), 473–81.
112. Noori, S., (2012). An overview of oxidative stress and antioxidant defensive system. *Scientific Reports, 1*(8), 1–9.
113. O'Flaherty, C., De Lamirande, E., & Gagnon, C., (2006). Positive role of reactive oxygen species in mammalian sperm capacitation: triggering and modulation of phosphorylation events, *Free Radical Biology Medicine, 41*(4), 528–540.
114. O'Flaherty, C. M., Beorlegui, N. B., & Beconi, M. T., (1999). Reactive oxygen species requirements for bovine sperm capacitation and acrosome reaction. *Theriogenology, 52*(2), 289–301.
115. Oguntibeju, O. O., Meyer, S., Aboua, Y. G., & Goboza, M., (2016). *Hypoxis hemerocallidea* significantly reduced hyperglycaemia and hyperglycaemic-induced oxidative stress in the liver and kidney tissues of streptozotocin-induced diabetic male Wister rats. *Evidence-Based Complementary and Alternative Medicine, 2016,* 1–10.
116. Ozougwu, J. C., Obimba, K. C., Belonwu, C. D., & Unakalamba, C. B., (2013). The pathogenesis and pathophysiology of type-1 and type-2 diabetes mellitus. *Journal of Physiology and Pathophysiology, 4*(4), 46–57.
117. Padgett, L. E., Broniowska, K. A., Hansen, P. A., Corbett, J. A., & Tse, H. M., (2013). The role of reactive oxygen species and pro-inflammatory cytokines in type-1 diabetes pathogenesis. *Annals of the New York Academy of Sciences, 1281*(1), 16–35.
118. Pan, H. Z., Zhang, L., Guo, M. Y., Sui, H., Li, H., Wu, W. H., & Chang, D., (2010). The oxidative stress status in diabetes mellitus and diabetic nephropathy. *Acta. Diabetologica, 47*(1), 71–76.

119. Pantuck, A. J., Leppert, J. T., & Zomorodian, N., (2006). Phase II study of pomegranate juice for men with rising prostate-specific antigen following surgery or radiation for prostate cancer. *Clinical Cancer Research, 12*(13), 4018–4026.
120. Pazdro, R., & Burgess, J. R., (2010). The role of vitamin E and oxidative stress in diabetes complications. *Mechanisms of Ageing and Development, 131*(4), 276–286.
121. Peppa, M., Uribarri, J., & Vlassara, H., (2003). Glucose, advanced glycation end products, & diabetes complications: what is new and what works. *Clinical Diabetes, 21*(4), 186–187.
122. Pitocco, D., Zaccardi, F., Di Stasio, E., Romitelli, F., Santini, S. A., Zuppi, C., & Ghirlanda, G., (2010). Oxidative stress, nitric oxide, & diabetes. *The Review of Diabetic Studies Journal, 7*(1), 15–25.
123. Radoi, V., Llixandru, D., Mahora, M., & Virgolici, B., (2012). Advanced glycation end products in diabetes mellitus: mechanism of action and focused treatment. *Proc. Rom. Acad., Series B., 1*, 9–19.
124. Rahimi, R., Nikfar, S., Larijani, B., & Abdollahi, M., (2005). A review on the role of antioxidants in the management of diabetes and its complications. *Biomedicine & Pharmacotherapy, 59*(7), 365–373.
125. Rains, J. L., & Jain, S. K., (2011). Oxidative stress, insulin signaling, & diabetes. *Free Radical Biology and Medicine, 50*(5), 567–575.
126. Ramos, S., (2008). Cancer chemoprevention and chemotherapy: dietary polyphenols and signaling pathways. *Molecular Nutrition & Food Research, 52*(5), 507–526.
127. Rasineni, K., Bellamkonda, R., Singareddy, S. R., & Desireddy, S., (2010). Antihyperglycemic activity of *Catharanthus roseus* leaf powder in streptozotocin-induced diabetic rats. *Pharmacognosy Research, 2*(3), 195.
128. Raza, H., Ahmed, I., & John, A., (2004). Tissue specific expression and immunohistochemical localization of glutathione S-transferase in streptozotocin induced diabetic rats: modulation by *Momordica charantia (karela)* extract. *Life Sciences, 74*(12), 1503–1511.
129. Rolo, A. P., & Palmeira, C. M., (2006). Diabetes and mitochondrial function: Role of hyperglycemia and oxidative stress. *Toxicology and Applied Pharmacology, 212*(2), 167–78.
130. Rosenblat, M., Hayek, T., & Aviram, M., (2006). Anti-oxidative effects of pomegranate juice (PJ) consumption by diabetic patients on serum and on macrophages. *Atherosclerosis, 187*(2), 363–371.
131. Rosenblat, M., Volkova, N., Coleman, R., & Aviram, M., (2006). Pomegranate by-product administration to apolipoprotein e-deficient mice attenuates atherosclerosis development as a result of decreased macrophage oxidative stress and reduced cellular uptake of oxidized low-density lipoprotein. *Journal of Agriculture and Food Chemistry, 54*(5), 1928–1935.
132. Roveri, A., Ursini, F., Flohe, L., & Maiorino, M., (2001). PHGPx and spermatogenesis. *Biofactors, 14*(1–4), 213–222.
133. Sakkas, D., Mariethoz, E., Mnicsirdi, G., Bizzaro, D., Bianchi, P. G., & Bianchi, U., (1999). Origin of DNA damage in ejaculated human spermatozoa. *Reviews of Reproduction, 4*(1), 31–37.

134. Saleh, R. A., Agarwal, A., Sharma, R. K., Nelson, D. R., & Thomas, A. J., (2002). Effect of cigarette smoking on levels of seminal oxidative stress in infertile men: a prospective study. *Fertility Sterility*, *78*(3), 491–499.
135. Salib, J. Y., Michael, H. N., & Eskande, E. F., (2013). Antidiabetic properties of flavonoid compounds isolated from *Hyphaene thebaica* epicarpo on alloxan induced diabetic rats. *Pharmacognosy Research*, *5*(1), 22–29.
136. Samy, R. P., & Gopalakrishnakone, P., (2007). Current status of herbal and their future perspectives. *Nature precedings*, *10101*(1176.1), 1–13.
137. Sawyer, D. E., Roman, S. D., & Aitken, R. J., (2001). Relative susceptibilities of mitochondrial and nuclear DNA to damage induced by hydrogen peroxide in two mouse germ cell lines. *Redox Reports, 6*(3), 182–184.
138. Schmitz-Peiffer, C., & Biden, T. J., (2008). Protein kinase C function in muscle, liver, & β-cells and its therapeutic implications for type-2 diabetes. *Diabetes, 57*(7), 1774–1783.
139. Schubert, S. Y., Lansky, E. P., & Neeman, I., (1999). Antioxidant and eicosanoid enzyme inhibition properties of pomegranate seed oil and fermented juice flavonoids. *Journal of Ethnopharmacology*, *66*(1), 11–17.
140. Schwarzlander, M., Murphy, M. P., Duchen, M. R., Logan, D. C., Fricker, M. D., Halestrap, A. P., & Muller, F. L., (2012). Mitochondrial flashes: A radical concept refined. *Trends in Cell Biology, 22*(10), 503–508.
141. Sen, S., Chakraborty, R., Sridhar, C., Reddy, Y. S. R., & De, B., (2010). Free radicals, antioxidants, diseases and phytomedicines: Current status and future prospect. *International Journal of Pharmaceutical Sciences Review and Research, 3*(1), 91–100.
142. Sesti, G., (2006). Pathophysiology of insulin resistance. *Best Practice & Research Clinical Endocrinology & Metabolism, 20*(4), 665–679.
143. Sesti, G., Federici, M., Lauro, D., Sbraccia, P., & Lauro, R., (2001). Molecular mechanism of insulin resistance in type-2 diabetes mellitus: role of the insulin receptor variant forms. *Diabetes/Metabolism Research and Reviews*, *17*(5), 363–373.
144. Shabbir, M., Khan, M. R., & Saeed, N., (2013). Assessment of phytochemicals, antioxidant, anti-lipid peroxidation and anti-hemolytic activity of extract and various fractions of *Maytenus royleanus* leaves. *BMC complementary and alternative medicine*, *13*(1), 1.
145. Sharma, R. K., & Agarwal, A., (1996). Role of reactive oxygen species in male infertility, *Urology*, *48*(6), 835–850.
146. Sies, H., & Masumoto, H., (1996). *Ebselen* as a glutathione peroxidase mimic and as a scavenger of peroxynitrite. *Advances in pharmacology*, *31*(38), 229–46.
147. Sies, H., (1985). Oxidative Stress. *Elsevier, 82*(2), 291–295.
148. Sikka, S. C., (2001). Relative impact of oxidative stress on male reproductive function, *Current Medical Chemistry, 8*(7), 851–862.
149. Singh, R., Barden, A., Mori, T., & Beilin, L., (2001b). Advanced glycation end-products: a review. *Diabetologia*, *44*(2), 129–146.
150. Singh, S. N., Vats, P., Suri, S., Shyam, R., Kumria, M. M., Ranganathan, S., & Sridharan, K., (2001). Effect of an antidiabetic extract of *Catharanthus roseus* on enzymic activities in streptozotocin induced diabetic rats. *Journal of Ethnopharmacology, 76*(3), 269–277.

151. Soldatos, G., & Cooper, M. E. (2008). Diabetic nephropathy: important pathophysiologic mechanisms. *Diabetes Research and clinical Practice*, *200*(82), S75–S79.
152. Spiropoulos, J., Turnbull, D. M., & Chinnerry, P. F., (2002). Can mitochondrial DNA mutations cause sperm dysfunctions? *Molecular Human Reproduction*, *8*(8), 719–721.
153. Srinivasan, D., Nathan, S., Suresh, T., & Perumalsamy, P. L., (2001). Antimicrobial activity of certain Indian medicinal plants used in folkloric medicine. *Journal of Ethnopharmacology*, *74*(3), 217–220.
154. Srinivasan, K., Viswanad, B., Asrat, L., Kaul, C. L., & Ramarao, P., (2005). Combination of high fat diet-fed and low dose streptozotocin-treated rat: a model for type-2 diabetes and pharmacological screening. *Pharmacological Research*, *52*(4), 313–320.
155. Storgaard, H., Song, X. M., Jensen, C. B., Madsbad, S., Bjornholm, M., Vaag, A., & Zierath, J. R., (2001). Insulin signal transduction in skeletal muscle from glucose-intolerant relatives with type-2 diabetes. *Diabetes*, *50*(12), 2770–2778.
156. Suresh, S., Prithiviraj, E., Lakshmi, N. V., Ganesh, M. K., Ganesh, L., & Prakash, S., (2013). Effect of *Mucuna pruriens* (Linn.) on mitochondrial dysfunction and DNA damage in epididymal sperm of streptozotocin induced diabetic rat. *Journal of Ethnopharmacology*, *145*(1), 32–41.
157. Tiganis, T., (2011). Reactive oxygen species and insulin resistance: the good, the bad and the ugly. *Trends in Pharmacological Sciences*, *32*(2), 82–89.
158. Tiong, S. H., Looi, C. Y., Arya, A., Wong, W. F., Hazni, H., Mustafa, M. R., & Awang, K., (2015). Vindogentianine, a hypoglycemic alkaloid from *Catharanthus roseus* (L.) G., Don (Apocynaceae). *Fitoterapia*, *102*, 182–188.
159. Tiong, S. H., Looi, C. Y., Hazni, H., Arya, A., Paydar, M., Wong, W. F., Cheah, S. C., Mustafa, M. R., & Awang, K., (2013). Antidiabetic and antioxidant properties of alkaloids from *Catharanthus roseus* (L.) G., Don. *Molecules*, *18*(8), 9770–9784.
160. Toi, M., Bando, H., Ramachandran, C., Melnick, S. J., Imai, A., Fife, R. S., Carr, R. E., Oikawa, T., & Lansky, E. P., (2003). Preliminary studies on the anti-angiogenic potential of pomegranate fractions in vitro and in vivo. *Angiogenesis*, *6*(2), 121–128.
161. Torre, L. A., Bray, F., Siegel, R. L., Ferlay, J., Lortet-Tieulent, J., & Jemal, A., (2015). Global cancer statistics, 2012. *CA: A Cancer Journal for Clinicians*, *65*(2), 87–108.
162. Turk, G., Sonmez, M., & Aydin, M., (2008). Effects of pomegranate juice consumption on sperm quality, spermatogenic cell density, antioxidant activity, & testosterone level in male rats. *Clinical Nutrition*, *27*(2), 289–296.
163. Twigg, J. P., Irvine, D. S., & Aitken, R. J., (1998). Oxidative damage to DNA in human spermatozoa does not preclude pronucleus formation at intracytoplasmic sperm injection, *Human Reproduction*, *13*(7), 1864–1871.
164. Udenta, E. A., Obizoba, I. C., & Oguntibeju, O. O., (2014). Antidiabetic effects of Nigerian indigenous plant foods /diet. Chapter 3, In: *Antioxidant-Antidiabetic Agents and Human Health*, Oguntibeju, O. O., ed., *In. Tech. Croatia*, pp. 59–93.
165. Valko, M., Leibfritz, D., Moncol, J., Cronin, M. T., Mazur, M., & Telser, J., (2007). Free radicals and antioxidants in normal physiological functions and human disease. *The International Journal of Biochemistry & Cell Biology*, *39*(1), 44–84.
166. Vasconcelos, L. C., Sampaio, M. C., Sampaio, F. C., & Higino, J. S., (2003). Use of *Punica granatum* as an antifungal agent against candidosis associated with denture stomatitis. *Mycoses*, *46*(5–6), 192–196.

167. Verma, S., & Singh, S. P., (2008). Current and future status of herbal medicines. *Veterinary world, 1*(11), 347–350.
168. Veskoukis, A. S., Tsatsakis, A. M., & Kouretas, D., (2012). Dietary oxidative stress and antioxidant defense with an emphasis on plant extract administration. *Cell Stress and Chaperones, 17*(1), 11–21.
169. Voravuthikunchai, S. P., & Limsuwan, S., (2006). Medicinal plant extracts as anti-*Escherichia coli* O157:H7 agents and their effects on bacterial cell aggregation. *Journal of Food Prot., 69*(10), 2336–2341.
170. Wagener F. A. D. T. G., Dekker, D., Berden, J. H., Scharstuhl, A., & Van der Vlag, J., (2009). The role of reactive oxygen species in apoptosis of the diabetic kidney. *Apoptosis, 14*(12), 1451–8.
171. Wang, X., Sharma, R. K., Sikka, S. C., Thomas, A. J., Falcone, Jr. T., & Agarwal, A., (2003). Oxidative stress is associated with increased apoptosis leading to spermatozoa DNA damage in patients with male factor infertility. *Fertility Sterility, 80*(3), 531–535.
172. Wathes, D. C., Abayasekara, D. R., & Aitken, R. J., (2007). Polyunsaturated fatty acids in male and female reproduction. *Biology Reproduction, 77*(2), 190–201.
173. World Health Organization, (2016). Global reports on Diabetes, 6–35.
174. Young, I. S., & Woodside, J. V., (2001). Antioxidants in health and disease. *Journal of Clinical Pathology, 54*(3), 176–186.
175. Zhang, L., Wei, G., Liu, Y., Zu, Y., Gai, Q., & Yang, L., (2016). Antihyperglycemic and antioxidant activities of total alkaloids from *Catharanthus roseus* in streptozotocin-induced diabetic rats. *Journal of Forestry Research, 27*(1), 167–74.
176. Zhang, L., & Eisenbarth, G. S., (2011). Prediction and prevention of type-1 diabetes Mellitus. *Journal of Diabetes, 3*(1), 48–57.

PART IV

MEDICINAL PLANTS AND MANAGEMENT OF HYPERTENSION

CHAPTER 12

GINGER AND TURMERIC SUPPLEMENTED DIET AS A NOVEL DIETARY APPROACH FOR MANAGEMENT OF HYPERTENSION: A REVIEW

AYODELE JACOB AKINYEMI, GANIYU OBOH, and MARIA ROSA CHITOLINA SCHETINGER

CONTENTS

12.1 INTRODUCTION

Hypertension is a part of the metabolic syndrome in which the blood pressure is elevated. Persistent hypertension has been linked to several cardiovascular diseases (CVDs) such as stroke, heart attack, and heart failure [56]. The oxidative stress (OS) associated with these diseases can result in inflammation of the heart [29]. Evidence has shown that these diseases are caused by free radicals [38]. Oxidative stress results from either a decrease in natural cell antioxidant capacity or an increase in the amount of reactive oxygen species (ROS) in organisms.

Worldwide hypertension is estimated to cause 7.1 million premature deaths and its prevalence in developing countries is as high as those in developed countries [65]. This pathologic condition can be caused by an increase in the cardiac output volume or by an increase in peripheral resistance. However, in 90% of patients with high blood pressure, the cause of hypertension is unknown, and these individuals are termed as carriers of essential hypertension [65].

Hypertension is grouped into two main categories: primary and secondary hypertension. Primary hypertension is also known as essential hypertension and it affects 95% of persons suffering from this disease. The primary causes of hypertension are not yet known; however, factors such as age, high salt intake, low potassium diet, sedentary lifestyle, stress as well as genes have been found to be contributors to hypertension [20]. High blood pressure occurring as a result of a consequence of another disorder or a side effect of medication is referred to as secondary high blood pressure. Such disorders may include renal failure or reno-vascular disease. This type of blood pressure is evident in about 5 to 10% of cases [20].

The prevention of high blood pressure is recognized as the controlling key to hypertension especially in developing countries. The treatment of hypertension in developing countries is unaffordable for the average worker. This is due to the fact that, the lowest treatment pharmacologically is recorded to be 7.5–12% of the monthly income of the average worker in developing countries [20]. In effect, it is impossible for a better treatment pharmacologically. Thus, the need for understanding the disease and controlling it with preventive measures is a key to reduce high prevalence in developing country. However, consumption of foods rich in antioxidant phytochemicals may help fight degenerative diseases caused by oxidative stress of the heart by improving the body's antioxidant status [20].

Although many antihypertensive drugs have been discovered such as captopril, enalapril, atenolol, etc., recent research studies have focused on finding cheap and affordable therapeutic agents with minimal side effects for effective treatment of hypertension, due to high cost and side effects associated with these drugs. During the last two decades, many studies have focused on the dietary prevention of hypertension development, with a particular interest in plant foods with promising potential but the mechanisms still remain to be explored.

Ginger (*Zingiber officinale*) rhizome is a perennial plant in the family Zingiberaceae and is a large biennial herb that grows abundantly in South Asia. It is consumed as delicacy, medicine, or spice and used in folk medicine for the treatment of several diseases such as stomach ache, diarrhea, asthma, and gastrointestinal and respiratory diseases [58]. It contains several hundred valuable compounds and new constituents are still being found [1]. The hypotensive effect of aqueous extract of ginger rhizomes has been reported by Ghayur and Gilani [25]. It is also recommended by the traditional healers in South Asia for use in cardiopathy, high blood pressure, palpitations and to improve the circulation for its use as a vasodilator [1]. The most famous traditional medicinal use of *Z. officinale* is to promote the blood circulation for removing blood stasis and the mechanism is related to anti-platelet aggregation activity [33]. In traditional medicine, it is used as a therapy against hypertension and several cardiovascular diseases with limited scientific basis for their action.

Another notable member of ginger family is turmeric (*Curcuma longa*), which is a rhizomatous herbaceous perennial plant in the ginger family, employed as a dye source and food colorant due to its characteristic yellow color [21]. Turmeric is one of the main ingredients for curry powder and can be used as a drink to treat colds and stomach ache [21]. Curcuminoid compounds are major phytochemicals of the turmeric and they are responsible for the characteristic yellow color and have been reported to exert several medicinal properties [9, 21]. Turmeric rhizomes have been reported to reduce the uptake of cholesterol from the gut [9]. Arun and Nalini [10] have reported the hypoglycemic properties of turmeric in diabetic albino rats. In addition, curcumin from turmeric rhizomes have been shown to possess anti-inflammatory properties and potential therapeutic effect against neurodegenerative and cardiovascular diseases [32].

In this review, the authors examined recent research on mechanisms of action by which ginger and turmeric rhizomes are currently being used in traditional medicine for the treatment of hypertension and several cardiovascular diseases.

12.2 ROLE OF REACTIVE OXYGEN SPECIES IN HYPERTENSION

Literature has shown that excessive generation of reactive oxygen species (ROS) contributes to hypertension and that scavenging of ROS decreases blood pressure. Nakazono et al. [41] showed that bolus administration of a modified form of superoxide dismutase (SOD) acutely lowered blood pressure in hypertensive rats. Membrane-targeted forms of SOD and SOD mimetics (such as tempol) can lower blood pressure and decrease renovascular resistance in hypertensive animal models [4, 52]. There is ample evidence suggesting that ROS not only contribute to hypertension but that the reduced nicotinamide adenine dinucleotide phosphate (NADPH) oxidase is their major source. Components of this enzyme system are upregulated by hypertensive stimuli, and NADPH oxidase enzyme activity is increased by these same stimuli. Hypertension is associated with increased ROS formation in multiple organs, including the brain, the vasculature, and the kidney, all of which could contribute to hypertension.

12.3 EFFECT OF GINGER AND TURMERIC RHIZOMES ON ANGIOTENSIN-1 CONVERTING ENZYME (ACE) ACTIVITY

The renin-angiotensin aldosterone system (RAS) plays a major physiologic role in the regulation of blood pressure and maintaining sodium homeostasis [8]. This system has been suggested to play important role in pathologic conditions such as hypertension and other cardiovascular diseases [60]. Angiotensin-1 converting enzyme (ACE) is a zinc metallopeptidase, which plays a vital role in regulation of blood pressure by converting the inactive peptide, angiotensin I (Ang I) into vasoconstrictor and trophic angiotensin II (Ang II) [18]. ACE exists both as a membrane-bound enzyme in various organs such as heart, blood vessels, kidney and in a freely soluble form in plasma [8, 60]. Several reports have demonstrated that RAS is altered in a model of chronic administration of L-NAME [2, 5]. It has been shown that blockade of the RAS with angiotensin-converting enzyme inhibitor or with an angiotensin II (Ang II) receptor antagonist prevents the elevation of blood pressure in L-NAME-treated animals [2]. Recent studies by Akinyemi et al. [5] revealed that oral administration of L-NAME resulted in the activation of renin–angiotensin system via increased ACE activity. However, dietary supplementation with both ginger and turmeric rhizomes caused a significant reduction in serum and kidney ACE activity. The decrease in ACE activity as a result of introduction of the dietary rhizomes has been linked to the synergistic effect of the phenolic compounds such as caffeic acid, gallic acid, quercetin, curcumin, etc. that are present in the rhizomes. This was in agreement with an in vitro study reported by Akinyemi et al. [5, 8], where aqueous extracts from ginger and turmeric rhizomes inhibited ACE in a dose-dependent manner.

Phenolic compounds such as curcumin from turmeric rhizomes have been reported to inhibit ACE activity either as a single compound or in synergy with other compounds [13, 43]. This approach has been used in various practices of traditional medicine, where mixture of plant constituents is commonly prescribed for the treatment/management of hypertension [35, 67].

12.4 EFFECTS OF GINGER AND TURMERIC RHIZOMES ON ENDOTHELIAL NITRIC OXIDE SYNTHASE (ENOS) AND ARGINASE ACTIVITIES

Angiotensin II produced in RAS from angiotensin I by the action of ACE is a vasoconstrictor in renal vessels and has been implicated in hypertension [50]. However, its effect under pathological conditions is counteracted by nitric oxide (NO), which serves as a potent vasodilator and plays an important role in maintaining vascular tone [50]. In the kidney, NO is synthesized primarily by endothelial nitric oxide synthase (eNOS) and plays crucial role in vasodilation [44]. L-arginine is primarily viewed as a substrate for NO formation, but L-arginine is involved in multiple biochemical pathways, and the availability of L-arginine for NO formation depends upon the concentration of plasma L-arginine and the relative activity of competing intracellular pathways [40]. Arginase converts L-arginine to form urea and ornithine, and several studies have suggested that there is an intracellular competition between eNOS and arginase enzymes for their common substrate L-arginine [5, 40].

Akinyemi et al. [5] observed that there was an increase in arginase activity in hypertensive rats treated with L-NAME. However, dietary ginger and turmeric rhizomes inhibited arginase activity in hypertensive rats. Previous studies have reported up-regulation of arginase activity and decreased NO in hypertension [11, 37]. The decrease in arginase activity in ginger and turmeric rhizomes in hypertensive rat has been linked to their inhibitory effect on arginase activity in a dose-dependent manner in vitro. However, several authors have reported that dietary plant phenolics exhibited an inhibitory effect on arginase activity [31, 36].

Enhanced arginase activity can impair endothelium-dependent vasorelaxation by decreasing L-arginine availability to endothelial nitric oxide synthase (eNOS), thereby reducing NO production and uncoupling eNOS function. Nitric oxide (NO) is essential to normal cardiovascular function and blood pressure control.

12.5 EFFECTS OF GINGER AND TURMERIC RHIZOMES ON INFLAMMATORY CYTOKINES

Low grade inflammation has been recognized to play a crucial pathophysiological role in hypertension and other cardiovascular diseases [32]. There is evidence indicating that innate and adaptive immune systems, and in particular T-cells, are involved. A balance between T-effector lymphocytes and T-regulatory lymphocytes represents a crucial regulatory mechanism that, when altered, favors blood pressure elevation and organ damage development [62]. Inflammation participates in many processes that contribute to the development of elevated blood pressure (BP) by enhancing the proliferation of smooth muscle cells vascular remodeling [62, 63]. Moreover, the release of pro-inflammatory cytokines modifies the normal state of vasodilatation mainly because of a low availability of nitric oxide (NO) [62, 63]. The immune system is essential to host defenses comprising an interactive network of lymphoid organs and immune cells [22].

Thus, agents that can modify the immune function have been suggested as a therapeutic approach in improving the outcomes in hypertension and cardiovascular diseases associated with inflammation. This approach has been discovered to be a novel means in the treatment of hypertension and other cardiovascular diseases [12, 47].

The immune and inflammatory processes can be altered by extracellular adenine nucleotides and nucleosides such as ATP and adenosine, which have been recognized as key components of the purinergic system [16]. The ATP acting through specific purinergic receptors on cell surface and is involved in pro-inflammatory actions such as lymphocyte stimulation and proliferation as well as cytokine release including IL-2, IFN-γ, IL-1β, and TNF-α [12, 45, 47]. The breakdown product, adenosine, exhibits potent anti-inflammatory and immunosuppressive actions by inhibiting both proliferation of T-cells and secretion of pro-inflammatory cytokines, such as: TNF-α and IFN-γ [28].

Thus, current specific hypertension treatments have focused on targeting inflammatory cytokines, and some studies demonstrated the correlation between levels of circulating inflammatory mediators and patient survival [30, 39]. In a more recent study by Akinyemi et al. [6], dietary supplementation with ginger and turmeric rhizomes prevented alterations in the inflammatory cytokines in hypertensive rats. This effect has been attributed to their immunomodulatory potential by causing a decrease in pro-inflammatory cytokines with a concomitant increase in anti-inflammatory cytokines under hypertensive state. A previous study has reported the use of anti-inflammatory/immunomodulatory agents (thiazolidinedinones, rapamycin, cyclosporine, and STAT3 inhibitors) as an emerging therapy for the treatment of hypertension in animal models [56].

12.6 EFFECTS OF GINGER AND TURMERIC RHIZOMES ON PLATELETS PURINERGIC SYSTEMS

Hypertension is a risk factor for athero-thrombosis events, in which platelets play a crucial role [64]. Platelets are one of the most important blood components that participate in maintaining vascular integrity promoting the primary and secondary hemostasis that occur after vessel damage [64].

Altered platelet morphology and function have been reported in patients with hypertension and may be associated with an increased risk of developing vascular disease [59, 61]. Furthermore, it has been demonstrated that platelets of these patients exhibit a greater tendency toward spontaneous aggregation and are highly hypersensitive to agonists such as ADP [26, 53]. Platelets express a multienzymatic complex on their surface, which is responsible for extracellular nucleotide hydrolysis. This complex includes enzymes: ecto-nucleoside triphosphate phosphohydrolase (NTPDase), ecto-5′-nucleotidase, and ecto-adenosine deaminase (ADA) [24]. NTPDases hydrolyses ATP and ADP to AMP [46], while ecto-5'-nucleotidase hydrolyses the resulting AMP to adenosine [57]. ADA enzyme catalyzes the irreversible deamination of adenosine to inosine [15]. Together, these ecto-enzymes constitute a highly organized enzymatic cascade that is able to regulate the extracellular concentrations of adenine nucleotides and nucleosides and play an important role in the maintenance of normal hemostasis and thrombo-genesis, mainly by regulating the platelet aggregation status [68].

Akinyemi et al. [7] demonstrated that platelet NTPDase, 5′-nucleotidase and ADA activities were altered in hypertensive rats. They reported an increase in ATP hydrolysis with a concomitant decrease in ADP and AMP hydrolysis in the L-NAME hypertensive rats, while ADA activity was significantly increased when compared with the control. Previous studies have demonstrated an alteration in NTPDase and 5'-nucleotidase activities in platelets of hypertensive rats [19, 34, 51].

Extracellular nucleotides such as ATP and ADP and their nucleoside adenosine are known to regulate the vascular response to endothelial damage by exerting a variety of effects on platelets [14]. The metabolism of these extracellular nucleotides of adenine in platelets occurs by the action of the surface located enzyme cascade constituted by NTPDase, 5'-nucleotidase and ADA [14, 24]. These enzymes have an important role in thrombo-regulation process, and alterations in their activities have been observed in hypertension, suggesting that this could be an important physiological and pathological parameter [19, 34, 51]. However, treatment with ginger and turmeric supplemented diet prevented an increase in platelets ADP and AMP hydrolysis in L-NAME hypertensive rat. Based on this finding, the authors suggest that treatment with ginger and turmeric rhizomes can maintain a high level of adenosine in the extracellular environment, which promotes vasodilation and has an important protective role under hypertensive state. This increase in adenosine production in hypertensive rats by both rhizomes in platelet will prevent platelet aggregation, which may suggest possible mechanism for their action in lowering systolic blood pressure.

12.7 EFFECTS OF GINGER AND TURMERIC RHIZOMES ON ANTIOXIDANT STATUS

It is well established that there is a link between hypertension and oxidative stress [17, 27]. The relationship between the development of hypertension and the increased bioavailability of ROS, decreased antioxidant capacity, or both, has been demonstrated in several experimental models of hypertension, as well as in human hypertension [3, 4, 27]. The increased levels of ROS such as superoxide anion, hydrogen peroxide and lipid peroxides have been reported in hypertensive patients [17]. Arterial hypertension is often associated with pathologies related to oxidative stress and may be considered as the result of oxygen free radicals systemic damage in different target tissues. Furthermore, growing evidence from animal studies suggests that oxidative stress in the kidney could be a key factor in the development and persistence of hypertension [66].

Intracellular defense against active oxygen species is performed by antioxidant enzymes (superoxide dismutase and catalase) and non-enzymatic antioxidants such as reduced glutathione (GSH) and vitamin C [48]. The loss of the balance between oxidation and antioxidation may lead to promoting the generation of OH^-, which is a powerful oxidant for many compounds.

In a recent work by Akinyemi and colleagues but not yet published, oral administration of L-NAME caused a decrease in the antioxidant enzymes (superoxide dismutase and catalase) and non-enzymatic antioxidants (total and non-protein thiols and vitamin C) with a concomitant increase in serum reactive oxygen species (ROS) level in L-NAME hypertensive rats. However, pre-treatment with both ginger and turmeric rhizomes were able to restore the levels of both enzymatic and non-enzymatic antioxidants with a concomitant decrease in serum ROS.

Furthermore, δ-Aminolevulinic acid dehydratase (δ-ALA-D), also known as porphobilinogen synthase (PBGS, EC 4.2.1.24), catalyzes the asymmetric condensation of two molecules of 5-aminolevulinate to produce porphobilinogen, which is the precursor of porphyrins [49]. It is a metalloenzyme containing sulfhydryl (-SH) groups and zinc, which are essential for its activity [49]. Evidence has indicated that δ-ALA-D enzyme is highly sensitive to the presence of a variety of pro-oxidant elements, which oxidize SH^- groups of this enzyme impairing its activity [18, 42]. The inhibition of δ-ALA-D activity may prejudice heme biosynthesis and can result in the accumulation of aminolevulinic acid that, under physiologically conditions, can have pro-oxidant effects contributing to the oxidative stress [18]. In the same study by Akinyemi and colleagues but unpublished, the activity of δ-ALA-D was significantly decreased in the whole blood, hepatic and renal tissues of L-NAME hypertensive rats. The results are in accordance with data found in human and experimental pathological conditions [23, 54]. Thus, the decrease in δ-ALA-D activity found in this study may be linked to the significant reduction in the antioxidant defenses in renal tissues of L-NAME hypertensive rats, especially in non-protein thiol (NPSH) content, which is responsible for preventing the oxidation of the sulfhydryl groups necessary for the activity of this enzyme [42].

The treatment with ginger and turmeric rhizomes was able to prevent the inhibition in δ-ALA-D activity in whole blood, liver and kidney of hypertensive rats suggesting that both rhizomes can prevent the oxidation of essential SH^- groups located at its active site of δ-ALA-D and consequently its inhibition. Indeed, in this study ginger and turmeric rhizomes prevented the reduction of NPSH levels in renal tissues in L-NAME induced hypertensive rats. Therefore, it could be expected to protect other endogenous thiols such as those found in δ-ALA-D enzyme. Consequently, the prevention of a decrease in NPSH content, as well as a decrease in oxidative stress in hypertensive rats by both rhizomes, could be associated in prevention of decrease in δ-ALA-D activity. Furthermore, this study revealed that the inhibition of δ-ALA-D enzyme is closely related to the development of hypertension in rats, pointing out the importance of antioxidants to minimize deleterious effects of hypertension.

12.8 SUMMARY

Ginger and turmeric rhizomes have shown to exhibit inhibitory effects on ACE activity, antioxidant properties (by improving antioxidant status and suppressing oxidative stress), anti-inflammatory properties (by preventing release of pro-inflammatory agents such as IL-1, IL-6, IFN-γ and TNF-α that can mediate inflammation), and modulatory effect on platelet ecto-nucleotidase (NTPDase, 5'-nucleotidase and ADA) activities resulting in an increase adenosine levels in hypertensive rats. Therefore, the prevention of alterations in the enzyme activities linked to rennin and purinergic systems in hypertensive rats by both rhizomes could suggest some possible mechanism of action for their antihypertensive benefits in traditional medicine. However, the observed effect could be attributed to the phenolic compounds present in the rhizomes.

KEYWORDS

- **ACE activity**
- **ADA**
- **angiotensin**
- **antihypertensive benefits**
- **antioxidant**
- **ginger**
- **hypertension**
- **inflammation**
- **NTPDase**
- **oxidative stress**
- **phenolic compounds**
- **purinergic system**
- **traditional medicine**
- **turmeric rhizomes**

REFERENCES

1. Abdullahi, M., (2011). Biopotency role of culinary spices and herbs and their chemical constituents in health and commonly used spices in Nigerian dishes and snacks. *African Journal of Food Science*, *5*, 111–124.
2. Ackermann, A., Fernandez-Alfonso, M. S., & Gonzalez, C., (1998). Modulation of angiotensin converting enzyme by nitric oxide. *British Journal of Pharmacology*, *124*, 291–298.
3. Addabbo, F., Montagnani, M., & Goligorsky, M. S., (2009). Mitochondria and reactive oxygen species. *Hypertension*, *53*, 885–892.
4. Adeagbo, A. S., Zhang, X., & Patel, D., (2005). Cyclo-oxygenase-2, endothelium and aortic reactivity during deoxycorticosterone acetate salt-induced hypertension. *Journal of Hypertension*, *23*(5), 1025–1036.
5. Akinyemi, A. J., Thome, G. R., Morsch, V. M., Stefanello, N., Goularte, J. F., Bello-Klein, A., Oboh, G., & Schetinger, M. R. C., (2015). Effect of dietary supplementation of ginger and turmeric rhizomes on angiotensin-1 converting enzyme (ACE) and arginase activities in L-NAME induced hypertensive rats. *Journal of Functional Foods*, *17*, 792–801.
6. Akinyemi, A. J., Thome, G., Morsch, V. M., Bottari, N. B., Baldissarelli, J., de Oliveira, L. S., Goularte, J. F., Belló-Klein, A., Duarte, T., Duarte, M., Akindahunsi, A. A., Oboh, G., & Schetinger, M. R., (2016a). Effect of ginger and turmeric rhizomes on inflammatory cytokines levels and enzyme activities of cholinergic and purinergic systems in hypertensive rats. *Planta Medica*, *82*, 612–620.
7. Akinyemi, A. J., Thome, G., Morsch, V. M., Bottari, N. B., Baldissarelli, J., de Oliveira, L. S., Goularte, J. F., Belló-Klein, A., Oboh, G., & Schetinger, M. R., (2016b). Dietary supplementation of ginger and turmeric rhizomes modulates platelets ecto-nucleotidase and adenosine deaminase activities in normotensive and hypertensive rats. *Phytotherapy Research*, *30*(7), 1156–1163.
8. Akinyemi, A. J., Ademiluyi, A. O., & Oboh, G., (2013). Aqueous extracts of two varieties of ginger (*Zingiber officinale*) inhibit angiotensin I–converting enzyme, Iron (II) and Sodium nitroprusside-induced lipid peroxidation in the rat heart in vitro. *Journal of Medicinal Food*, *16*(7), 641–646.
9. Arafa, H. M., (2005). Curcumin attenuates diet-induced hypercholesterolemia in rats. *Medical Science Monitor*, *11*, 228–234.
10. Arun, N., & Nalini, N., (2002). Efficacy of turmeric on blood sugar and polyol pathway in diabetic albino rats. *Plant Foods for Human Nutrition*, *57*, 41–52.
11. Bagnost, T., Ma, L., Da Silva, R. F., Rezakhaniha, R., Houdayer, C., & Stergiopulos, N., (2010). Cardiovascular effects of arginase inhibition in spontaneously hypertensive rats with fully developed hypertension. *Cardiovascular Research*, *87*, 569–577.
12. Bertoncheli, C. M., Zimmermann, C. E., Jaques, J. A., Leal, C. A., Ruchel, J. B., Rocha, B. C., Pinheiro, K. V., Souza, V. C., Stainki, D. R., Luz, S. C., Schetinger, M. R., & Leal, D. B., (2012). Increased NTPDase activity in lymphocytes during experimental sepsis. *Scientific World Journal*, 941906.

13. Bhullar, K. S., Lassalle-Claux, G., Touaibi, M., & Rupasinghe, H. P. V., (2014). Antihypertensive effect of caffeic acid and its analogs through dual rennin-angiotensin–aldosterone system inhibition. *European Journal of Pharmacology*, *730*, 125–132.
14. Birk, A. V., Broekman, J., & Gladek, E. M., (2002). Role of a novel soluble nucleotide phosphohydrolase from sheep plasma in inhibition of platelet reactivity: hemostasis, thrombosis, & vascular biology. *Journal of Laboratory and Clinical Medicine*, *139*, 116–124.
15. Blackburn, M. R., & Kellems, R. E., (2005). Adenosine deaminase deficiency: metabolic basis of immune deficiency and pulmonary inflammation. *Advance Immunology*, *86*, 1–41.
16. Bours, M. J., Swennen, E. L., Di Virgilio, F., Cronstein, B. N., & Dagnelie, P. C., (2006). Adenosine 5'-triphosphate and adenosine as endogenous signaling molecules in immunity and inflammation. *Pharmacology Therapy*, *112*, 358–404.
17. Briones, A. M., & Touyz, R. M., (2010). Oxidative stress and hypertension: current concepts. *Current Hypertension Reports*, *12*, 135–142.
18. Brito, V. B., Folmer, V., Soares, J. C., Silveira, I. D., & Rocha, J. B. T., (2007). Long-term sucrose and glucose consumption decreases the delta-aminolevulinate dehydratase activity in mice. *Nutrition*, *23*, 818–826.
19. Cardoso, A. M., Bagatini, M. D., & Martins, C. C., (2012). Exercise training prevents ectonucleotidases alterations in platelets of hypertensive rats. *Molecules and Cell Biochemistry*, *371*, 147–156.
20. Carretero, O. A., & Oparil, S., (2000). Essential hypertension. Part I: definition and etiology. *Circulation*, *101*(3), 329–335.
21. Chan, E. W. C., Lim, Y. Y., Wong, S. K., Lim, K. K., Tan, S. P., Lianto, F. S., & Yong, M. Y., (2009). Effects of different drying methods on the antioxidant properties of leaves and tea of ginger species. *Food Chemistry*, *113*(1), 166–172.
22. Delves, P. J., & Roitt, I. M., (2000). The immune system. Second of two parts. *N. Engl. J. Med.*, *343*, 108–117.
23. Fernández-Cuartero, B., Rebollar, J. L., Batlle, A., & De Salamanca, E. R., (1999). Delta aminolevulinate dehydratase (ALA-D) activity in human and experimental diabetes mellitus. *International Journal of Biochemistry and Cell Biology*, *31*, 479–488.
24. Furstenau, C. R., Trentin, D. S., & Gossenheimer, A. N., (2008). Ectonucleotidase activities are altered in serum and platelets of L-NAME-treated rats. *Blood Cells Mol Dis*, *41*, 223–229.
25. Ghayur, M. N., & Gilani, A. H., (2005). Ginger lowers blood pressure through blockade of voltage-dependent calcium channels. *Journal of Cardiovascular Pharmacology*, *45*, 74–80.
26. Haouari, M. E., & Rosado, J. A., (2008). Platelet signalling abnormalities in patients with type 2 diabetes mellitus: A review. *Blood Cells Mol. Dis.*, *41*(1), 119–123.
27. Harrison, D. G., Gongora, M. C., Guzik, T. J., & Widder, J., (2007). Oxidative stress and hypertension. *Journal of American Society of Hypertension*, *1*, 30–44.
28. Hasko, G., Linden, J., Cronstein, B., & Pacher, P., (2008). Adenosine receptors: therapeutic aspects for inflammatory and immune diseases. *National Review*, *7*, 759–770.

29. Huang, W., & Glass, C. K., (2010). Nuclear receptors and inflammation control molecular mechanisms and pathophysiological relevance. *Arterioscler Thromb Vasc Biol*, *30*, 1542–1549.
30. Kherbeck, N., Tamby, M. C., Bussone, G., Dib, H., Perros, F., Humbert, M., & Mouthon, L., (2013). The role of inflammation and autoimmunity in the pathophysiology of pulmonary arterial hypertension. *Clinical Review on Allergy and Immunology*, *44*, 31–38.
31. Kim, S. W., Cuong, T. D., Hung, T. M., Ryoo, S., Lee, J. H., & Min, B. S., (2013). Arginase II inhibitory activity of flavonoid compounds from *Scutellaria indica*. *Archives of Pharmaceutical Research*, *36*, 922–926.
32. Kowluru, R. A., & Kanwar, M., (2007). Effects of curcumin on retinal oxidative stress and inflammation in diabetes. *Nutrition and Metabolism*, *4*, 1–8.
33. Liao, Y. R., Leu, Y. L., Chan, Y. Y., Kuo, P. C., & Wu, T. S., (2012). Anti-platelet aggregation and vasorelaxing effects of the constituents of the rhizomes of *Zingiber officinale*. *Molecules*, *17*, 8928–8937.
34. Lunkes, G., Lunkes, D., & Morsch, V. M., (2004). NTPDase and 5'-nucleotidase activities in rats alloxan-induced diabetes. *Diabetes Research and Clinical Practice*, *65*, 1–6.
35. Luo, J., Xu, H., & Chen, K. J., (2013). Potential benefits of Chinese herbal medicine for elderly patients with cardiovascular diseases. *Journal of Geriatric Cardiology*, *10*, 305–309.
36. Manjolin, L. C., Dos Reis, M. B., Maquiaveli, C., Santos-Filho, O. A., & Da Silva, E. R., (2013). Dietary flavonoids fisetin, luteolin and their derived compounds inhibit arginase, a central enzyme in Leishmania (Leishmania) amazonensis infection. *Food Chemistry*, *141*, 2253–2262.
37. Maquiaveli, C. C., Da Silva, E. R., Rosa, L. C., Francescato, H. D., Lucon Júnior, J. F., & Silva, C. G., (2014). *Cecropia pachystachya* extract attenuated the renal lesion in 5/6 nephrectomized rats by reducing inflammation and renal arginase activity. *Journal of Ethnopharmacology*, *158*, 49–57.
38. Martinon, F., (2010). Signaling by ROS drives inflammasome activation. *Eur. J. Immunol*, *40*, 616–619.
39. Meloche, J., Renard, S., Provencher, S., & Bonnet, S., (2013). Anti-inflammatory and immunosuppressive agents in PAH., *Handbook on Experimental Pharmacology*, *218*, 437–476.
40. Morris, S. M., Jr., (2009). Recent advances in arginine metabolism: Roles and regulation of the arginases. *British Journal of Pharmacology*, *157*, 922–930.
41. Nakazono, K., Watanabe, N., & Matsuno, K., (1991). Does superoxide underlie the pathogenesis of hypertension? *Proc. Natl. Acad. Sci. U S A*, *88*(22), 10045–10048.
42. Nogueira, C. W., Soares, F. A., Nascimento, P. C., Muller, D., & Rocha, J. B., (2003). T-2,3-Dimercaptopropane-1-sulfonic acid and meso-2, 3-dimercaptosuccinic acid increase mercury and cadmium-induced inhibition of delta-aminolevulinate dehydratase. *Toxicology*, *184*, 85–95.
43. Pang, X. F., Zhang, L. H., Bai, F., Wang, N. P., Shah, A., & Garner, R., (2015). Dual ACE-inhibition and angiotensin II AT1 receptor antagonism with curcumin attenuate maladaptive cardiac repair and improve ventricular systolic function after myocardial infarction in rat heart. *European Journal of Pharmacology*, *746*, 22–30.

44. Patzak, A., Steege, A., Lai, E. Y., Brinkmann, J. O., Kupsch, E., & Spielmann, N., (2008). Angiotensin II response in afferent arterioles of mice lacking either the endothelial or neuronal isoform of nitric oxide synthase. *American Journal of Physiology. Regulatory, Integrative and Comparative Physiology*, *294*, 429–437.
45. Polachini, C. R., Spanevello, R. M., Casali, E. A., Zanini, D., Pereira, L. B., Martins, C. C., Baldissareli, J., Cardoso, A. M., Duarte, M. F., Da Costa, P., Prado, A. L., Schetinger, M. R., & Morsch, V. M., (2014). Alterations in the cholinesterase and adenosine deaminase activities and inflammation biomarker levels in patients with multiple sclerosis. *Neuroscience*, *266*, 266–274.
46. Robson, S. C., Sevigny, J., & Zimmermann, H., (2006). The E-NTPDase family of ectonucleotidases: Structure function relationships and pathophysiological significance. *Purinergic Signaling*, *2*, 409–430.
47. Rodrigues, R., Debom, G., Soares, F., Machado, C., Pureza, J., Peres, W., De Lima, G. G., Duarte, M. F., Schetinger, M. R., Stefanello, F., Braganhol, E., & Spanevello, R., (2014). Alterations of ectonucleotidases and acetylcholinesterase activities in lymphocytes of Down syndrome subjects: relation with inflammatory parameters. *Clin. Chim. Acta.*, *433*, 105–110.
48. Romero, D., & Roche, E., (1996). High blood pressure, oxygen radicals and antioxidants: Etiological relationships. *Medical Hypotheses*, *46*, 414–420.
49. Sassa, S., (1998). Delta-aminolevulinic acid porphyria. *Seminars in Liver Disease*, *18*, 95–101.
50. Satoh, M., Fujimoto, S., Arakawa, S., Yada, T., Namikoshi, T., & Haruna, Y., (2008). Angiotensin II type1 receptor blocker ameliorates uncoupled endothelial nitric oxide synthase in rats with experimental diabetic nephropathy. *Nephrology Dialysis Transplant*, *23*, 3806–3813.
51. Schmatz, R., Schetinger, M. R., & Spanevello, R. M., (2009). Effects of resveratrol on nucleotide degrading enzymes in streptozotocin-induced diabetic rats. *Life Science*, *84*, 345–350.
52. Schnackenberg, C. G., & Wilcox, C. S., (1999). Two-week administration of tempol attenuates both hypertension and renal excretion of 8-Iso prostaglandin f2 alpha. *Hypertension*, *33*(1: Part 2), 424–428.
53. Sobol, A. B., & Watala, C., (2000). The role of platelets in diabetes-related vascular complications. *Diabetes Research and Clinical Practice*, *50*(1), 1–16.
54. Souza, J. B., Rocha, J. B. T., Nogueira, C. W., Borges, V. C., Kaizer, R. R., Morsch, V. M., Dressler, V. L., Martins, A. F., Flores, E. M., & Schetinger, M. R., (2007). Delta-aminolevulinate dehydratase (delta-ALA-D) activity in diabetes and hypothyroidism. *Clinical Biochemistry*, *40*, 321–325.
55. Sowers, J. R., & Epstein, M., (1995). Diabetes mellitus and associated hypertension, vascular disease, & nephropathy: an update. *Hypertens*, *26*, 869–879.
56. Stenmark, K. R., & Rabinovitch, M., (2010). Emerging therapies for the treatment of pulmonary hypertension. *Pediatr. Crit. Care Med.*, *11*, 85–90.
57. Strater, N., (2006). Ecto-5'-nucleotidase: structure function relationships. *Purinergic Signaling*, *2*, 343–350.
58. Tang, W., & Eisenbrand, G., (1992). *Chinese Drugs of Plant Origin*. Springer-Verlag, Berlin, Germany, pp. 1011–1052.

59. Tousoulis, D., Paroutoglou, L. P., Papageorgiou, N., Charakida, M., & Stefanadis, C., (2010). Recent therapeutic approaches to platelet activation in coronary artery disease. *Pharmacology Therapeutics*, *127*, 108–120.
60. Udenigwe, C. C., Lin, Y. S., Hou, W. C., & Aluko, R. E., (2009). Kinetics of inhibition of renin and angiotensin I-converting enzyme by flaxseed protein hydrolysate fractions. *Journal of Functional Foods*, *1*, 199–207.
61. Vilahur, G., & Badimon, L., (2013). Antiplatelet properties of natural products. *Vascular Pharmacology*, *59*, 67–75.
62. Virdis, A., Dell'Agnello, U., & Taddei, S., (2014). Impact of inflammation on vascular disease in hypertension. *Maturitas*, *78*, 179–183.
63. Virdis, A., & Schiffrin, E. L., (2003). Vascular inflammation: a role in vascular disease in hypertension. *Curr. Opin. Nephrol. Hypertens.*, *12*, 181–187.
64. Vorchheimer, D. A., & Becker, R., (2006). Platelets in atherothrombosis. *Mayo. Clin. Proc.*, *81*, 59–68.
65. WHO, (1985). *Diabetes Mellitus:* World Health Organization Study Group Technical Report. Series 727, Geneva: World Health Organization.
66. Wilcox, C. S., (2005). Oxidative stress and nitric oxide deficiency in the kidney: a critical link to hypertension? *American Journal of Physiology, Regulatory, Integrative and Comparative Physiology*, *289*, 13–35.
67. Xiong, X. J., & Wang, J., (2014). Chinese classical formulas for treatment of essential hypertension. *China Journal of Chinese Materia Medica*, *39*, 929–933.
68. Yegutkin, G. G., (2008). Nucleotide- and nucleoside-converting ectoenzymes: important modulators of purinergic signalling cascade. *Biochimica. et. Biophysica. Acta. (BBA) – Molecules and Cell Research*, *1783*, 673–694.

GLOSSARY OF TECHNICAL TERMS

Abscess is a swollen area within body tissue, containing an accumulation of pus.

Adenylate cyclase is an enzyme with key regulatory roles in essentially all cells, catalyzing the conversion of adenosine triphosphate (ATP) to 3',5'-cyclic AMP (cAMP) and pyrophosphate.

Agar diffusion assay is a test of the antibiotic sensitivity of bacteria.

Allopathy is the scientific, evidence-based approach to healing, such as the use of conventional medicines for treatment.

Alpha-amylase is an enzyme that hydrolyses alpha bonds of large, alpha-linked polysaccharides, such as starch and glycogen, yielding glucose and maltose.

Alpha-glucosidase is an enzyme that breaks down starch and disaccharides to glucose.

Ameliorate refers to improve or make an unpleasant situation better, more bearable or more satisfactory.

Ankylosing spondylitis is a type of autoimmune arthritis which causes chronic inflammation in the joints of the spine and pelvis.

Antidiabetic refers to drugs that treat diabetes mellitus by lowering glucose levels in the blood.

Antimicrobial is an agent that kills microorganisms or inhibits their growth.

Antioxidant is a substance that inhibits oxidation, especially one used to counteract the deterioration of stored food products.

Apoptosis is the pre-programmed cell death, but can occur due to the presence of molecules that trigger events leading to characteristic cell changes (morphology) and death. These changes include cell shrinkage, nuclear fragmentation, chromatin condensation, chromosomal DNA fragmentation, and global mRNA decay.

Autoimmune inflammatory diseases are a group of diseases arising from abnormal immune responses to a normal body part.

Axenic amastigote is a culture in which only a single type of protist cell that does not have visible external flagella or cilia and at a certain phase in the life-cycle of trypanosome protozoans is present and entirely free of all other contaminating organisms.

Ayurveda is a traditional Hindu system of medicine based on the idea of balancing body systems through diet, herbal treatment and yoga.

Bambara groundnut (BGN) refers to grain legume grown mainly by subsistence farmers in sub-Sahara Africa.

Bioactive compound is a compound that has an effect on a living organism, tissue/cell.

Bioactivity is the effect of a drug upon a living organism or living tissue.

Bioavailability refers to the degree and rate at which a drug is absorbed and becomes available at the site of physiological activity.

Biomarker refers to a naturally occurring molecule, gene, or characteristic by which a particular pathological or physiological process, disease, can be identified.

Cancer is a group of diseases involving abnormal cell growth with the potential to invade or spread to other parts of the body.

Candidiasis is a fungal infection due to any type of Candida (a type of yeast). When it affects the mouth, it is commonly called thrush.

Capacitation is the penultimate step in the maturation of mammalian spermatozoa and is required to render them competent to fertilize an oocyte.

Carbuncle is a severe abscess or multiple boil in the skin, typically infected with Staphylococcus bacteria.

Carcinogenesis is the formation of a cancer, whereby normal cells are transformed into cancer cells.

Chalcones are aromatic ketone and an enone that forms the central core for a variety of important biological compounds, which are known collectively as chalcones or chalconoids.

Chromatography is an analytical technique that is used to separate mixture of substances into its components based on movement of two phases (one stationary phase and one mobile phase) moving at definite direction to achieve separation.

Chronic diseases are long medical conditions that are generally progressive.

Cisplatin is a chemotherapy agent within a class of platinum-containing anti-cancer drugs.

Crohn's disease is a chronic inflammatory disease of the gastrointestinal tract.

Cytokines are a category of small proteins (~5 to 20 kDa) that are important in cell signaling, involved in autocrine signaling, paracrine signaling and endocrine signaling as immunomodulating agents. Cytokines also refers to a group of proteins made by the immune system that act as chemical messengers.

Cytotoxicity is the quality of being toxic to cells.

Decoction is an extract prepared by suspending the material to be extracted in cold water.

Dermatomycosis is a superficial fungal infection of the skin or its appendages.

Diabetes is a group of metabolic diseases in which there are high blood sugar levels over a prolonged period.

Diabetes mellitus is a group of metabolic disorders of carbohydrate metabolism in which glucose produced is being underutilized, thereby resulting in excess glucose in the blood stream.

Diuresis refers to increased or excessive production of urine.

Enzyme is a substance produced by a living organism which acts as a catalyst to bring about a specific biochemical reaction.

Epimastigote is a growth stage in the life cycle of some Trypanosomatids.

Essential oils are natural oils obtained by distillation of plant materials.

Ethnobotany is the study of regions plants and their practical uses through the traditional knowledge.

Ethnopharmacology refers to a study or comparison of the traditional medicine practiced by various ethnic groups, and especially by indigenous peoples.

Flavonoids are a large group of plant or fungal secondary metabolites which contain a 15 carbon skeleton which consists of 2 phenyl rings and a heterocyclic ring.

Folklore refers to the traditional beliefs, myths, tales, and practices of a people, which have been disseminated in an informal manner from generations to genrations.

Free radical is an uncharged molecule (typically highly reactive and short-lived) having an unpaired valency electron. Free radical is any atom or molecule that has a single unpaired electron in an outer shell.

GC-MS is gas chromatography–mass spectrometry; an analytical method that combines the features of gas-chromatography and mass spectrometry to identify different substances within a test sample.

Genus is a taxonomic rank used in the biological classification of living and fossil organisms in biology.

Glycation is the result of the typically covalent bonding of a protein or lipid molecule with a sugar molecule, such as glucose, without the controlling action of an enzyme.

Hydroxyl radical (•OH) is the neutral form of the hydroxide ion (OH^-). Hydroxyl radicals are highly reactive and thus generally short lived.

Hyperglycemia is the abnormal increase of glucose in the blood.

Hypertension is also known as high blood pressure, is a long-term medical condition in which the blood pressure in the arteries is persistently elevated.

IC_{50} is the half maximal inhibitory concentration, representing the concentration of a drug that is required for 50% inhibition in vitro.

IDDM is a type of diabetes mellitus whereby there is relative or absolute secretion of insulin due to pancreatic beta cells destructions. It is otherwise referred to as type 1 DM.

Infertility refers to an inability of a couple, living together, to achieve conception after a period of regular and unprotected sexual intercourse.

Inflammation refers to a localized physical condition in which part of the body becomes reddened, swollen, hot and often painful, especially as a reaction to injury or infection.

Infusion is the process of extracting chemical compounds or flavors from plant material in hot water by allowing the material to remain suspended in the solvent over time.

LC_{50} is the concentration of extract required to produce mortality in 50% of the test subjects.

Lipid peroxidation refers to the process under which oxidants such as free radicals or non-radical species attack lipids containing carbon-carbon double bond(s), especially polyunsaturated fatty acids (PUFAs) that involve hydrogen abstraction from a carbon, with oxygen insertion resulting in lipid peroxyl radicals and hydroperoxides.

Lipopolysaccharides are large molecules consisting of a lipid and a polysaccharide composed of O-antigen, outer core and inner core joined by a covalent bond, and which elicit strong immune responses in animals.

Medicinal plants are naturally occurring substances that possess potent compounds that can be utilized for therapeutic reasons or synthetically modified to pharmacological drugs.

Minimum inhibitory concentration (MIC) is the lowest concentration of a chemical that prevents visible growth of a bacterium.

Multiple sclerosis is a chronic, progressive autoimmune inflammatory disease which results in damage to nerve sheath, resulting in a myriad of symptoms.

Neurodegenerative diseases is defined as a deterioration, often irreversible, of the intellectual and cognitive faculties and it is generally associated with ageing and/or AD, PD, stroke.

Neuroprotection refers to the strategies and relative mechanisms able to defend the central nervous system (CNS) against neuronal injury due to both acute (e.g. stroke or trauma) and chronic neurodegenerative disorders (e.g. Alzheimer's disease, AD, and Parkinson's disease, PD).

Non-communicable disease is a medical condition or disease that is not caused by infectious agents (non-infectious or non-transmissible organisms.

Non-insulin dependent diabetes mellitus is a type of diabetes mellitus whereby the insulin produced or secreted by pancreatic beta cells is resistant to body cell for glucose uptake. It is maturity onset. Otherwise referred to as type 2 DM.

Nutritional refers to the process of nourishing or being nourished.

Nutritional therapy is a healing system using functional foods and nutraceuticals as therapeutics. This complementary therapy is based on the assumption that food is not only a source of nutrients and energy, but can also provide health benefits. In particular, the reported health-promoting effects of plant foods and beverages can be ascribed to the numerous bioactive chemicals present in plant tissues.

Oligonucleotide is a polynucleotide, whose molecules contain a relatively small number of nucleotides.

Oxidation is the loss of electrons or an increase in oxidation state by a molecule, atom, or ion.

Oxidative stress is caused by reactive oxygen species (ROS). It is known to result in the oxidation of biomolecules, thereby leading to cellular damage and it plays a key pathogenic role in the aging process.

Paclitaxel is a taxane drug used to treat ovarian, breast, lung, pancreatic and other cancers.

Pathogen is a bacterium, virus, or other microorganism that can cause disease.

Pathogenesis refers to the origin and development of a disease. It involves the biological mechanism(s) that leads to the diseased state.

Peroxyl radical is a type of free radical derived from peroxide.

Per-partum occurs during the last month of gestation or the first few months after delivery, with reference to the mother.

Pharmacopeia is a book containing directions for the identification of compound medicines.

Phytochemical refers to any of various biologically active compounds found in plants.

Platelets are also called thrombocytes are a component of blood whose function is to stop bleeding by clumping and clotting blood vessel injuries.

Pleurisy refers to inflammation of the membranes (pleurae) that surround the lungs and line the chest cavity.

Pluripotent activity is any extract or medicinal property that functions via multiple mechanisms.

Polymenorrhea refers to the menstrual cycle abnormality.

Polymerase chain reaction (PCR) is a technique used in molecular biology to amplify a single copy or a few copies of a piece of DNA across several orders of magnitude, generating thousands to millions of copies of a particular DNA sequence.

Proliferation is cell growth used in the contexts of biological cell development and cell division.

Promastigote is the flagellate stage of a trypanosomatid protozoan in which the flagellum arises from a kinetoplast in front of the nucleus and emerges from the anterior end of the organism.

Pro-oxidants are chemicals that induce oxidative stress, either by generating reactive oxygen species or by inhibiting antioxidant systems.

Protozoa is a diverse group of single celled eukaryotic organisms, which may often be pathogenic.

Reactive oxygen species (ROS) are chemically reactive compounds containing oxygen (e.g., peroxides, superoxide, hydroxyl radical, singlet oxygen).

Resorption refers to the process or action by which something is reabsorbed.

Reverse transcription is the process used to generate complementary DNA (cDNA) from an RNA template, using an enzyme called reverse transcriptase.

Rheumatoid arthritis is a painful autoimmune inflammatory disease, which results in chronic inflammation in joints (especially in the fingers, wrists, feet and ankles) resulting in immobility.

Ringworm is a contagious itching skin disease occurring in small circular patches, caused by any of a number of fungi and affecting chiefly the scalp or the feet. The commonest form is athlete's foot.

Signal transduction is a set of chemical reactions that occur in cells when a molecule (e.g. hormone) binds to a receptor and initiates a cascade of reactions that amplify the response until eventually reaching a target molecule or reaction.

Superoxide is a compound, which contains the superoxide anion (O_2^-). It is a subclass of free radical.

Tannins are astringent polyphenolic plant secondary metabolites, which have roles in protection of plants from predation.

Terpenoids (also called isoprenoids) are a large class of organic compounds produced by plants derived from five carbon isoprene units. Most are multicyclic and include monoterpenes, diterpenes, and sesquiterpenes. They have the general formula $(C_5H_8)n$.

Therapeutic relates to the healing disease.

Tincture is typically an alcoholic extract of plant or animal material or solution.

Traditional Chinese medicine (TCM) is a style of traditional Asian medicine, which uses herbal medicine, acupuncture, massage, exercise and dietary therapy.

Trophozoite is a growth stage in the life cycle of some sporozoan parasites. This is the stage in which the absorb nutrients in the host.

Trypomastigote is a developmental stage of the genus *Trypanosoma*. This is the freely circulating form of the protozoan parasite.

Tumor is a swelling of a part of the body, generally without inflammation, caused by an abnormal growth of tissue, whether benign or malignant.

Vasoactive refers to an endogenous agent or pharmaceutical drug that has the effect of either increasing or decreasing blood pressure and/or heart rate through its vascular activity (effect on blood vessels).

Wart is raised bumps on the skin caused by the human papillomavirus (HPV).

***Xhosa* people** are a Bantu ethnic group of Southern Africa mainly found in the Eastern Cape, South Africa.

Zone of inhibition is the clear region without microbial growth around the disc (or well) in disc (or well) diffusion assays.

α-amylase is a protein enzyme that hydrolyses alpha bonds of large, alpha-linked polysaccharides, such as starch and glycogen, yielding glucose and maltose.

β-lactam antibiotic is a class of broad-spectrum antibiotics that contain a β-lactam ring in their structure. This class includes penams (penicillin derivatives), cephalosporins, monobactams and carbapenems. They generally work by inhibiting bacterial cell wall biosynthesis.

INDEX

C

D

E

H

M

Q

R

T

U

V

W

X

Y

Z